Locomotives & Coaching Stock 1990

The Complete Guide to all BR Locomotives and Coaching Stock Vehicles (excluding Departmental Stock)

Peter Fox

PLATFORM 5

ISBN 1 872524 06 0

CONTENTS

This book lists details of all BR locomotives and coaching stock vehicles in service 1st January 1990, together with vehicles on order which may be delivered during 1990. Departmental coaching stock vehicles are excluded. In addition, information is provided on vehicles of light railways and metros (other than London Underground). The information on locomotives and DMUs has been provided by Metro Enterprises Ltd.

1.LOCOMOTIVES

The following notes are applicable to locomotives:

LOCOMOTIVE CLASS DETAILS

Principal details and dimensions are given for each class in metric units. Imperial equivalents are also given for power. Maximum speeds are still quoted in miles per hour since the operating department of BR still uses imperial units. Where the present maximum permissible speed is lower than the design speed, the latter is shown in parentheses after the present maximum. Many classes of locomotive are now subjected to such speed limits, nominally to make savings on maintenance costs. However, such limits are often raised when necessary to suit operating conditions and thus most speed limits which are less than the original design speed can be regarded as purely arbitrary.

Standard abbreviations used are:

BR	British Railways	kW	kilowatts
BREL	British Rail Engineering Ltd.	lbf	pounds force
hp	horse power	mph	miles per hour
kN	kilonewtons	RA	Route availability
kV	kilovolts	t	tonnes

LOCOMOTIVE DETAIL DIFFERENCES

Detail differences which affect the areas and types of train which locos work are shown. Where detail differences occur within a class or part class of locos., these are shown against the individual loco number. If all members of a class or part class are similar, then the characteristics are shown at the end of the class details. Standard abbreviations used are:

a	Train air brakes only
c	Driver to shore radio-telephone fitted.
e	Fitted with electric heating apparatus (ETH).
g	Fitted with driver–guard telephone.
p	Fitted with tdm push/pull & multiple working equipment.
o	No train heating apparatus
r	Fitted with radio electronic token block equipment.
s	Slow speed control fitted (and operable).
v	Train vacuum brakes only.
x	Dual train brakes (air & vacuum).
y	ETH equipped but equipment isolated.
+	Additional fuel tank capacity compared with others in class.

NAMES AND ALLOCATIONS

All official names are shown as they appear on the locomotive i.e. all upper case or upper & lower case lettering. (S) denotes stored serviceable and (U) stored unserviceable. Last known allocations of stored locomotives are shown, but readers should note that locomotives may not necessarily be stored at their home depots.

After the locomotive number is shown the livery code, followed by the sector code, then notes regarding braking, heating etc.

Thus the layout is as follows:

No.	*Old No.*	*Liv.*	*Sector*	*Notes*	*Depot*	*Name*
47595	(47268)	**M**	IEDA	xec	ED	Confederation of British Industry

Old numbers and names are included where appropriate.

GENERAL INFORMATION ON BRITISH RAILWAYS' LOCOMOTIVES

CLASSIFICATION & NUMBERING

Initially BR diesel locomotives were allocated numbers in the 1xxxx series, with electrics allotted numbers in the 2xxxx series. Around 1957 diesel locomotives were allocated new four digit numbers with 'D' prefixes. Diesel electric shunters in the 13xxx series had the '1' replaced by a 'D', but diesel mechanical shunters were completely renumbered. Electric locomotives retained their previous numbers but with an 'E' prefix. When all standard gauge steam locomotives had been withdrawn, the prefix letter was removed.

In 1972, the present TOPS numbering system was introduced whereby the loco number consisted of a two-digit class number followed by a serial number. In some cases the last two digits of the former number were generally retained (classes 20, 37, 50), but in other classes this is not the case. In this book former TOPS numbers carried by converted locos. are shown in parentheses. Full renumbering information is to be found in the second edition of 'Diesel & Electric loco Register'.

Diesel locomotives are classified as "types" depending on their engine horsepower as follows:

Type	*Engine hp.*	*Old Number Range*	*Current Classes*
1	800–1000	D 8000–D 8999	20
2	1001–1499	D 5000–D 6499/D 7500–D 7999	26, 31.
3	1500–1999	D 6500–D 7499	33, 37.
4	2000–2999	D 1 –D 1999	47, 50.
5	3000+	D 9000–D 9499	56, 58, 59, 60.
Shunter	Under 300	D 2000–D 2999	03.
Shunter	300–799	D 3000–D 4999	08, 09.

Class 14 (650 hp diesel hydraulics) were numbered in the D95xx series.

Electric and electro-diesel locomotives are classified according to their supply system. Locomotives operating on a d.c. system are allocated classes 70–79, whilst a.c. locomotives are allocated classes starting at 80.
Departmental locomotives which remain self propelled or which are likely to move around on a day to day basis are classified 97.

WHEEL ARRANGEMENT

For main line diesel and electric locomotives the system whereby the number of driven axles on a bogie or frame is denoted by a letter (A=1, B=2, C=3 etc.) and the number of undriven axles is noted by a number is used. The letter 'o' after a letter indicates that each axle is individually powered and a + sign indicates that the bogies are intercoupled.

For shunters and steam locomotives the Whyte notation is used. The number of leading wheels are given, followed by the number of driving wheels and then the trailing wheels. Suffix 'T' on a steam locomotive indicates a tank locomotive, and 'PT' a pannier tank loco.

For shunters and steam locomotives the Whyte notation is used. The number of leading wheels are given, followed by the number of driving wheels and then the trailing wheels. Suffix 'T' on a steam locomotive indicates a tank locomotive, and 'PT' a pannier tank loco.

HAULING CAPABILITY OF DIESEL LOCOS

The hauling capability of a diesel locomotive depends basically upon three factors:

1. Its adhesive weight. The greater the weight on its driving wheels, the greater the adhesion and thus more tractive power can be applied before wheel slip occurs.

2. The characteristics of its transmission. In order to start a train the locomotive has to exert a pull at standstill. A direct drive diesel engine cannot do this, hence the need for transmission. This may be mechanical, hydraulic or electric. The current BR standard for locomotives is electric transmission. Here the diesel engine drives a generator or alternator and the current produced is fed to the traction motors. The force produced by each driven wheel depends on the current in its traction motor. In other words the larger the current, the harder it pulls.

As the locomotive speed increases, the current in the traction motors falls hence the *Maximum Tractive Effort* is the maximum force at its wheels that the locomotive can exert at a standstill. The electrical equipment cannot take such high currrents for long without overheating. Hence the *Continuous Tractive Effort* is quoted which represents the current which the equipment can take continuously.

3. The power of its engine. Not all of this power reaches the rail as electrical machines are approximately 90% efficient. As the electrical energy passes through two such machines (the generator/alternator and the traction motors), the *Power At Rail* is about 81% (90% of 90%) of the engine power, less a further amount used for auxiliary equipment such as radiator fans, traction motor cooling fans, air compressors, battery charging, cab heating, ETH, etc. The power of the locomotive is proportional to the tractive effort times the speed. Hence when on full power there is a speed corresponding to the continuous tractive effort.

HAULING CAPABILITY OF ELECTRIC LOCOS

Unlike a diesel locomotive, an electric locomotive does not develop its power on board and its performance is determined only by two factors, namely its weight and the characteristics of its electrical equipment. Whereas a diesel locomotive tends to be a constant power machine, the power of an electric locomotive varies considerably. Up to a certain speed it can produce virtually a constant tractive effort. Hence power rises with speed according to the formula given in section 3 above, until a maximum speed is reached at which tractive effort falls, such that the power also falls. Hence the power at the speed corresponding to the maximum tractive effort is lower than the maximum.

BRAKE FORCE

The brake force is a measure of the braking power of a locomotive. This is shown on the locomotive data panels so that railway staff can ensure that sufficient brake power is available on freight trains.

TRAIN HEATING EQUIPMENT

Electric train heating (ETH) is now the standard system in use on BR for loco-hauled trains. Locomotives which were equipped to provide steam heating have had this equipment removed or rendered inoperable (isolated). Electric heat is provided from the locomotive by means of a separate alternator on the loco., except in the case of 47401–18 and classes 33 and 50 which have a d.c. generator. The *ETH Index* is a measure of the electrical power available for train heating. All electrically heated coaches have an ETH index and the total of these in a train must not exceed the ETH power of a locomotive.

ROUTE AVAILABILITY

This is a measure of a railway vehicle's axle load. The higher the axle load of a vehicle, the higher the RA number on a scale 1 to 10. Each route on BR has an RA number and in theory no vehicle with a higher RA number may travel on that route without special clearance. Exceptions are made, however.

MULTIPLE WORKING

Multiple working between diesel locomotives on BR has usually been provided by means of an electro-pneumatic system, with special jumper cables connecting the locos. A coloured symbol is painted on the end of the locomotive to denote which system is in use. Class 47/7 used a time-division multiplex (tdm) system which utilised the existing RCH (an abbreviation for the former railway clearing house, a pre-nationalisation standards organisation) jumper cables which have in the past only been used for train lighting control, and more recently for public address (pa) and driver–guard communication. A new standard tdm system is now being fitted to electric locomotives.

1.1 DIESEL LOCOMOTIVES

CLASS 03 BR SHUNTER 0–6–0

Built: 1960 at BR Doncaster Works (03079) or 1962 at BR Swindon Works (03179).
Engine: Gardner 8L3 of 152 kW (204 hp) at 1200 rpm.
Transmission: Mechanical. Fluidrive type 23 hydraulic coupling to Wilson-Drewry CA5R7 gearbox with SCG type RF11 final drive.
Max. Tractive Effort: 68 kN (15300 lbf).
Brake Force: 13 t.
Weight: 31 t.
Max. Speed: 28 mph.
Length over Buffers: 7.92 m.
Wheel Diameter: 1092 mm.
RA: 1.

Formerly numbered 2079–2179.

03079		FSAE	vo	RY
03179	**N**	FSAE	xo	RY

CLASS 08 BR SHUNTER 0–6–0

Built: 1953–62 by BR at Crewe, Darlington, Derby, Doncaster or Horwich Works.
Engine: English Electric 6KT of 298 kW (400 hp) at 680 rpm.
Main Generator: English Electric 801.
Traction Motors: Two English Electric 506.
Max. Tractive Effort: 156 kN (35000 lbf).
Cont. Tractive Effort: 49 kN (11100 lbf) at 8.8 mph.
Power At Rail: 194 kW (260 hp).
Brake Force: 19 t.
Weight: 50 t.
Max. Speed: 15 (20) or 20* mph.
Length over Buffers: 8.92 m.
Wheel Diameter: 1372 mm.
RA: 5.

Non standard liveries:

08011/556/647/772/869 are BR green.
08601 is London Midland & Scottish black.
08604 is Great Western Railway green with the number 604 being on a cast numberplate.
08641/765 are rail blue with yellow cabsides.
08642 is London & South Western Railway black and also carries its former number D 3809.
08691/786 have large bodyside numbers.
08793 is London & North Eastern Railway apple green.
08833 is Great Eastern Railway blue.
08867/730 are BR black.
08892 is Southern Railway Maunsell green.
08907 is London & North Western Railway black.
08938 is grey and red.
08944 is GWR green.

Formerly numbered in series 3000–4192.
n–Waterproofed for working at Oxley Carriage Depot.

z–Fitted with buckeye coupling at nose end for HST depot shunting.

CLASS 08/0. Standard Design.

08011	**0**	DCBS	vo	BY
08308		FSHT	vo	TI
08309	**F**	FSGH	vo	HO
08375		FMSK	vo*	CF
08388	**F**	FPZA	ao	IM
08389		FSFN	ao	NL
08390		FHTA	ao	LE
08393	**D**	FSJS	ao	SF
08397		FPZA	ao	IM
08399		RXLS	ao	DY
08401		FPZA	ao	IM
08402		FSMA	ao	AN
08405		FPZA	ao	IM
08407	**F**	FSJS	ao	SF
08410		FSQA	ao	BR
08411		FMZA	ao	TE
08413	**D**	FSJS	ao	SF
08414		FSJS	ao*	SF
08415		FSMA	xo	AN
08416		FHWN	ao	TO
08417	**D**	FSJS	ao	SF
08418	**F**	FSID	ao	DR
08419		FSLC	ao	CL
08421		FHUB	ao	BL
08428		RXLS	ao	DY
08434		FSHT	vo	TI
08436		FSHT	vo	TI
08440		FSJS	ao	SF
08441		FHUB	ao	BL
08442	**F**	FHUB	ao	BL
08445		FPZA	ao	IM
08447		FSLC	ao	CL
08448		RXLS	ao	DY
08449		FPZA	ao	IM
08451		IWNB	xo	WN
08454		FSSO	xo	OC
08460	**F**	FSJS	ao	SF
08466	**FO**	FSGB	ao	BS
08472		FSNC	ao	CD
08479		FMSK	vo	CF
08480		NWRB	aoz	OC
08481		FMSK	xo	CF
08482		FSMA	ao	AN
08483		FSQA	ao	BR
08484	**F**	DBMS	ao	ZN
08485	**F**	FSMA	ao	AN
08489	**F**	FSEG	ao	GD
08492		FSHT	ao	TI
08493		FMSK	ao	CF
08495		FSKC	xo	CA
08496		FSKC	ao	CA
08498		FSJS	ao	SF
08499	**F**	DCEB	ao	YK
08500		DCEB	xo	YK
08506		FMZA	ao	TE
08507		FSEG	ao	GD
08508		FSHT	vo	TI
08509	**F**	FSHT	ao	TI
08510		FSID	ao	DR
08511	**F**	FHWN	ao	TO
08512	**F**	FHUB	ao	BL
08514		FSID	ao	DR
08515		FSEG	ao	GD
08516	**D**	FSGH	ao	HO
08517		FSGB	ao	BS
08519		FMZA	ao	TE
08521		FSEG	ao	GD
08523		NWRB	xo	OC
08525	**F**	DCEB	xo	YK
08526		FSKC	xo	CA
08527		FSJS	xo	SF
08528	**D**	FSKC	xo	CA
08529		FSKC	xo	CA
08530		FSJS	xo	SF
08531		FSJS	xo	SF
08532		FSMA	xo	AN
08533		FSJS	xo	SF
08534		FSLC	xo	CL
08535	**FO**	FSGB	xo	BS
08536		RXLS	xo	DY
08537	**FO**	FPZA	xo	IM
08538		FSKC	xo	CA
08539		FSKC	xo	CA
08540		FSKC	xo	CA
08541		FSJS	xo	SF
08542	**F**	FSJS	xo	SF
08543		FPZA	xo	IM
08544		FSEG	xo	GD
08556	**0**	IWNB	vo*	WN
08561		FHZR	xo	AY
08562		FSID	xo	DR
08565		FSBM	xo	ML
08567		DCEB	xo	YK
08568		FSBM	xo	ML
08569		FSMA	xo	AN
08570	**IO**	FSDH	xo	HA
08571		FSDH	xoz	HA
08573		FSOB	xo	OC

08575		FMZA	xo	TE
08576		IWRB	xo	LA
08577		FSEG	xo	GD
08578		FSEG	xo	GD
08580		FSKC	xo	CA
08581		FSBM	xo	ML
08582		FMZA	ao	TE
08583		FPZA	xo	YK
08584		IWRB	xo	LA
08585		FSNC	xo	CD
08586		FHZR	ao	AY
08587		FSEG	xo	GD
08588		FMZA	xo	TE
08589		FMSK	xo	CF
08590		FSFN	xo	NL
08591		FHZR	xo	AY
08593		FSJS	xo	SF
08594		FSKC	xo	CA
08595		FSID	xo	DR
08597		FHWN	xo	TO
08599		FSNC	xo	CD
08600	**N**	DCSC	ao	AF
08601	**0**	FSGB	xo	BS
08603		FSGB	xo	BS
08604	**0**	FSGB	xo	BS
08605		DCEB	xo	YK
08607		FSID	xo	DR
08609		IWNB	xo	WN
08610		FSGB	xo	BS
08611		DCLS	xo	LO
08612		FSSO	xo	OC
08613		FSNC	xo	CD
08614		IWNB	xo	WN
08615		FSMA	xo	AN
08616		FSGB	xo	BS
08617		IWNB	xo	WN
08618		FSEG	xo	GD
08619		DCLS	xo	LO
08620		FSDH	xo	HA
08622		FSBM	xo	ML
08623		FHWN	xo	TO
08624		DCLS	xo	LO
08625		IWNB	xo	WN
08627		FSJS	ao	SF
08628		DCBS	xo	BY
08629		DBMS	xo	ZN
08630		FSCG	xo	GM
08631	**N**	FSKC	xo	CA
08632		FPZA	xo	IM
08633		FSNC	xo	CD
08634		NWRB	xo	OC
08635		FSNC	xo	CD
08637		FMSK	xo	CF
08638		FSKC	xo	CA
08641	**0**	IWRB	xoz	LA
08642	**0**	DBMS	xo*	ZG
08643		FSQA	xoz	BR
08644	**M**	IWRB	xoz	LA
08645		IWRB	xoz	LA
08646	**F**	FHTA	xo	LE
08647	**0**	DBMS	xo	DL
08648		IWNB	xo*	WN
08649		FMSK	xo	CF
08651		NWRB	xoz	OC
08652		FMSK	xo	CF
08653		NWRB	xo*	OC
08654		FMSK	xo	CF
08655	**F**	FSJS	xo*	SF
08656		DCBS	ao	BY
08657		DCEB	ao	YK
08658		IANB	xo	NC
08659		FSGH	ao	HO
08660		FMSK	vo	CF
08661		FSFN	ao	NL
08662		DCEB	xo	YK
08663		IWRB	ao	LA
08664		FMSK	xo	CF
08665		FPZA	xo	IM
08666		DCLS	xo	LO
08667		FSFN	xoz	NL
08668		FMSK	xo	CF
08670		FSSO	ao	OC
08672		FSGB	ao	BS
08673	**I0**	DCLS	xo	LO
08675		FHZR	xo	AY
08676		DCLS	xo	LO
08677		IWNB	xo	WN
08680		IABB	xo	AC
08682		FSID	xo	DR
08683		IWNB	xo	WN
08685		IANB	xo	NC
08686		FSMA	ao	AN
08688		FSMA	ao	AN
08689		FSJS	ao	SF
08690		FSLC	xo	CL
08691	**0**	FSHT	xo	TI
08692		FSGB	xo	BS
08693		FHZR	xo	AY
08694		FSMA	xo	AN
08695		FSNC	xo	CD
08696		IWNB	ao	WN
08697		RXLS	xo	TO
08698		FSJS	ao	SF
08699		FSNC	xo	CD

08700		FSGB	ao	BS
08701		FSEG	xo	GD
08702		FSNC	xo	CD
08703		FSMA	ao	AN
08704		DCBS	xo	BY (U)
08705		FSKC	ao	CA
08706		FSGH	xo	HO
08707		FSGH	ao	HO
08708		FSJS	ao	SF
08709		FSJS	xo	SF
08710		FSDH	xo	HA
08711		FSKC	xo	CA
08712		DCHB	xo	PH
08713		FSKC	ao	CA
08714		FSKC	xo	CA
08715		FSJS	vo	SF
08717		IISB	xo	IS
08718		FSDH	xo	HA
08719		DCEB	xo	YK
08720	**P**	FSDH	ao	HA
08721	**O**	DCLS	xo	LO
08723		FSID	xo	DR
08724		FSJS	xo	SF
08725		FSCG	xo	GM
08727		FHZR	xo	AY
08729		FSID	ao	DR
08730	**O**	FSDH	xo	HA
08731		FSCG	xo	GM
08732		FSBM	xo	ML
08733		FSBM	xo	ML
08734		FPZA	xo	IM
08735		FHZR	xo	AY
08737	**F**	FSNC	xo	CD
08738	**D**	FSBM	xo	ML
08739		FSNC	xo	CD
08740	**F**	FSJS	xo	SF
08741		FSEG	xo	GD
08742		FSNC	xo	CD
08743		FPZA	xo	IM
08744		FHVW	xo	SP
08745		FSFN	xo	NL
08746		FSGB	xo	BS
08747		FSEG	xo	GD
08748		FSJS	xo	SF
08749		FSHT	xo	TI
08750		FSJS	xo	SF
08751		FPZA	xo	IM
08752		FSJS	xo	SF
08753		IISB	xo	IS
08754		IISB	xo	IS
08755		FSDH	xo	HA
08756		FHTA	xo	LE
08757		FSKC	xo	CA
08758		FSJS	xo	SF
08759		FSGB	xo	BS
08760	**F**	FSAE	xo*	EH
08761	**P**	FSDH	xo	HA
08762		DCHB	xo	PH
08765	**O**	FSGB	xon	BS
08766		FSFN	xoz	NL
08767		FSJS	xo	SF
08768		FSLC	xo	CL
08770	**F**	FMZA	ao	TE
08771		FSEG	xo	GD
08772	**O**	FSJS	xo	SF
08773		FSGH	xo	HO
08775		FSJS	xo	SF
08776		FSGH	ao	HO
08777		DCEB	xo	YK
08778		FSPG	xo	GL
08780		FHTA	xo	LE
08781		FSPG	xo	GL
08782		FSGH	ao	HO
08783		FSGH	xo	HO
08784		FSNC	xo	CD
08786	**O**	FMZA	ao	TE
08787		FMSK	xo	CF
08788		RXLS	xo	DY
08789		DCBS	ao	BY
08790		DCLS	xo	LO
08791	**F**	FSDH	ao	HA
08792		IWRB	xo	LA
08793	**O**	FSDH	ao	HA
08794		FSFN	xoz	NL
08795		FSPG	xo	GL
08796		DCSC	ao	AF
08797		FSEG	xo	GD
08798		FHTA	xo	LE
08799		FSPG	xo	GL
08800		FSQA	xo	BR
08801		IWRB	xo	LA
08802		FSEG	xo	GD
08803		NWRB	xo	RG
08804		NWRB	xo	OC
08805	**FO**	FSGB	xo	BS
08806	**F**	FSGH	ao	HO
08807		DCBS	xo	BY
08808		FSLC	xo	CL
08809		FSMA	xo	AN
08810		IANB	ao	NC
08811		FSSO	ao*	OC
08813		FSID	ao	DR
08814		RXLS	ao	DY
08815		FSMA	xo	AN

08817		FMZA	xo	TE
08818		FMSK	xo	CF
08819		FSQA	xo	BR
08820		DCLS	ao	LO
08821		NWRB	xo	OC
08822		FMSK	xo	CF
08823		FSNC	ao	CD
08824		FSID	ao	DR
08825		FSSO	ao	OC
08826		FSLC	ao	CL
08827		FSMA	ao	AN
08828		FSJS	ao	SF
08829		FHWN	ao	TO
08830		FSAE	xo*	EH
08831		FSAE	xo*	EH
08832	**FO**	FSGB	xo	BS
08833	**O**	FSJS	xo*	SF
08834	**F**	FSJS	xo	SF
08835		FMSK	xoz	CF
08836		FMSK	xo	CF
08837		NWRB	xo*	RG
08838		RXLS	xo	DY
08839		IWRB	xo	LA
08840		IWRB	xo	LA
08841		FSGB	xon	BS
08842		RXLS	xo	DY
08843		DCLS	xo	LO
08844		FSLC	xo	CL
08845		FSAE	xo*	EH
08847		FSAE	xo*	EH
08848		FMSK	xo	CF
08849		IWRB	xo	LA
08850		FHTA	xo	LE
08851		FSBM	xo	ML
08853		FSBM	xo	ML
08854		DCSB	xo*	SU
08855		IAAB	xo	AC
08856		FSMA	xo	AN
08857		FSHT	xo	TI
08858		FSMA	xo	AN
08859		IANB	xo	NC
08865		FSKC	xo	CA
08866		FSID	xo	DR
08867	**O**	FMZA	xo	TE
08868		IANB	xo	NC
08869	**O**	IANB	xo	NC
08870		FSHT	xo	TI
08871		FSHT	xo	TI
08872		FSEG	xo	GD
08873	**M**	FSJS	xo	SF
08874		FSFN	xoz	NL
08875		FSFN	xo	NL

08876		FSID	xo	DR
08877		FPZA	xo	IM
08878		FSHT	xo	TI
08879		FSHT	xo	TI
08880		FSHT	xo	TI
08881		DBMS	xo	ZH
08882		IAAB	xo	AC
08883		DBMS	xo	ZH
08884		FSMA	xo	AN
08885		FSID	xo	DR
08886		FSEG	xo	GD
08887		FSSO	xo	OC
08888		FSEG	xoz	GD
08889		FSKC	xo	CA
08890	**D**	IWNB	xo	WN
08891		DCLS	xo	LO
08892	**O**	DBMS	xo*	ZG
08893	**FO**	FSGB	xo	BS
08894		FSMA	xo	AN
08895		FHTA	xo	LE
08896		FHTA	xo	LE
08897	**D**	FHTA	xo	LE
08899		RXLS	xo	DY
08900		FSQA	xo	BR
08901		FSGB	xon	BS
08902		FSMA	xo	AN
08903		FSID	xo	DR
08904		FSSO	xo	OC
08905		FSSO	xo	OC
08906		FMZA	xo	TE
08907	**O**	FSNC	xo	CD
08908		FSFN	xo	NL
08909		DCBS	xo	BY
08910		FSLC	xo	CL
08911		FSLC	xo	CL
08912		FSLC	xo	CL
08913		FSMA	xo	AN
08914		DCBS	xo	BY
08915	**F**	DCLS	xo	LO
08916		FSMA	xo	AN
08917		FSMA	xo	AN
08918	**D**	FSMA	xo	AN
08919		FSHT	xo	TI
08920	**F**	FSGB	xo	BS
08921		FSNC	xo	CD
08922		DCBS	xo	BY
08923	**F**	FSJS	xo	SF
08924		FSMA	xo	AN
08925		FSMA	xo	AN
08926		IWNB	xo	WN
08927		DCBS	xo	BY
08928	**FO**	FSGB	xo	BS

08929	NWRB	xo*	OC
08930	FSJS	xo	SF
08931	FSEG	xo	GD
08932	FMSK	xo	CF
08933	FSAE	xo*	EH
08934	IWNB	xo	WN
08935	FSOA	xo	BR
08936	IANB	xo	NC
08937	IWRB	xo	LA
08938 **0**	FSBM	xor	ML
08939	FSMA	xo	AN
08940 **F**	FMSK	xo	CF
08941	IWRB	xo	LA
08942	FMSK	xo	CF
08944 **0**	NWRB	xo	OC
08945	IWRB	xo	LA
08946	NWRB	xo	RG
08947	FSSO	xo	OC
08948	NWRB	xo	OC
08949	FSOA	xo	BR
08950 **F**	DCSC	xo	AF
08951	FSOA	xo	BR
08952	FSBM	xo	ML
08953 **D**	IWRB	xo	LA
08954 **F**	IWRB	xo	LA
08955	IWRB	xo	LA
08956	FSJS	xo	SF
08957	FSJS	xo	SF
08958	FSJS	xo	SF

Names:

08562	The Doncaster Postman
08631	Eagle C.U.R.C.
08647	Crimpsall
08691	ESCAFELD
08772	CAMULODNUM
08869	The Canary

Class 08/9. Fitted with cut-down cab and headlight for Cwmmawr branch.

08993 (08592)		FHTA	xo	LE	ASHBURNHAM
08994 (08462)	**FO**	FHTA	ao	LE	GWENDRAETH
08995 (08687)	**F**	FHTA	ao	LE	KIDWELLY

CLASS 09 BR SHUNTER 0–6–0

Built: 1959–62 by BR at Darlington or Horwich Works.
Engine: English Electric 6KT of 298 kW (400 hp) at 680 rpm.
Main Generator: English Electric 801.
Traction Motors: English Electric 506.
Max. Tractive Effort: 111 kN (25000 lbf).
Cont. Tractive Effort: 39 kN (8800 lbf) at 11.6 mph.
Power At Rail: 201 kW (269 hp).
Brake Force: 19 t.
Length over Buffers: 8.92 m.
Weight: 50 t.
Wheel Diameter: 1372 mm.
Max. Speed: 27 mph.
RA: 5.
Formerly numbered 3665–71, 3719–21, 4099–4114.
Non standard liveries:

09011 is Network South East blue.

09001	FMSK	xo	CF
09002	DCSC	xo	AF
09003	DCSB	xo	SU
09004	DCSB	xo	SU
09005	DCSB	xo	SU
09006	DCSB	xo	SU
09007	DCSB	xo	SL
09008	FMMT	xo	TI
09009	DCSB	xo	SU
09010	DCSB	xo	SU
09011 **0**	FSAE	xo	EH
09012 **M**	DCSB	xo	SU
09013	FMMT	xo	TI
09014	DCSB	xo	SU
09015	FMSK	xo	CF
09016	DCSB	xo	SU
09018	DCSC	xo	AF
09019	DCSC	xo	AF

09020	DCSB	xo	SU	09024	DCSC	xo	AF
09021	DCSC	xo	AF	09025	FSAE	xo	EH
09022	DCSC	xo	AF	09026 **D**	FSAE	xo	EH
09023	DCSB	xo	SU				

Names:

09012 Dick Hardy
09026 William Pearson

CLASS 20 ENGLISH ELECTRIC TYPE 1 Bo–Bo

Built: 1957–68 by English Electric Company at Vulcan Foundry, Newton le Willows or Robert Stephenson & Hawthorn, Darlington. 20004–128 were originally built with disc indicators whilst 20129–228 were built with four character headcode panels.
Engine: English Electric 8SVT Mk. II of 746 kW (1000 hp) at 850 rpm.
Main Generator: English Electric 819/3C.
Traction Motors: English Electric 526/5D (20001–48) or 526/8D (others).
Max. Tractive Effort: 187 kN (42000 lbf).
Cont. Tractive Effort: 111 kN (25000 lbf) at 11 mph.
Power At Rail: 574 kW (770 hp).
Brake Force: 35 t. **Length over Buffers**: 14.25 m.
Weight: 73t (20004–48). 74t (others). **Wheel Diameter**: 1092 mm.
Max. Speed: 60 (75) mph. **RA**: 5.
Multiple Working: Blue Star Coupling Code.
Non-standard Liveries: 20028/70/172/3 are BR blue with red solebar. 20030/64 are Brunswick green with red solebar. 20172 also has large numbers. 20901–7 are in Hunslet–Barclay two-tone grey livery with red lettering.

Formerly numbered in series 8004–8199, 8302–8327.

m Fitted with remote control multiple working.
Note: 20023/59/134/68/72/3/94/6 were numbered 20301–8 during 1986.

CLASS 20/1. BR-owned Locomotives.

20004	FEFN	xos	TO	20035	DCMM	xo	TO
20006	FEGN	xos	TO	20040	FEFN	xos	TO
20007	FEFN	xos	TO	20042	DCMC	xo	TO
20010 **FR**	FEGN	xos	TO	20043	DCMC	xo	TO
20013	FEGN	xos	TO	20045	FEGN	xos	TO
20016	FEGN	xos	TO	20046	FMYI	xo	IM
20019	FEGN	xos	TO	20047	FEFN	xos	TO
20020	FEGN	xos	TO	20048	DCMM	xo	TO
20021	FEGN	xos	TO	20051	FGYX	xos	TO (U)
20023 **FR**	FEGN	xos	TO	20052	FEFN	xos	TO
20025	DCEA	xo	IM	20053	FEFN	xos	TO
20026	FEFN	xos	TO	20055	FEGN	xos	TO
20028 **0**	FEGN	xos	TO	20056	DCMN	xos	TO
20029	DCMN	xo	TO	20057	FEGN	xos	TO
20030 **0**	DCEA	xo	IM	20058	DRTC	xosm	RT
20031	DCMN	xo	TO	20059 **FR**	FEFN	xos	TO
20032	DCMN	xos	TO	20061	FMYI	xo	IM
20034	DCMC	xos	TO	20063	DCMN	xo	TO

20064	**0**	DCEA	xo	IM	
20065		FGYX	xos	TO	(U)
20066		FPAE	xo	ED	
20069		DCMC	xo	TO	
20070	**0**	DCMB	xos	TO	
20071		FEGN	xos	TO	
20072		DCMB	xos	TO	
20073		FEFN	xos	TO	
20075		FEFN	xos	TO	
20078		FEFN	xos	TO	
20080		FEGN	xos	TO	
20081		FEFN	xos	TO	
20082		FEGN	xos	TO	
20084		FEFN	xos+	TO	
20085		FEFN	xos	TO	
20087		DRTC	xosm	RT	
20088	**F**	FEFN	xos	TO	
20090	**FR**	FEGN	xos	TO	
20092		DCMC	xo	TO	
20093		FMYI	xo	IM	
20094		FEFN	xos	TO	
20095		DCMN	xo	TO	
20096		DCEA	xo	IM	
20098		FMYI	xo	IM	
20099		FGYX	xos	TO	(U)
20102		DCMC	xo	TO	
20103		FEFN	xos	TO	
20104	**FR**	FEFN	xos	TO	
20105		FEFN	xos	TO	
20106		FEGN	xos	TO	
20107		FMYI	xo	IM	
20108	**FR**	FEFN	xos	TO	
20110		DCMB	xo	TO	
20112		FMYI	xo	IM	
20113		FEGN	xos	TO	
20114		DCMB	xo	TO	
20117		FEGN	xos	TO	
20118	**FR**	FCTY	xo	TE	
20119		FCTY	xo	TE	
20120		FEGN	xos	TO	
20121		FEGN	xos	TO	
20122	**FR**	FCTY	xo	TE	
20124		FCTY	xo	TE	
20127		DCMB	xo	TO	
20128		FEGN	xos	TO	
20129		FEFN	xos	TO	
20130		FEGN	xos	TO	
20131		FEGN	xos	TO	
20132	**FR**	FEGN	xos	TO	
20133		FEGN	xos	TO	
20135		FEGN	xos	TO	
20136		FEFN	xos	TO	
20137	**FR**	FCTY	xo	TE	
20138	**FR**	FCTY	xo†	TE	
20139		DCMM	xo	TO	
20140		FEGN	xos	TO	
20141	**FR**	FEGN	xos	TO	
20142		FEFN	xos	TO	
20143		FEFN	xos	TO	
20144		FCTY	xo	TE	
20145		DCMB	xo	TO	
20148		FPAE	xo	ED	
20151		FEFN	xos	TO	
20154		FEGN	xos	TO	
20156	**FR**	FCTY	xo	TE	
20157		FEFN	xos	TO	
20159		FEGN	xos	TO	
20160		DCMM	xos	TO	
20163	**FR**	FEFN	xos	TO	
20165	**FR**	FCTY	xo	TE	
20166		FEFN	xos	TO	
20168		FEGN	xos	TO	
20169		FEGN	xos	TO	
20170	**FR**	FEFN	xos	TO	
20172	**0**	FEGN	xos	TO	
20173	**0**	FGYX	xos	TO	(U)
20175	**FR**	FEGN	xos	TO	
20176		DCMN	xo	TO	
20177		FEFN	xos	TO	
20182		FEGN	xos	TO	
20183		DCMA	xos	TO	(U)
20185		FPAE	xo	ED	
20186		FEFN	xos	TO	
20187		FEFN	xos	TO	
20188		DCML	xos	TO	
20189		DCMN	xo	TO	
20190		FEFN	xos	TO	
20194		FEGN	xos	TO	
20195		FEGN	xos	TO	
20196		FEFN	xos	TO	
20197		FEGN	xos	TO	
20198		FGTE	xos	ED	
20199		DCMB	xos	TO	
20206		DMEA	xos	ED	
20208		FGYX	xos	TO	(U)
20210		FEFN	xos	TO	
20211		FGTE	xos	ED	
20212		FGTE	xos	ED	
20213		DMEA	xos	ED	
20214		FEFN	xos	TO	
20215	**FR**	FEFN	xos	TO	
20224		DCMA	xos	TO	(U)
20227	**FR**	DCML	xo	TO	
20228		DCMB	xo	TO	

Names:

20118	Saltburn by the Sea	20137	Murray B. Hofmeyr
20122	Cleveland Potash	20165	Henry Pease

CLASS 20/9. Privately-owned by Hunslet–Barclay Ltd.

20901 (20041)	**0**	CYPD	xo	HB (S)	NANCY
20902 (20060)	**0**	CYPD	xo+	HB (S)	LORNA
20903 (20083)	**0**	CYPD	xo+	HB (S)	ALISON
20904 (20101)	**0**	CYPD	xo	HB (S)	JANIS
20905 (20225)	**0**	CYPD	xo	HB (S)	IONA
20906 (20219)	**0**	CYPD	xo	HB (S)	GEORGINA

The above locos are stored for the winter.

CLASS 26 BRCW TYPE 2 Bo–Bo

Built: 1958–59 by the Birmingham Rly Carriage & Wagon Co., Smethwick.
Engine: Sulzer 6LDA28 of 870 kW (1160 hp) at 750rpm.
Main Generator: Crompton Parkinson CG391A1.
Traction Motors: Crompton Parkinson C171A1 (26/0), C171D3 (26/1).
Max. Tractive Effort: 187 kN (42000 lbf).
Cont. Tractive Effort: 133 kN (30000 lbf) at 11.25 mph.
Power At Rail: 671 kW (900 hp).
Brake Force: 35 t. **Length over Buffers**: 15.47 m.
Weight: 75 t. **Wheel Diameter**: 1092 mm.
Max. Speed: 60 (80) mph. **RA**: 5 or 6.
Multiple Working: Blue Star Coupling Code.
Formerly numbered 5301–46 (except 26007 (5300), 26020 (5307) & 26028 (5320).

CLASS 26/0. RA6.

26001	**F**	FEPE	xos	ED
26002	**F**	FEPE	xos	ED
26003	**F**	FEPE	xos	ED
26004	**F**	FEPE	xos	ED
26005	**F**	FEPE	xos	ED
26006	**F**	FEPE	xos	ED
26007	**F**	FEPE	xos	ED
26008	**F**	FEPE	xo	ED
26010	**FR**	DMEA	xo	ED
26011		DCHA	xo	ED
26014		DCHA	xo	ED
26015		DCHA	xo	ED

CLASS 26/1. RA5.

26021		DCHA	xo	ED
26023		DCHA	xo	ED
26024		DCHA	xo	ED
26025	**FR**	FPAE	xo	ED
26026	**FR**	DCHA	xo	ED
26027		DCHA	xo	ED
26028		DCHA	xo	ED
26032	**FR**	FPAE	xo	ED
26035	**FR**	DCHA	xo	ED
26036		DCHA	xo	ED
26037	**FR**	FPAE	xo	ED
26038	**FR**	FPAE	xo	ED
26039		DCHA	xo	ED
26040	**FR**	FPAE	xo	ED
26041	**FR**	FPAE	xo	ED
26042		DMEA	xo	ED
26043		DMEA	xo	ED
26046		DMEA	xo	ED

CLASS 31 BRUSH TYPE 2 A1A–A1A

Built: 1957–62 by Brush Traction at Loughborough. 31102/5–8/10/25/34/44/

418/44/50/61 retain two headcode lights. Others have roof-mounted headcode boxes. 31215 has headcode ,box one end only.
Engine: English Electric 12SVT of 1100 kW (1470 hp) at 850 rpm.
Main Generator: Brush TG160-48.
Traction Motors: Brush TM73-68.
Max. Tractive Effort: 160 kN (35900 lbf) (190 kN (42800 lbf)*).
Cont. Tractive Effort: 83 kN (18700 lbf) at 23.5 mph. (99 kN (22250 lbf) at 19.7 mph *.)
Power At Rail: 872 kW (1170 hp). **ETH Index (Class 31/4):** 66.
Brake Force: 49 t. **Length over Buffers:** 17.30 m.
Max. Speed: 90 or 60 (90 or 80*) mph. **RA:** 5 or 6.
Weight: 108–111 t.
Wheel Diameter: Driving 1092 mm. Centre 1003 mm.
Multiple Working: Blue Star Coupling Code.
Formerly numbered 5518/20–5862 (not in order).

Non standard liveries:

31281/83 have large bodyside numbers.
31327 is as 'FR' but has large bodyside numbers.
31413 is BR blue with yellow cabsides and a light blue stripe along the lower bodyside.
31430 is BR blue with grey roof & carries its number on a cast plate.
31970 is British Rail Research red/blue/white.

CLASS 31/1. Standard Design. RA5. 60 (90 or 80*) mph.

31101		DMEA	xo	BS
31102	**FO**	FGOT	xo	TI
31105	**FR**	DMEA	xo*	BS
31106		DCEA	xo*	IM
31107	**FR**	DCMA	xo	BS
31108	**FO**	FGOT	xo	TI
31110	**FO**	FGOT	xo	TI
31112	**FR**	DCMW	xo*	BS
31113	**FO**	FGOT	xo	TI
31116	**FO**	FAGS	xo	SF
31119	**FR**	DCMA	xo	BS
31120	**F**	FHAC	xo	CD
31123		FALY	xo	TE
31124		FALY	xo	TE
31125	**FO**	FGOT	xo	TI
31126	**FO**	FGOT	xo	TI
31128	**FO**	FAGS	xo	SF
31130	**F**	FHAC	xo	CD
31132	**FO**	FGOT	xo	TI
31134	**FO**	FAGS	xo	SF
31135	**FR**	DCAA	xo	SF
31142	**FO**	FGOT	xo	TI
31144	**FO**	FGOT	xo	TI
31145	**FO**	FGOT	xo	TI
31146	**FO**	FGOT	xo	TI
31147	**FO**	FGOT	xo	TI
31149	**FR**	FAGS	xo	SF
31154	**FO**	FPCI	xo	IM
31155	**F**	FAGS	xo	SF
31156		FPCI	xo	IM
31158	**FO**	FGOT	xo	TI
31159	**FR**	FGOT	xo	TI
31160	**F**	FGOT	xo	TI
31162		DCMN	xo	BS
31163	**FO**	FGOT	xo	TI
31164	**FO**	FGOT	xo	TI
31165	**FO**	DCAA	xo	SF
31166	**FO**	FGOT	xo	TI
31168		FPFS	xo	SF
31170		FPCI	xo	IM
31171	**FO**	FGOT	xo	TI
31174	**FO**	FGOT	xo	TI
31178	**FR**	DCMW	xo	BS
31180	**FR**	FAGS	xo	SF
31181	**FR**	DCAA	xo	SF
31184	**FO**	FALY	xo	TE
31185	**F**	FPCI	xo	IM
31186	**FR**	DCAA	xo	SF
31187	**FR**	DCAA	xo	SF
31188	**FR**	FPCI	xo	IM
31190	**FR**	DCAA	xo	SF
31191	**FR**	DCAA	xo	SF
31196	**FR**	DCAA	xo	SF
31198	**FO**	FAGS	xo	SF

31199	**FO**	FPCI	xo	IM
31200	**F**	FHAC	xo	CD
31201	**F**	FPCI	xo	IM
31203	**FO**	FPCI	xo	IM
31205	**FR**	FPCI	xo	IM
31206	**FO**	FCCC	xo	CD
31207	**F**	FPCI	xo	IM
31208		DCEA	xo	IM
31209	**F**	FAGS	xo	SF
31210	**FO**	FPCI	xo	IM
31212		DCEA	xo	IM
31215	**FO**	FALY	xo	TE
31217	**F**	FHAC	xo	CD
31219	**FR**	DCAA	xo	SF
31221		DCEA	xo	IM
31223		DCEA	xo	IM
31224	**FR**	DCAA	xo	SF
31229	**FO**	FALY	xo	TE
31230	**FO**	DCEA	xo*	IM
31231		FPFS	xo	SF
31232	**FR**	FCCC	xo	CD
31233	**F**	FPCI	xo	IM
31234	**FO**	FAGS	xo	SF
31235	**FO**	DCMA	xo	BS
31237	**FR**	DCMC	xo	BS
31238	**F**	FPCI	xo	IM
31240	**FO**	FAGS	xo	SF
31242	**FR**	FCCC	xo	CD
31243	**FO**	FPCI	xo	IM
31247	**FR**	FAGS	xo	SF
31248	**FO**	FGOT	xo	TI
31249		DCEA	xo	IM
31250	**FR**	DCAA	xo	SF
31252	**FO**	DCEA	xo	IM
31255	**FO**	FPCI	xo	IM
31257		FGOT	xo	TI
31260		DMEA	xo	TE
31263	**FR**	DCAA	xo	SF
31264		DMEA	xo	TE
31268	**FR**	DCAA	xo	SF
31270	**F**	FHAC	xo	CD
31271	**F**	FAGS	xo	SF
31272	**FR**	DMEA	xo	BS
31273	**F**	FPCI	xo	IM
31275	**F**	FHAC	xo	CD
31276	**F**	FHAC	xo	CD
31282	**FR**	DCEA	xo	IM
31285	**FR**	DCEA	xo	IM
31286		FGOT	xo	TI
31288		FPCI	xo	IM
31289		FAGS	xo	SF
31290	**FO**	DMEA	xo	BS
31292		DMEA	xo	BS
31293		FAGS	xo	SF
31294	**F**	FAGS	xo	SF
31296	**F**	FAGS	xo	SF
31299	**FO**	FPCI	xo	IM
31301	**FR**	FAGS	xo	SF
31302	**FO**	FPCI	xo	IM
31304	**FO**	FPCI	xo	IM
31305		DMEA	xo	TE
31306	**FO**	FAGS	xo	SF
31308	**FO**	FAGS	xo	SF
31309		FGOT	xo	TI
31312	**F**	FHAC	xo	CD
31317	**FO**	DCMW	xo	BS
31319	**FO**	FPCI	xo	IM
31320		DCEA	xo	TE
31324	**F**	FHAC	xo	CD
31327	**O**	FAGS	xo	SF

Names:

31102 Cricklewood
31201 Fina Energy
31276 Calder Hall Power Station
31296 Amlwch Freighter/Trên Nwyddau Amlwych
31327 Phillips-Imperial

CLASS 31/4. Equipped with Train Heating. RA6. 90 mph.

31400 (31161)		RXLB	xe	CD
31402		DMEA	xe	BS
31403		DMEA	xe	BS
31404		RXLB	xe	CD
31405		DCMW	xe	BS
31406		DCMA	xe	BS
31407		DCMC	xe	BS
31408		RXLB	xe	CD
31409		DCMN	xe	BS
31410		RXLB	xe	CD
31411		DCMW	xe	BS
31412	**D**	DCQA	xe	CD
31413	**O**	RXLB	xe	CD
31414		DCQA	xe	CD
31415		DMEA	xe	BS
31416		DCQA	xe	CD

31417	D	DMEA	xe	BS
31418		RXLB	xe	CD
31419		DCMN	xe	BS
31420 (31172)		DCMW	xe	BS
31421 (31140)		RXLB	xe	CD
31422 (31310)		DCMA	xe	BS
31423 (31197)	M	DCMC	xe	BS
31424 (31157)	D	DCMW	xe	BS
31425 (31274)		RXLB	xe	CD
31426 (31193)		DCQA	xe	CD
31427 (31194)		RXLB	xe	CD
31428 (31211)		RXLB	xe	CD
31429 (31269)		RXLB	xe	CD
31430 (31265)	O	DCMA	xe	BS
31431 (31253)		PCDB	xe	CD
31432 (31153)		RXLB	xe	CD
31433 (31236)		DCMA	xe	BS
31434 (31258)		DCMW	xe	BS
31435 (31179)		DCMC	xe	BS
31437 (31182)		DCMN	xe	BS
31438 (31139)		RXLB	xe	CD
31439 (31239)		RXLB	xe	CD
31441 (31220)		DCEA	xe	IM
31442 (31251)		RXLB	xe	CD
31444 (31137)		DCEA	xe	IM
31445 (31300)		DCMA	xe	BS
31446 (31316)		DCMC	xe	BS
31447 (31295)		DCEA	xe	IM
31448 (31148)		DCMA	xe	BS
31449 (31307)		PCDB	xe	CD
31450 (31133)		RXLB	xe	CD
31451 (31318)	D	DCMC	xe	BS
31452 (31279)		DCEA	xe	IM
31453 (31114)	D	DCEA	xe	IM
31454 (31228)	D	DCMC	xe	BS
31455 (31246)	D	DCMA	xe	BS
31456 (31291)		DCEA	xe	IM
31457 (31169)	D	DMEA	xe	BS
31458 (31303)		DCEA	xe	IM
31459 (31256)		DMEA	xe	BS
31460 (31266)		PCDB	xe	CD
31461 (31129)	D	DMEA	xe	BS
31462 (31315)	D	DMEA	xe	BS
31463 (31297)		PCDB	xe	CD
31464 (31325)		DMEA	xe	BS
31465 (31213)	D	DMEA	xe	BS
31466 (31115)		DCEA	xe	BS
31467 (31216)		DCEA	xe	BS
31468 (31321)	D	DCMW	xe	BS
31469 (31277)		PCDB	xe	CD

Names:

31413 Severn Valley Railway
31428 The North Yorkshire Moors Railway
31430 Sister Dora
31444 Keighley and Worth Valley Railway

CLASS 31/1. Standard Design. RA5. 90 mph. Owned by British Rail Research.

31970 (97204)	O	DRTC	xo	CD

CLASS 33 BRCW TYPE 3 Bo–Bo

Built: 1960–62 by the Birmingham Railway Carriage & Wagon Company, Smethwick.
Engine: Sulzer 8LDA28 of 1160 kW (1550 hp) at 750 rpm.
Main Generator: Crompton Parkinson CG391B1.
Traction Motors: Crompton Parkinson C171C2.
Max. Tractive Effort: 200 kN (45000 lbf).
Cont. Tractive Effort: 116 kN (26000 lbf) at 17.5 mph.
Power At Rail: 906 kW (1215 hp). **ETH Index**: 48
Brake Force: 35 t. **Length over Buffers**: 15.47 m.
Weight: 78t (79t 33/1). **Wheel Diameter**: 1092 mm.
Max. Speed: 60 (85) mph. **RA**: 6.
Multiple Working: Blue Star Coupling Code.
Non standard livery: BR green.
Formerly numbered in series 6500–97 but not in order.

Class 33/0. Standard Design.

33002	**D**	DCSA	xy	EH
33004		FASB	xe	SL
33006		DCSA	xe	SL
33008	**0**	DCSA	xy	EH
33009		FASB	xe	SL
33012		FASB	xe	SL
33019		FASB	xe	SL
33020		FASB	xe	SL
33021	**F**	FASB	xe	SL
33023		FASB	xe	SL
33025		DMEA	xe	SL
33026	**D**	DCSA	xe	EH
33027		FASB	xe	SL
33029		FASB	xe	SL
33030		DMEA	xe	SL
33033	**F**	FASB	xe	SL
33040		FASB	xe	SL
33042	**F**	FASB	xy	SL
33046		FASB	xy	SL
33047		FASB	xy	SL
33048		FASB	xes	SL
33050	**F**	FASB	xe	SL
33051	**F**	FASB	xe	SL
33052		FASB	xe	SL
33053	**F**	FASB	xe	SL
33056	**F**	FASB	xy	SL
33057		FASB	xys	SL
33058		FASB	xe	SL
33060		FASB	xe	SL
33063	**F**	FASB	xys	SL
33064	**F**	FASB	xe	SL
33065	**D**	DCSA	xe	EH

Class 33/1. Fitted with Buckeye Couplings, SR Multiple WorkingEquipment and flashing light adaptor for Weymouth Quay line. Work with SR e.m.u.s, TC stock & class 73. Note: SR Multiple Working Equipment is isolated on 33107.

33101		DCSA	xe	EH
33102		DCSA	xe	EH
33103	**D**	DCSA	xe	EH
33106		DCSA	xe	EH
33108		DCSA	xe	EH
33109		DCSA	xe	EH
33110		DCSA	xe	EH
33111		DCSA	xe	EH
33113		DCSA	xe	EH
33114		DCSA	xe	EH
33116		DCSA	xe	EH
33117		DCSA	xe	EH
33118		DCSA	xe	EH

Class 33/2. Built to Former Loading Gauge of Hastings Line.

33201	**D**	DCSA	xes	SL
33202	**F**	FASB	xes	SL
33203	**F**	FGWB	xes	SL
33204		FASB	xes	SL
33205	**F**	FGWB	xes	SL
33206	**F**	FGWB	xes	SL
33207	**F**	FASB	xes	SL
33208		DCSA	xes	SL
33211		DCSA	xys	SL

Note: 33205 was numbered 33302 for a period during 1988.

Names:

33008	Eastleigh
33025	Sultan
33050	Isle of Grain
33051	Shakespeare Cliff
33052	Ashford
33056	The Burma Star
33207	Earl Mountbatten of Burma

CLASS 37 ENGLISH ELECTRIC TYPE 3 Co–Co

Built: 1960–5 by English Electric Company at Vulcan Foundry, Newton le Willows or Robert Stephenson & Hawthorn, Darlington. 37001–119 were built with box-type route indicators, the remainder with central headcode panels. The former have had these removed on refurbishment, together with 37054/65/100 following accident damage.

Engine: English Electric 12CSVT of 1300 kW (1750 hp) at 850 rpm.
Main Generator: English Electric 822/10G.
Traction Motors: English Electric 538/A.
Max. Tractive Effort: 245 kN (55500 lbf).
Cont. Tractive Effort: 156 kN (35000 lbf) at 13.6 mph.
Power At Rail: 932 kW (1250 hp). **ETH Index (37/4)**: 38
Brake Force: 50 t. **Length over Buffers**: 18.75 m.
Weight: 103–108 t. **Wheel Diameter**: 1092 mm.
Max. Speed: 80 (90) mph. **RA**: 5 or 7.
Multiple Working: Blue Star Coupling Code.
Formerly numbered 6600–8, 6700–6999 (not in order).

Class 37/0. Unrefurbished Locos. Technical details as above. RA5.
Non-standard Liveries: 37038/145/203 are BR blue, but with large numbers. 37146 is 'R' but with small logo. 37250 is 'FR' but with small logo.

37003			FGET	xo+	TI
37004		**F**	FMTY	xo	TE
37008	(37352)	**F**	FGET	xo+	TI
37009			FGET	xo+	TI
37010			FMGM	xo	ML
37012		**R**	FGET	xo	TI
37013			FGET	xo+	TI
37015		**F**	FGET	xo+	TI
37019		**F**	FGET	xo	TI
37023			FMTY	xo	TE
37025		**D**	DCHA	xo	ED
37026	(37320)	**R**	FMGM	ao+	ML
37029		**F**	FGET	xo+	TI
37031		**F**	FGET	xo+	TI
37032	(37353)	**FR**	FGET	xo+	TI
37035		**F**	FPAE	xo	ED
37037	(37321)	**F**	FMGM	ao	ML
37038			FGET	xo	TI
37040			FMGM	xo	ML
37042		**F**	FMTY	xo+	TE
37046			FGTY	xo	TE
37047		**F**	FGET	xo	TI
37048		**F**	FMTY	xo	TE
37049		**F**	FMGM	ao	ML
37051		**F**	FMGM	xo	ML
37053		**F**	FGET	xo	TI
37054			FMTY	xo	TE
37055		**F**	FGET	xo	TI
37057		**R**	FGET	xo	TI
37058		**F**	FMTY	xo+	TE
37059		**F**	FGET	xo+	TI
37063		**F**	FGET	xo+	TI
37065		**F**	FGET	xo+	TI
37066			FGET	xo+	TI
37068	(37356)	**FR**	FGET	xo+	TI
37069		**F**	FMTY	xo+	TE
37070		**F**	FMTY	xo	TE
37071		**D**	DCEA	xo+	IM
37072			FGET	xo+	TI
37073			FGET	xo+	TI
37074		**F**	FGET	xo	TI
37075		**F**	FMTY	xo	TE
37077			FGET	xo	TI
37078		**F**	FPEK	xo+	CF
37080		**F**	FPAE	xo	ED
37083			FMTY	xo+	TE
37087			FGET	xo	TI
37088	(37323)	**0**	FMGM	ao	ML
37092			FMGM	xo	ML
37095			FGET	xo+	TI
37096			FGYX	xo+	TI (U)
37097		**F**	FMGM	xo	ML
37098			FMTY	xo+	TE
37099	(37324)	**F**	FMGM	ao	ML
37100			FMTY	xo	TE
37101		**F**	FGET	xo+	TI
37104		**D**	DCEA	xo	IM
37106		**F**	FMYI	xo+	IM
37107			FGET	xo	TI
37108	(37325)	**R**	FMGM	ao	ML
37109		**F**	FMTY	xo	TE
37110		**F**	FMTY	xo	TE
37111	(37326)	**F**	FMGM	ao	ML
37113		**F**	FGUV	xo	IS
37114		**F**	FGUV	xor	IS
37116		**R**	FGET	xo	TI
37128		**R**	FGUV	xo	IS
37131			FQCK	xo+	CF
37133			DCWA	xo	CF
37137	(37312)	**F**	FMGM	ao	ML
37138		**F**	FPCI	xor	IM
37139			FQCK	xo+	CF

37140			DCAA	xo	SF
37141			DCWA	xo	CF
37142		**D**	DCWA	xo	CF
37144			FAGS	xor	SF
37145	(37313)	**0**	FMGM	ao	ML
37146		**0**	DCWA	xo	CF
37152	(37310)	**R**	FMGM	ao	ML
37153		**R**	DCHA	xo	ED
37154			FGUV	xo	IS
37156	(37311)	**R**	FMGM	ao	ML
37158			DCWA	xo	CF
37162			FQCK	xo+	CF
37165	(37374)	**F**	FEPE	xo+	ED
37167		**F**	FQCK	xo+	CF
37170			DCHA	xo	ED
37174		**R**	DCWA	xo	CF
37175		**R**	DCHA	xo	ED
37178			FGET	xo	TI
37184		**F**	FPBC	xo	IM
37185			FGET	xo+	TI
37188		**F**	FPAE	xo	ED
37190	(37314)	**F**	FMGM	ao	ML
37191		**R**	FPAE	xo	ED
37194		**F**	FGET	xo+	TI
37196		**FR**	DCHA	xo	ED
37197			FMAK	xo+	CF
37198		**F**	FGET	xo+	TI
37201		**F**	FMGM	xo	ML
37202		**F**	FMTY	xo	TE
37203		**0**	FMTY	xo	TE
37207		**D**	DCWA	xo	CF
37209		**R**	FGET	xo	TI
37211		**F**	FAGS	xo	SF
37212		**F**	FQCK	xo+	CF
37213		**F**	FQCK	xo+	CF
37214			FQCK	xo+	CF
37215		**F**	FPEK	xo	CF
37216		**R**	DCAA	xor+	SF
37217			FQCK	xo+	CF
37218			FAGS	xo	SF
37219			FAGS	xor	SF
37220		**F**	FPEK	xo+	CF
37221			FPEK	xo	CF
37222			FQCK	xo+	CF
37223		**F**	FQCK	xo+	CF
37225			FMYI	xo+	IM
37227		**F**	FMTY	xo+	TE
37229			FEPE	xo+	ED
37230			DCWA	xo+	CF
37232			DCEA	xo	IM
37235		**F**	FQCK	xo+	CF
37238			FGET	xo	TI
37239		**F**	FQCK	xo+	CF
37240			FMTY	xo+	TE
37241		**F**	FMTY	xo	TE
37242		**F**	FGET	xo+	TI
37244			FQCK	xo+	CF
37245			FPAE	xo	ED
37248		**F**	FPEK	xo+	CF
37250		**0**	FMTY	xo+	TE
37251		**D**	DCEA	xo+	IM
37252		**F**	FGET	xo	TI
37254			FMAK	xo+	CF
37255		**FR**	FMTY	xo+	TE
37258			FMTY	xo+	TE
37261		**R**	FGUV	xor	IS
37262		**D**	FGUV	xor	IS
37263		**D**	DCWA	xo	CF
37264		**R**	DCWA	xo	CF
37271	(37303)	**F**	FGET	xo+	TI
37272	(37304)	**F**	FGET	xo+	TI
37273	(37306)		FPEK	xo+	CF
37274	(37308)	**F**	FQCK	xo+	CF
37275		**F**	FMYI	xo+	IM
37278		**F**	FMAK	xo+	CF
37280		**F**	FPEK	xo+	CF
37285		**F**	FGET	xo+	TI
37293			FMAK	xo+	CF
37294		**F**	FPEK	xo+	CF
37298			FGET	xo+	TI

Names:

37026 Shap Fell
37037 Gartcosh
37049 Imperial
37059 Port of Tilbury
37068 Grainflow
37069 Thornaby T.M.D.
37088 Clydesdale
37099 Clydebridge
37108 Lanarkshire Steel
37111 Glengarnock
37113 Radio Highland
37114 Dunrobin Castle
37137 Clyde Iron
37152 British Steel Ravenscraig
37156 British Steel Hunterston
37188 Jimmy Shand
37190 Dalzell
37235 The Coal Merchants' Association of Scotland
37261 Caithness

37262	Dounreay		37275	Stainless Pioneer	

Class 37/3. Unrefurbished locos fitted with regeared (CP7 bogies). Details as Class 37/0 except:
Max. Tractive Effort: 250 kN (56180 lbf).
Cont. Tractive Effort: 184 kN (41250 lbf) at 11.4 mph.
Non-standard Livery: 37350 is BR green.

37350	(37119)	**0**	FPEK	xo+	CF	
37351	(37002)	**FR**	FMYI	xo+	IM	
37354	(37043)	**F**	FAGS	xo	SF	
37355	(37045)	**F**	FGET	xo+	TI	
37357	(37079)	**F**	FGET	xo+	TI	
37358	(37091)	**R**	FGET	xo	TI	P & O Containers
37359	(37118)	**F**	FPAE	xo	ED	
37370	(37127)	**FR**	FACM	xo	ML	
37371	(37147)	**FR**	FPEK	xo+	CF	
37372	(37159)	**FR**	DCWA	xo	CF	
37373	(37160)	**FR**	FACM	xo	ML	
37375	(37193)	**F**	FEPE	xo+	ED	
37376	(37199)	**F**	FEPE	xo+	ED	
37377	(37200)	**FR**	FMYI	xo+	IM	
37378	(37204)	**FO**	FGET	xo+	TI	
37379	(37226)	**FR**	FACM	xo	ML	
37380	(37259)	**F**	FEPE	xo	ED	
37381	(37284)	**F**	FMYI	xo+	IM	

Class 37/4. Refurbished locos fitted with train heating. Main generator replaced by alternator. Regeared (CP7) bogies. Details as class 37/0 except:

Main Alternator: Brush BA1005A.
Max. Tractive Effort: 256 kN (57440 lbf).
Cont. Tractive Effort: 184 kN (41250 lbf) at 10.2 mph.
Power At Rail: 935 kW (1254 hp).

37401	(37268)	**M**	FGTE	xer+	ED	Mary Queen of Scots
37402	(37274)	**R**	FGTE	xer+	ED	Oor Wullie
37403	(37307)	**F**	FGTE	xer+	ED	Glendarroch
37404	(37286)	**M**	FGTE	xer+	ED	Ben Cruachan
37405	(37282)	**M**	FGTE	xer+	ED	Strathclyde Region
37406	(37295)	**M**	FGTE	xer+	ED	The Saltire Society
37407	(37305)	**R**	DCEA	xe+	IM	Loch Long
37408	(37289)	**R**	DCEA	xe+	IM	Loch Rannoch
37409	(37270)	**M**	FGTE	xer+	ED	Loch Awe
37410	(37273)	**R**	FGTE	xer+	ED	Aluminium 100
37411	(37290)	**R**	FABT	xe+	TI	The Institution of Railway Signal Engineers
37412	(37301)	**F**	FJLL	xe+	LA	
37413	(37276)	**R**	FGTE	xer+	ED	Loch Eil Outward Bound
37414	(37287)	**F**	FJLL	xer+	LA	
37415	(37277)	**R**	FGUV	xer+	IS	
37416	(37302)	**R**	FGUV	xer+	IS	
37417	(37269)	**M**	FGUV	xer+	IS	Highland Region
37418	(37271)	**R**	FGUV	xer+	MO	An Comunn Gaidhealach

37419	(37291)	**R**	FGUV	xer+	MO	
37420	(37297)	**R**	FGUV	xer+	MO	The Scottish Hosteller
37421	(37267)	**R**	FGUV	xer+	IS	
37422	(37266)	**F**	FABT	xer+	TI	
37423	(37296)	**M**	FGTE	xer+	ED	Sir Murray Morrison 1873–1948 Pioneer of British Aluminium Industry
37424	(37279)	**R**	FGTE	xer+	ED	
37425	(37292)	**F**	FABT	xer+	TI	Sir Robert McAlpine/Concrete Bob
37426	(37299)	**R**	DCEA	xerc+	IM	Y Lein Fach/Vale of Rheidol
37427	(37288)	**R**	PCFA	xerc+	CF	Bont Y Bermo
37428	(37281)	**R**	PCFA	xerc+	CF	David Lloyd George
37429	(37300)	**R**	PCFA	xerc+	CF	Eisteddfod Genedlaethol
37430	(37265)	**R**	FPBC	xerc+	IM	Cwmbrân
37431	(37272)	**R**	PCFA	xerc+	CF	Sir Powys/County of Powys

Class 37/5. Refurbished locos. Main generator replaced by alternator. Regeared (CP7) bogies. Details as class 37/4 except:

Max. Tractive Effort: 248 kN (55590 lbf).
Non-standard Liveries:
37501 is BSC light blue.
37504/508/513 are as 'FO' but with large bodyside numbers.
37511 is as 'F' but with large bodyside numbers.

37501	(37005)	**F**	FMTY	xo+	TE	Teesside Steelmaster
37502	(37082)	**F**	FMTY	xo+	TE	British Steel Teesside
37503	(37017)	**F**	FMTY	xo+	TE	British Steel Shelton
37504	(37039)	**F**	FMTY	xo+	TE	British Steel Corby
37505	(37028)	**F**	FMTY	xo+	TE	British Steel Workington
37506	(37007)	**0**	FMTY	xo+	TE	British Steel Skinningrove
37507	(37036)	**F**	FMTY	xo+	TE	Hartlepool Pipe Mill
37508	(37090)	**0**	FMTY	xo+	TE	
37509	(37093)	**F**	FMTY	xo+	TE	
37510	(37112)	**F**	FMTY	xo+	TE	
37511	(37103)	**0**	FMTY	xo+	TE	Stockton Haulage
37512	(37022)	**FO**	FMTY	xo+	TE	Thornaby Demon
37513	(37056)	**0**	FMTY	xo+	TE	
37514	(37115)	**0**	FMTY	xos+	TE	
37515	(37064)	**0**	FMTY	xos+	TE	
37516	(37086)	**0**	FMTY	xos+	TE	
37517	(37018)	**0**	FMTY	xos+	TE	
37518	(37076)	**0**	FMTY	xo+	TE	
37519	(37027)	**0**	FMTY	xo+	TE	
37520	(37041)	**0**	FMTY	xo+	TE	
37521	(37117)	**F**	FMTY	xo+	TE	
37667	(37151)	**F**	FMTY	xos+	TE	Wensleydale
37668	(37257)	**F**	FMTY	xos+	TE	Leyburn
37669	(37129)	**FR**	FJLL	xo+	LA	
37670	(37182)	**F**	FJLL	xo+	LA	
37671	(37247)	**F**	FJLL	xo+	LA	Tre Pol and Pen
37672	(37189)	**FR**	FJLL	xos+	LA	Freight Transport Association
37673	(37132)	**F**	FJLL	xo+	LA	
37674	(37169)	**FR**	FJLL	xo+	LA	

37675	(37164)	**FR**	FJLL	xos+	LA	William Cookworthy
37676	(37126)	**FR**	FABT	xo+	TI	
37677	(37121)	**FR**	FABT	xo+	TI	
37678	(37256)	**FR**	FABT	xo+	TI	
37679	(37123)	**FR**	FABT	xo+	TI	
37680	(37224)	**FR**	FABT	xo+	TI	
37681	(37130)	**F**	FABT	xo+	TI	
37682	(37236)	**FR**	FABT	xo+	TI	
37683	(37187)	**FR**	FABT	xo+	TI	
37684	(37134)	**F**	FABT	xo+	TI	
37685	(37234)	**FR**	FABT	xo+	TI	
37686	(37172)	**FR**	FABT	xo+	TI	
37687	(37181)	**F**	FABT	xo+	TI	
37688	(37205)	**F**	FABT	xo+	TI	Great Rocks
37689	(37195)	**F**	FHBK	xos+	CF	
37690	(37171)	**FO**	FHBK	xo+	CF	
37691	(37179)	**FO**	FHBK	xo+	CF	
37692	(37122)	**FO**	FHBK	xos+	CF	
37693	(37210)	**O**	FHBK	xos+	CF	
37694	(37192)	**F**	FEPE	xos+	ED	
37695	(37157)	**FO**	FHBK	xos+	CF	
37696	(37228)	**FO**	FHBK	xos+	CF	
37697	(37243)	**FO**	FHBK	xos+	CF	
37698	(37246)	**F**	FHBK	xos+	CF	Coedbach
37699	(37253)	**FO**	FHBK	xo+	CF	

Class 37/7. Refurbished locos. Main generator replaced by alternator. Regeared (CP7) bogies. Ballast weights added. Details as class 37/4 except:
Main Alternator: GEC G564AZ (37796–803) Brush BA1005A (others).
Max. Tractive Effort: 276 kN (62000 lbf).
Weight: 120 t. **RA**: 7.

37701	(37030)	**FO**	FEKK	xos+	CF	
37702	(37020)	**FO**	FEKK	xos+	CF	Taff Merthyr
37703	(37067)	**FO**	FEKK	xos+	CF	
37704	(37034)	**FO**	FEKK	xos+	CF	
37705	(37060)	**F**	FPFS	xo+	SF	
37706	(37016)	**F**	FPFS	xo+	SF	
37707	(37001)	**F**	FPAE	xo+	ED	
37708	(37089)	**F**	FPAE	xo+	ED	
37709	(37014)	**F**	FPFS	xo+	SF	
37710	(37044)	**F**	FMHK	xo+	CF	
37711	(37085)	**F**	FMHK	xo+	CF	Tremorfa Steel Works
37712	(37102)	**F**	FMHK	xo+	CF	The Cardiff Rod Mill
37713	(37052)	**F**	FMHK	xo+	CF	
37714	(37024)	**F**	FMHK	xo+	CF	
37715	(37021)	**F**	FMHK	xo+	CF	
37716	(37094)	**F**	FMHK	xo+	CF	
37717	(37050)	**F**	FMHK	xo+	CF	
37718	(37084)	**F**	FMHK	xo+	CF	
37719	(37033)	**F**	FMHK	xo+	CF	
37796	(37105)	**FO**	FEKK	xos+	CF	

37797 (37081) **FO** FEKK xos+ CF
37798 (37006) **FO** FEKK xos+ CF
37799 (37061) **FO** FEKK xos+ CF Sir Dyfed/County of Dyfed
37800 (37143) **FO** FEKK xos+ CF Glo Cymru
37801 (37173) **FO** FEKK xos+ CF Aberthaw/Aberddawan
37802 (37163) **FO** FEKK xos+ CF
37803 (37208) **FO** FEKK xos+ CF
37883 (37176) **F** FMHK xo+ CF
37884 (37183) **F** FMHK xo+ CF
37885 (37177) **F** FMHK xo+ CF
37886 (37180) **F** FMHK xo+ CF
37887 (37120) **F** FEKK xos+ CF
37888 (37135) **F** FPFS xo+ SF Petrolea
37889 (37233) **F** FEKK xo+ CF
37890 (37168) **F** FPFS xo+ SF
37891 (37166) **F** FPFS xo+ SF
37892 (37149) **F** FPFS xo+ SF Ripple Lane
37893 (37237) **F** FPFS xo+ SF
37894 (37124) **FO** FEKK xos+ CF
37895 (37283) **FO** FEKK xos+ CF
37896 (37231) **FO** FEKK xos+ CF
37897 (37155) **FO** FEKK xos+ CF
37898 (37186) **FO** FEKK xos+ CF
37899 (37161) **FO** FEKK xos+ CF

Class 37/9. Refurbished Locos. fitted with manufacturers prototype power units and ballast weights. Main generator replaced by alternator. Details as class 37/0 except:
Engine: Mirrlees MB275T of 1340 kW (1800 hp) at 1000 rpm (37901–4), Ruston RK270T of 1340 kW (1800 hp) at 900 rpm (37905–6).
Main Alternator: Brush BA1005A (GEC G564, 37905/6).
Max. Tractive Effort: 279 kN (62680 lbf).
Cont. Tractive Effort: 184 kN (41250 lbf) at 11.4 mph.
Weight: 120 t. **RA**: 7.
Tractive effort figures not yet available.

37901 (37150) **FO** FMCK xo+ CF Mirrlees Pioneer
37902 (37148) **F** FMCK xo+ CF
37903 (37249) **F** FMCK xo+ CF
37904 (37125) **F** FMCK xo+ CF
37905 (37136) **FO** FMCK xos+ CF Vulcan Enterprise
37906 (37206) **F** FMCK xos+ CF

CLASS 43 HST POWER CAR Bo–Bo

Built: 1976–82 by BREL Crewe Works. Formerly numbered as coaching stock but now classified as locomotives. Include luggage compartment. Built to coaching stock lot numbers as shown.
Engine: Paxman Valenta 12RP200L of 1680 kW (2250 hp) at 1500 rpm. (Mirrlees MB190 of 1680 kW (2250 hp)*.
Main Alternator: Brush BA1001B.
Traction Motors: Brush TMH68–46 or GEC G417AZ (43124–151/180). Frame mounted.

Max. Tractive Effort: 80 kN (17980 lbf).
Cont. Tractive Effort: 46 kN (10340 lbf) at 64.5 mph.
Power At Rail: 1320 kW (1770 hp). **ETH Index**:
Brake Force: **Length over Buffers**: 17.79 m.
Weight: 70 t. **Wheel Diameter**: 1020 mm.
Max. Speed: 125 mph. **RA**: 5.
Multiple Working: With one other similar vehicle.

† Retains guard's equipment.
§ Modified to be able to remotely control a class 91 locomotive and to be remotely controlled by a class 91 locomotive.

43002	**I**	IWRP	ae	OO	
43003	**I**	IWRP	ae	LA	
43004	**I**	IWRP	ae	LA	
43005	**I**	IWRP	aec	LA	
43006	**I**	IWRP	aec	LA	
43007	**I**	IWRP	aec	LA	
43008	**I**	IWRP	aec	LA	
43009	**I**	IWRP	aec	LA	
43010	**I**	IWRP	aec	LA	
43011	**I**	IWRP	aec	LA	
43012	**I**	ICCP	aec	LA	
43013	**I**	IECP	aec§	BN	
43014	**I**	IECP	aec§	BN	
43015	**I**	IWRP	aec	LA	
43016	**I**	ICCP	aec	LA	
43017	**I**	IWRP	ae	PM	
43018	**I**	IWRP	aec	PM	
43019	**I**	ICCP	aec	LA	
43020	**I**	IWRP	aec	LA	
43021	**I**	IWRP	aec	LA	
43022	**I0**	IWRP	aec	LA	
43023	**I**	IWRP	aec	LA	County of Cornwall
43024	**I0**	IWRP	aec	LA	
43025	**I0**	IWRP	aec	LA	
43026	**I0**	IWRP	aec	LA	City of Westminster
43027	**I0**	IWRP	aec	LA	Westminster Abbey
43028	**I0**	IWRP	aec	LA	
43029	**I**	IWRP	aec	LA	
43030	**I0**	IWRP	aec	PM	
43031	**I0**	IWRP	ae	PM	
43032	**I**	IWRP	aec	PM	The Royal Regiment of Wales
43033	**I0**	IWRP	ae	PM	
43034	**I0**	IWRP	aec	PM	
43035	**I**	ICCP	aec	PM	
43036	**I0**	ICCP	aec	PM	
43037	**I**	ICCP	aec	PM	
43038	**I**	IMLP	aec	NL	
43039	**I**	IMLP	aec	NL	
43040	**I0**	IECP	aec	EC	
43041	**I0**	IECP	aec	EC	
43042	**I0**	IECP	aec	EC	

43043	**I**	IMLP	aec	NL	
43044	**IO**	IMLP	aec†	NL	
43045	**IO**	IMLP	aec	NL	The Grammar School Doncaster AD1350
43046	**I**	IMLP	aec	NL	
43047	**I**	IMLP	aec	NL	
43048	**I**	IMLP	aec	NL	
43049	**IO**	IMLP	aec	NL	Neville Hill
43050	**I**	IMLP	aec	NL	
43051	**I**	IMLP	aec	NL	
43052	**IO**	IECP	aec	NL	City of Peterborough
43053	**IO**	IECP	aec	NL	County of Humberside
43054	**I**	IECP	aec	NL	
43055	**IO**	IMLP	aec	NL	
43056	**IO**	IECP	aec	BN	University of Bradford
43057	**I**	IECP	aec	BN	
43058	**I**	IECP	aec	NL	
43059	**IO**	IMLP	aec	NL	
43060	**IO**	IMLP	aec	NL	County of Leicestershire
43061	**I**	IMLP	aec	NL	City of Lincoln
43062	**IO**	IECP	aec	BN	
43063	**IO**	IECP	aec	BN	
43064	**I**	IMLP	aec	NL	
43065	**I**	IECP	aec§	BN	
43066	**I**	IMLP	aec	NL	
43067	**I**	IECP	aec§	BN	
43068	**I**	IECP	aec§	BN	
43069	**IO**	IECP	aec	EC	
43070	**I**	IECP	aec	EC	
43071	**IO**	IECP	aec	EC	
43072	**I**	IMLP	aec	NL	
43073	**I**	IMLP	aec	NL	
43074	**I**	IECP	aec	BN	
43075	**I**	IECP	aec	BN	
43076	**I**	IMLP	aec	NL	
43077	**I**	IMLP	aec	NL	
43078	**IO**	IECP	aec	EC	Shildon County Durham
43079	**IO**	IECP	aec	EC	
43080	**I**	IECP	ae	BN	
43081	**I**	IECP	aec	BN	
43082	**IO**	IECP	aec	BN	
43083	**I**	IECP	aec	BN	
43084	**I**	IECP	aec§	BN	County of Derbyshire
43085	**I**	IECP	aec§	BN	
43086	**I**	IECP	aec	EC	
43087	**IO**	IECP	aec	EC	
43088	**I**	IECP	aec	EC	
43089	**IO**	IECP	aec	EC	
43090	**I**	IECP	aec	EC	
43091	**I**	IECP	aec	EC	
43092	**I**	IECP	aec	EC	
43093	**I**	IECP	aec	EC	York Festival '88
43094	**I**	IECP	aec	EC	

43095	**I0**	IECP	ae	HT	Heaton
43096	**I**	IECP	aec†	HT	
43097	**I0**	IECP	ae	HT	The Light Infantry
43098	**I**	IECP	ae	HT	
43099	**I0**	IECP	aec†	HT	
43100	**I**	IECP	aec	EC	
43101	**I**	IECP	aec	EC	
43102	**I**	IECP	aec	HT	
43103	**I**	IECP	ae	HT	John Wesley
43104	**I0**	IECP	ae+	HT	County of Cleveland
43105	**I**	IECP	ae	HT	
43106	**I**	IECP	aec	BN	Songs of Praise
43107	**I**	IECP	aec§	BN	
43108	**I**	IMLP	aec	NL	
43109	**I**	IMLP	aec†	NL	Yorkshire Evening Press
43110	**I**	IECP	aec	NL	
43111	**I**	IECP	aec	NL	
43112	**I**	IMLP	aec	NL	
43113	**I**	IECP	ae	HT	
43114	**I**	IECP	aec	NL	National Garden Festival Gateshead 1990
43115	**I**	IECP	aec	EC	Yorkshire Cricket Academy
43116	**I**	IECP	aec†	NL	
43117	**I**	IECP	aec	NL	
43118	**I**	IECP	aec	NL	Charles Wesley
43119	**I**	IECP	aec	NL	
43120	**I0**	IECP	aec	NL	
43121	**I**	IECP	aec	NL	
43122	**I0**	IECP	aec	NL	
43123	**I**	IECP	aec§	BN	
43124	**I**	IWRP	aec	PM	
43125	**I**	IWRP	aec	PM	
43126	**I**	ICCP	aec	PM	
43127	**I**	IWRP	aec	PM	
43128	**I0**	IWRP	aec	PM	
43129	**I**	IWRP	aec	PM	
43130	**I**	IWRP	aec	PM	
43131	**I**	IWRP	aec	PM	
43132	**I**	IWRP	aec	PM	Worshipful Company of Carmen
43133	**I**	IWRP	aec	PM	
43134	**I**	IWRP	aec	PM	
43135	**I**	IWRP	aec	PM	
43136	**I**	IWRP	aec	PM	
43137	**I0**	IWRP	ae	PM	
43138	**I0**	IWRP	aec	PM	
43139	**I**	IWRP	aec	OO	
43140	**I**	IWRP	aec	OO	
43141	**I**	IWRP	aec	OO	
43142	**I**	IWRP	aec	OO	
43143	**I0**	IWRP	aec	OO	
43144	**I0**	IWRP	aec	OO	
43145	**I**	IWRP	ae	OO	
43146	**I**	IWRP	aec	OO	

43147	I	IWRP	ae†	OO	Red Cross
43148	I	IWRP	aec†	OO	
43149	I	IWRP	aec	PM	B.B.C. Wales Today
43150	I	IWRP	ae	PM	Bristol Evening Post
43151	I	IWRP	aec	PM	
43152	I	IMLP	aec	NL	
43153	I	IMLP	aec	NL	
43154	I	IECP	aec	NL	
43155	I0	IECP	aec	NL	B.B.C. Look North
43156	I0	IECP	aec	NL	
43157	I0	IMLP	aec	NL	Yorkshire Evening Post
43158	I0	IMLP	aec	NL	
43159	I0	IMLP	aec	NL	
43160	I0	IWRP	aec	LA	
43161	I	IWRP	aec	LA	
43162	I0	IMLP	aec	NL	Borough of Stevenage
43163	I	ICCP	ae	LA	
43164	I	ICCP	aec	LA	
43165	I	ICCP	aec	LA	
43166	I	ICCP	aec	LA	
43167	I	IWRP	aec*	PM	
43168	I	IWRP	aec*	PM	
43169	I	IWRP	aec*	PM	The National Trust
43170	I	IWRP	aec*	PM	
43171	I	ICCP	aec	PM	
43172	I0	ICCP	aec	PM	
43173	I0	ICCP	aec	PM	
43174	I0	ICCP	aec	PM	
43175	I0	ICCP	aec	PM	
43176	I0	IWRP	aec	PM	
43177	I0	ICCP	aec	LA	
43178	I	ICCP	aec	LA	
43179	I	ICCP	aec	LA	
43180	I	IWRP	aec	OO	
43181	I0	ICCP	aec	LA	
43182	I0	ICCP	aec	LA	
43183	I0	ICCP	aec	LA	
43184	I	ICCP	aec	LA	
43185	I	ICCP	aec	LA	
43186	I	ICCP	aec	LA	Sir Francis Drake
43187	I	IWRP	aec	LA	
43188	I	IWRP	aec	LA	
43189	I	IWRP	ae	LA	
43190	I	IWRP	aec	LA	
43191	I	IWRP	aec	LA	Seahawk
43192	I	IWRP	aec	LA	City of Truro
43193	I0	IMLP	aec	NL	Yorkshire Post
43194	I	IECP	aec	NL	
43195	I	IECP	aec	NL	
43196	I	IECP	aec	NL	
43197	I	IECP	aec	NL	
43198	I	IECP	aec	NL	

CLASS 47 BRUSH TYPE 4 Co–Co

Built: 1963–67 by Brush Traction, Loughborough or BR Crewe Works.
Engine: Sulzer 12LDA28C of 1920 kW (2580 hp) at 750 rpm.
Main Generator: Brush TG160-60 Mk2, TG 160-60 Mk4 or TM172-50 Mk1.
Traction Motors: Brush TM64-68 Mk1 or Mk1A (axle hung).
Max. Tractive Effort: 267 kN (60000 lbf).(245 kN (55000 lbf) 47401–18)
Cont. Tractive Effort: 133 kN (30000 lbf) at 26 mph.
Power At Rail: 1550 kW (2080 hp). **ETH Index (47/4 & 47/7)**: 66
Brake Force: 61 t. **Length over Buffers**: 19.38 m.
Weight: 112–125 t. **Wheel Diameter**: 1143 mm.
Max. Speed: various (95) mph. **RA**: 6 or 7.
Non-standard Liveries:
47361 is 'R' but with small logo and a yellow stripe along the bottom of the body.
47301/302/363 are as 'FR' but with large bodyside numbers. 47346 is as 'FO' but with large bodyside numbers. 47484/500/628 are Great Western Railway green with cast numberplates.
47522 is London & North Eastern Railway apple green with yellow cabsides.
47577 is as 'R' but with blue cabsides.
47814/822/828 are as 'R' but with small cabside numbers.
47973 is Midland Railway red with yellow cabsides.

Formerly numbered 1100–11, 1500–1999 not in order.

Class 47/0. Built with train heating boiler. RA6. Max Speed 75 (95) mph.

47002		FGAT	xoc	TI	
47003		FGAT	xoc	TI	
47004		FAME	xoc	ED	
47005		FGAT	xoc	TI	
47006	**F**	FAME	xoc	ED	
47007	**FO**	FAGS	xoc	SF	Stratford
47009		FGAT	xoc	TI	
47010	**F**	FPCI	xoc	IM	
47014	**FO**	FGAT	xo	TI	
47016	**FO**	FGAT	xoc	TI	The Toleman Group
47017		FGYX	xoc	TI (U)	
47018	**FO**	FGAT	xoc	TI	
47019	**FO**	FGBC	xoc	CD	
47033		FAWK	xoc	CF	
47049	**F**	FGAT	xoc	TI	
47050	**FO**	FGBC	xoc	CD	
47051		FGBC	xoc	CD	
47052	**F**	FGAT	xoc	TI	
47053		DCMA	xoc	CD	
47054	**F**	FPFS	xoc+	SF	
47060	**F**	FGAT	xoc	TI	Halewood Silver Jubilee 1988
47063	**F**	FAWK	xoc	CF	
47079	**F**	FAWK	xoc	CF	
47085	**F**	FPBC	xoc	CD	Conidae
47094	**F**	FPEK	xo	CF	

47095	**FO**	FGAT	xoc	TI	
47096		FGAT	xoc	TI	
47097		FGAT	xoc	TI	
47098		FGAT	xoc	TI	
47099	**FO**	FGAT	xoc	TI	
47100		FGAT	xoc	TI	
47102		FGAT	xoc	TI	
47105		FGAT	xoc	TI	
47107	**FO**	FGAT	xoc	TI	
47108		DCMA	xoc	BS	
47112	**FO**	FGAT	xoc	TI	
47114	**F**	FAME	xoc	ED	
47115		FPCI	xoc+	IM	
47116		FGAT	xoc	TI	
47117		FGAT	xoc	TI	
47118	**R**	FGAT	xoc	TI	
47119	**F**	FPCI	xoc	IM	
47120	**R**	FGAT	xoc	TI	
47121		FGAT	xoc	TI	
47123		FGAT	xoc	TI	
47125	**F**	FPBC	xoc	CD	Tonnidae
47142	**FR**	FGAT	xoc	TI	The Sapper
47144	**F**	FGAT	xoc	TI	
47145		FGAT	xo	TI	
47146		FGAT	xoc	TI	
47147	**F**	FGAT	xoc	TI	
47150	**F**	FGAT	xoc	TI	
47152	**F**	FGAT	xoc	TI	
47156	**F**	FGAT	xoc	TI	
47157	**F**	FGAT	xoc	TI	
47186	**FO**	FGBC	xoc	CD	
47187	**F**	FGBC	xoc	CD	
47188	**F**	FGAT	xoc	TI	
47190	**F**	FPBC	xoc	CD	Pectinidae
47193	**F**	FPBC	xoc	CD	Lucinidae
47194	**F**	FPBC	xoc	CD	Bullidae
47195	**F**	FPBC	xoc	CD	Muricidae
47196	**F**	FPBC	xoc	CD	Haliotidae
47197	**F**	FPEK	xoc	CF	
47200		FGAT	xoc	TI	
47201	**F**	FGAT	xoc	TI	
47204	**F**	FGAT	xoc	TI	
47205	**F**	FGAT	xoc	TI	
47206	**F**	FGAT	xoc	TI	
47207	**F**	FGAT	xoc	TI	Bulmers of Hereford
47209		FGAT	xoc	TI	
47210	**F**	FAME	xoc	ED	Blue Circle Cement
47211	**F**	FGAT	xoc	TI	
47212	**F**	FPCI	xoc+	IM	
47213	**F**	FGAT	xoc	TI	
47214	**FO**	FGAT	xoc	TI	Tinsley Traction Depot
47215	**FO**	FGYX	xoc	TI (U)	

47217		FGAT	xoc	TI	
47218		FGAT	xoc	TI	
47219		FGAT	xoc	TI	
47220	**FO**	FGAT	xoc	TI	
47221	**F**	FPCI	xoc+	IM	
47222		FPCI	xoc+	IM	Appleby-Frodingham
47223	**F**	FPCI	xoc+	IM	
47224	**F**	FPCI	xoc+	IM	
47225	**F**	FGAT	xoc	TI	
47226		FGAT	xoc	TI	
47227	**FR**	FGBC	xoc	CD	
47228		FGBC	xoc	CD	
47229		FGBC	xoc	CD	
47231	**F**	FGAT	xoc	TI	The Silcock Express
47233	**F**	FPBC	xoc	CD	Strombidae
47234		FGAT	xoc	TI	
47236	**F**	FGAT	xoc	TI	
47237	**FO**	FGAT	xoc	TI	
47238	**F**	FGAT	xoc	TI	Bescot Yard
47241	**F**	FGAT	xoc	TI	
47245		FGAT	xoc	TI	
47249	**FR**	FGAT	xoc	TI	
47256		FGAT	xoc	TI	
47258	**F**	FGAT	xoc	TI	
47270		FGAT	xoc	TI	
47276	**F**	FPCI	xoc+	IM	
47277	**F**	FPEK	xoc	CF	
47278	**F**	FPBC	xoc	CD	Vasidae
47279	**F**	FGAT	xoc	TI	
47280	**FO**	FGAT	xoc	TI	Pedigree
47281		FGAT	xoc	TI	
47283	**F**	FGAT	xoc	TI	Johnnie Walker
47284		FGAT	xoc	TI	
47285	**FO**	FGAT	xoc	TI	
47286	**F**	FGAT	xoc	TI	
47287	**F**	FGAT	xoc	TI	
47288	**F**	FGAT	xoc	TI	
47289	**F**	FGAT	xoc	TI	
47290	**F**	FGAT	xoc	TI	
47291	**R**	FGBC	xoc	CD	The Port of Felixstowe
47292	**F**	FGAT	xoc	TI	
47293	**F**	FGAT	xoc	TI	
47294	**F**	FPCI	xoc+	IM	
47295		FPCI	xoc+	IM	
47296	**F**	FGAT	xoc	TI	
47297	**FO**	FGAT	xoc	TI	
47298		FGAT	xoc	TI	
47299		FPCI	xoc+	IM	

Class 47/3. Built without Train Heat. RA6. Max Speed 75 (95) mph.

47301	**O**	FCTY	xosc	TE
47302	**O**	FCTY	xosc	TE
47303	**F**	FCTY	xosc	TE
47304		FCTY	xosc	TE

Class 03 No. 03179 in Network SouthEast livery stands at Ryde St. Johns on 13th July 1989. *Colin J. Marsden*

Cut-down-cab Class 08 No. 08994 'GWENDRAETH' in old Railfreight livery is pictured at Coedbach Washery on the Burry Port and Gwendraeth Valley line on 10th January 1989. *Hugh Ballantyne*

Class 20 locos Nos. 20045/159 in BR standard blue livery head an MGR train at Bickershaw on 7th June 1989. *Paul Shannon*

20904 propels an Oban–Fort William Chipman weedkiller special (headed by 20901) north of Crianlarich on 19th July 1989. The Class 20/9 are privately owned by Hunslet-Barclay and are painted in a special livery. *Douglas Young*

26010 in old Railfreight livery (P5 code 'FO') passes Dalmeny with an Inverkeithing–Millerhill ballast trip on 14th June 1989. *Douglas Young*

31201 'Fina Energy' and 31273, both in the new Railfreight livery with Petroleum sub-sector markings, are seen passing through Hartlepool on 27th March 1989.
John E. Augustson

Class 33/2 No. 33209 (now withdrawn) in standard blue livery is seen at Bristol Bath Road on 10th April 1989. These locos were built with narrow bodies for use on the Hastings line which used to be of sub-standard loading gauge.
Mike Goodfield

On 5th September 1988, Class 37/3 No. 37355 is hard at work attaching wagons onto a southbound Speedlink service at Stirling Yard. The loco is in the old Railfreight livery with a red bodyside stripe (P5 code 'FR'). *John E. Augustson*

HST power car No. 43048 in new InterCity livery is seen at the head of the 09.20 Leeds–St. Pancras at Sharnbrook on 21st May 1989. *A.J. Woof*

Class 47 No. 47714 in Network SouthEast livery leaves Exeter St. Davids with the 13.02 Paddington–Paignton on 24th June 1989. *John E. Augustson*

Class 50 No. 50033 'Glorious' in revised BR blue livery pauses at Reading on the 13.15 Paddington–Oxford service on 8th April 1989. This livery first appeared on the class in 1981. *Duncan Street*

Class 47/8 No. 47834 'FIRE FLY'heads the 07.07 Plymouth–Glasgow/Aberdeen at Stoke Works Jn on 17th August 1989. This loco has twin fuel tanks and is painted in the new InterCity livery with swallow motif. *R.J. Sweet*

Foster Yeoman's five Class 59s stand at Merehead on 25th June 1989. From left to right are 59001 'YEOMAN ENDEAVOUR', 59002 'YEOMAN ENTERPRISE', 59005 'KENNETH J. PAINTER', 59003 'YEOMAN HIGHLANDER' and 59004 'YEOMAN CHALLENGER'. *Colin J. Marsden*

Railfreight Coal-liveried 56030 is pictured after being named 'Eggborough Power Station' at its open day on 2nd September 1989. *B.J. Nicolle*

Brand new 60002 'Capability Brown', painted in Railfreight Petroleum livery, is star attraction at Brush's Falcon Works centenary open day at Loughborough on 12th August 1989. *Paul A. Biggs*

Old Railfreight liveried Class 58 No. 58040 'Cottam Power Station' hauls a southbound MGR train past the site of Stretton station between Clay Cross South Junction and Ambergate on 9th May 1989. *Brian Cuttell*

73001 is caught in Tonbridge Yard on 26th September 1989, its old blue BR livery contrasting with the old InterCity livery adorning 73129 'City of Winchester' behind.
Michael J. Collins

81010 is in charge of the Up Silcock Express car train at Madeley on 2nd February 1989.
Hugh Ballantyne

85018 leads a southbound Freightliner along the Up loop at Carlisle Kingmoor on 17th June 1989. *Dave McAlone*

BR mainline-liveried 86620 and 86416 'Wigan Pier' (in old InterCity livery) double-head a down Freightliner at Mill Meece on 3rd July 1989. *Hugh Ballantyne*

Class 87 No. 87006 'City of Glasgow' stands at Carlisle on 15th June 1989. This loco was named 'Glasgow Garden Festival' during 1988, but has since regained its original name.
John E. Augustson

All the better for receiving the new InterCity livery, 89001 'Avocet' poses at the Brush Falcon Works at Loughborough for the open day on 12th August 1989.
B.J. Nicolle

Class 90 No. 90017 in new InterCity livery stands at Euston on 11th August 1989.
Chris Wilson

Class 90 No. 90028, painted in Mainline livery, pauses at Carlisle with the 10.25 Inverness–Euston on 26th May 1989.
John E. Augustson

91001 'Swallow' propels the 13.10 from Leeds to London King's Cross. The train's Mark 4 carriages have been in revenue-earning service for just two days in this view on 3rd October 1989. *John F. Augustson*

47305	**F**	FCTY	xosc	TE
47306		FGAT	xosc	TI
47307		FGAT	xosc	TI
47308	**F**	FGAT	xosc	TI
47309	**F**	FGAT	xosc	TI
47310		FGAT	xosc	TI
47311	**F**	FGAT	xosc	TI
47312	**F**	FGAT	xos	TI
47313		FGAT	xosc	TI
47314		FGAT	xosc	TI
47315	**D**	DCWA	xosc	OC
47316		FGAT	xosc	TI
47317	**F**	FGAT	xosc	TI
47318	**FO**	FPEK	xosc	CF
47319	**F**	FPCI	xosc+	IM
47320	**FO**	FCCC	xosc	CD
47321	**F**	FGAT	xosc	TI
47322	**FR**	FCCC	xosc	CD
47323		FCCC	xosc	CD
47324	**F**	FPBC	xosc	CD
47325	**FO**	FAGS	xosc	SF
47326		FPEK	xos	CF
47327	**FO**	FPEK	xosc	CF
47328	**F**	FAGS	xosc	SF
47329	**D**	DCMA	xosc	CD
47330		FGBC	xosc	CD
47331	**FO**	DCEA	xosc	IM
47332		DCEA	xosc	IM
47333		DCML	xosc	CD
47334	**D**	DCWA	xosc	OC
47335	**F**	FGBC	xosc	CD
47336	**F**	FPCI	xosc+	IM
47337	**FO**	FGAT	xosc	TI
47338		FGAT	xosc	TI
47339	**FR**	DCMA	xosc	CD
47340	**FR**	DCMM	xos	CD
47341	**R**	DCWA	xosc	OC
47343	**D**	DCMA	xosc	CD
47344		DCEA	xosc	IM
47345	**FR**	FGBC	xosc	CD
47346	**O**	DCAA	xosc	SF
47347	**F**	FGBC	xosc	CD
47348	**FO**	DCEA	xosc	IM
47349	**F**	FGBC	xosc	CD
47350	**FO**	FGBC	xosc	CD
47351		FGBC	xosc	CD
47352	**D**	DCEA	xosc	IM
47353		DCMA	xosc	CD
47354	**F**	FCCC	xosc	CD
47355	**F**	FGBC	xosc	CD
47356	**FO**	DCMA	xosc	CD
47357	**F**	DCML	xosc	CD
47358	**FO**	DCMA	xosc	CD
47359	**F**	FGAT	xosc	TI
47360		FGAT	xosc	TI
47361	**F**	FCTY	xosc	TE
47362	**FR**	FCTY	xosc	TE
47363	**O**	FCTY	xosc	TE
47364	**D**	DCWA	xosc	OC
47365	**FO**	FCCC	xosc	CD
47366	**FO**	DCAA	xosc	SF
47367	**FR**	FAGS	xosc	SF
47368	**F**	FPBC	xosc	CD
47369		FPEK	xos	CF
47370	**FO**	FGAT	xosc	TI
47371	**FO**	FGAT	xosc	TI
47372		DCMA	xosc	CD
47373	**F**	FPCI	xosc+	IM
47374	**FO**	FPCI	xosc+	IM
47375	**F**	FGAT	xosc	TI
47376		FGAT	xosc	TI
47377	**F**	FGAT	xosc	TI
47378	**F**	FGAT	xosc	TI
47379	**F**	FPCI	xosc+	IM
47380	**F**	FPCI	xosc+	IM
47381	**F**	FPEK	xos	CF

Names:

47311 Warrington Yard
47317 Willesden Yard
47319 Norsk Hydro
47324 Glossidae
47337 Herbert Austin
47348 St. Christopher's Railway Home
47350 British Petroleum
47361 Wilton Endeavour
47363 Billingham Enterprise
47365 ICI Diamond Jubilee
47366 The Institution of Civil Engineers
47368 Neritidae

47379 Total Energy
47380 Immingham

Class 47/4. Equipped with train heating. RA7. Max Speed 95 mph.

47401	**R**	FGII	xec	IM	
47402		FGII	xec	IM	
47406	**IO**	IECA	xec	IM	
47407	**R**	IECA	xec	IM	
47413	**R**	IECA	xec	IM	
47417		IECA	xec	IM	
47418		IECA	xec	IM	
47421		RXLC	xec	CD	
47422	**R**	PCDA	xec	CD	
47423		FGBC	xec	CD	
47424	**R**	PCDA	xec	CD	The Brontës of Haworth
47425		FGBC	xec	CD	Holbeck
47426	**R**	DCMA	xec	CD	
47427	**R**	ICDA	xec	CD	
47430	**IO**	FGCS	xec	SF	
47431	**R**	DCMA	xec	CD	
47432	**R**	ICDA	xec	CD	
47433	**R**	RXLC	xec	CD	
47434	**R**	PCDA	xec	CD	Pride In Huddersfield
47435		RXLC	xec	CD	
47436	**R**	DCML	xec	CD	
47438	**R**	FCCC	xec	CD	
47439	**R**	FAGS	xec	SF	
47440	**R**	FCCC	xec	CD	
47441	**R**	FCCC	xec	CD	
47442	**R**	FCCC	xec	CD	
47443	**R**	PCDA	xec	CD	North Eastern
47444	**R**	PCDA	xec	CD	University of Nottingham
47445	**F**	FCCC	xec	CD	
47446	**R**	FCCC	xec	CD	
47447	**R**	FCCC	xec	CD	
47448	**R**	PCDA	xec	CD	Gateshead
47449	**R**	FCCC	xec	CD	
47450	**R**	DCMA	xec	CD	
47451	**R**	FGBC	xec	CD	
47452	**R**	FGBC	xec	CD	Aycliffe
47453	**R**	DCMA	xec	CD	
47454	**R**	FCCC	xec	CD	
47455	**R**	RXLC	xec	CD	
47456	**R**	ICDA	xec	CD	
47457	**R**	FGBC	xec	CD	Ben Line
47458	**R**	RXLD	xec	BR	
47459	**R**	ICDA	xec	CD	
47460	**R**	PISA	xec	IS	
47461	**PS**	PISA	xec	IS	Charles Rennie Mackintosh
47462		ICDA	xec	CD	
47463		RXLC	xec	CD	
47465	**R**	RXLC	xec	CD	

47466	**R**	RXLD	xec	BR	
47467	**R**	DCHA	xec	ED	
47468	**R**	RXLD	xec	BR	
47470	**M**	IEDA	xec	ED	University of Edinburgh
47471	**M**	ICDA	xec	CD	Norman Tunna G.C.
47472		DCWA	xec	OC	
47473	**R**	FCCC	xe	CD	
47474		RXLC	xec	CD	
47475	**P**	PCDA	xec	CD	
47476		RXLD	xec	BR	
47477		PCDA	xec	CD	
47478		ICDA	xec	CD	
47479	**R**	FCCC	xec	CD	
47481	**R**	RXLC	xec	CD	
47482	**R**	RXLD	xec	BR	
47483	**M**	ICDA	xec	CD	
47484	**O**	DWCC	xec	OC	ISAMBARD KINGDOM BRUNEL
47485	**R**	FCCC	xec	CD	
47488	**R**	PCDA	xec	CD	Rail Riders
47489	**R**	RXLD	xec	BR	
47490	**IO**	RXLD	xec	BR	Bristol Bath Road
47491		FCBC	xec	CD	Horwich Enterprise
47492	**IO**	PISA	xec	IS	The Enterprising Scot
47500	**O**	ICDA	xec	CD	GREAT WESTERN
47501	**IO**	RXLD	xec	BR	Craftsman
47503	**R**	PCDA	xec	CD	The Geordie
47508	**M**	ICDA	xec	CD	S.S. Great Britain
47509	**I**	IEDA	xec	ED	Albion
47512	**R**	ICDA	xec	CD	
47513	**O**	RXLD	xec	BR	Severn
47515	**M**	RXLD	xec	BR	Night Mail
47517	**R**	ICDA	xec	CD	Andrew Carnegie
47518	**R**	IEDA	xec	ED	
47519	**R**	ICDA	xec	CD	
47520	**M**	ICDA	xec	CD	
47521	**N**	NWRA	xec	OC	
47522	**O**	RXLC	xec	CD	Doncaster Enterprise
47523	**M**	RXLC	xec	CD	
47524	**M**	RXLD	xec	BR	
47525	**IO**	ICDA	xec	CD	
47526	**R**	RXLC	xe	CD	Northumbria
47527	**M**	ICDA	xec	CD	Kettering
47528	**M**	RXLD	xec	BR	The Queen's Own Mercian Yeomanry
47530	**N**	NWRA	xec	OC	
47531	**R**	DCHA	xec	ED	
47532	**R**	FCCC	xec	CD	
47533	**R**	RXLD	xec	OC	
47534	**R**	RXLD	xec	OC	
47535	**R**	RXLD	xec	OC	University of Leicester
47536	**R**	RXLD	xec	BR	
47537	**R**	RXLD	xec	BR	Sir Gwynedd – County of Gwynedd
47538	**R**	RXLD	xe	BR	

47539		**R**	RXLD	xec	BR	Rochdale Pioneers
47540		**R**	DCWA	xe	OC	
47541		**IO**	PISA	xec	IS	The Queen Mother
47543			RXLD	xec	OC	
47544			ICDA	xec	CD (U)	
47546		**R**	DCHA	xec	ED	Aviemore Centre
47547		**R**	NWRA	xec	OC	
47549		**IO**	NWRA	xec	BR	Royal Mail
47550		**M**	DCHA	xec	ED	University of Dundee
47555	(47126)	**IO**	ICDA	xec	CD	The Commonwealth Spirit
47556	(47020)		ICDA	xec	CD	
47557	(47024)	**M**	RXLD	xec	BR	
47558	(47027)	**M**	RXLD	xec	BR	Mayflower
47559	(47028)		RXLC	xec	CD	Sir Joshua Reynolds
47562	(47036)	**M**	IEDA	xec	ED	Sir William Burrell
47564	(47038)	**R**	RXLD	xec	BR	COLOSSUS
47565	(47039)	**M**	RXLD	xec	BR	
47566	(47043)	**M**	RXLD	xec	BR	
47567	(47044)		RXLC	xec	CD	Red Star
47568	(47045)	**M**	RXLD	xe	BR	
47569	(47047)	**R**	RXLD	xec	BR	
47572	(47168)	**R**	RXLD	xec	BR	Ely Cathedral
47573	(47173)	**N**	NWRA	xec	OC	The LONDON STANDARD
47574	(47174)	**R**	RXLD	xec	BR	Benjamin Gimbert G.C.
47575	(47175)		RXLD	xec	BR	City of Hereford
47576	(47176)	**N**	NNEA	xec	SF	King's Lynn
47578	(47181)	**R**	IEDA	xec	ED	The Royal Society of Edinburgh
47579	(47183)	**N**	NNEA	xec	SF	James Nightall G.C.
47580	(47167)	**R**	RXLD	xec	BR	County of Essex
47581	(47169)	**N**	NNEA	xec	SF	Great Eastern
47582	(47170)	**N**	NWRA	xec	OC	County of Norfolk
47583	(47172)	**N**	NWRA	xec	OC	County of Hertfordshire
47584	(47180)	**M**	RXLD	xe	BR	County of Suffolk
47585	(47184)	**R**	RXLD	xec	BR	County of Cambridgeshire
47586	(47042)	**I**	ICDA	xec	CD	Northamptonshire
47587	(47263)	**R**	NWRA	xec	OC	
47588	(47178)	**F**	FCCC	xec	CD	Carlisle Currock
47592	(47171)	**R**	RXLD	xec	BR	County of Avon
47593	(47272)	**IO**	IEDA	xec	ED	Galloway Princess
47594	(47035)	**F**	FCCC	xec	CD	
47595	(47268)	**M**	IEDA	xec	ED	Confederation of British Industry
47596	(47255)	**N**	NNEA	xec	SF	Aldeburgh Festival
47597	(47026)	**R**	PCDA	xec	CD	
47598	(47182)	**N**	NWRA	xec	OC	
47599	(47177)	**F**	FCCC	xec	CD	
47600	(47250)	**F**	FGBC	xec	CD	Dewi Sant/Saint David
47603	(47267)	**R**	RXLD	xec	BR	County of Somerset
47604	(47271)	**R**	DCHA	xec	ED	Women's Royal Voluntary Service
47605	(47160)	**F**	FCCC	xec	CD	
47606	(47081)	**IO**	ICDA	xec	CD	ODIN
47614	(47141)	**M**	ICDA	xec	CD	
47615	(47252)	**F**	FGBC	xec	CD	Castell Caerffili/Caerphilly Castle

47616	(47248)	**R**	RXLD	xec	BR	Y Ddraig Goch/The Red Dragon
47617	(47149)	**R**	IEDA	xec	ED	University of Stirling
47623	(47090)		ICDA	xe	CD	
47624	(47087)	**M**	RXLD	xec	BR	CYCLOPS
47625	(47076)	**M**	RXLD	xec	BR	
47626	(47082)	**M**	RXLD	xec	BR	ATLAS
47627	(47273)		RXLD	xec	BR	City of Oxford
47628	(47078)	**M**	RXLD	xec	BR	
47630	(47041)	**R**	PISA	xec	IS	
47631	(47059)	**R**	RXLD	xec	BR	
47633	(47083)	**R**	IEDA	xec	ED	
47634	(47158)	**R**	RXLD	xec	BR	Henry Ford
47635	(47029)	**R**	PISA	xec	IS	Jimmy Milne
47636	(47243)	**R**	IEDA	xec	ED	Sir John de Graeme
47639	(47064)	**R**	ICDA	xec	CD	
47640	(47244)	**R**	PISA	xec	IS	University of Strathclyde
47641	(47080)	**R**	IEDA	xec	ED	Fife Region
47642	(47040)	**IO**	PISA	xec	IS	Strathisla
47643	(47269)	**IO**	IEDA	xec	ED	
47644	(47246)	**R**	IEDA	xec	ED	The Permanent Way Institution
47645	(47075)	**R**	ICDA	xec	CD (U)	Robert F. Fairlie Locomotive Engineer 1831–1885
47646	(47074)	**R**	ICDA	xec	CD	
47648	(47151)	**R**	ICDA	xec	CD	

Class 47/7. Push & Pull fitted (RCH jumper system). RA6. 100 mph.

47701	(47493)	**PS**	PEDA	xegc+	ED	Saint Andrew
47702	(47504)	**PS**	PEDA	xegc+	ED	Saint Cuthbert
47703	(47514)	**PS**	PEDA	xegc+	ED	Saint Mungo
47704	(47495)	**PS**	PEDA	xegc+	ED	Dunedin
47705	(47554)	**N**	NWRA	xegc+	OC	
47706	(47494)	**PS**	PEDA	xegc+	ED	
47707	(47506)	**PS**	PEDA	xegc+	ED	Holyrood
47708	(47516)	**PS**	PEDA	xegc+	ED	Waverley
47709	(47499)	**PS**	PEDA	xegc+	ED	The Lord Provost
47710	(47496)	**PS**	PEDA	xegc+	ED	Sir Walter Scott
47711	(47498)	**PS**	PEDA	xegc+	ED	Greyfriars Bobby
47712	(47505)	**PS**	PEDA	xegc+	ED	Lady Diana Spencer
47714	(47511)	**N**	NWRA	xegc+	OC	
47715	(47502)	**PS**	PEDA	xegc+	ED	Haymarket
47716	(47507)	**PS**	PEDA	xegc+	ED	Duke of Edinburgh's Award
47717	(47497)	**PS**	PEDA	xegc+	ED	Tayside Region

Class 47/4 continued.

47801	(47551)	**R**	ILRA	xec+	BR	
47802	(47552)	**R**	ILRA	xec+	BR	
47803	(47553)	**IO**	ICDA	xec+	CD	Woman's Guild
47804	(47591)	**R**	ILRA	xec+	BR	
47805	(47650)	**R**	ILRA	xec+	BR	
47806	(47651)	**R**	ILRA	xec+	BR	
47807	(47652)	**R**	ILRA	xec+	BR	
47808	(47653)	**R**	ILRA	xec+	BR	

47809	(47654)	**R**	ILRA	xec+	BR	Finsbury Park
47810	(47655)	**R**	ILRA	xec+	BR	
47811	(47656)	**R**	ILRA	xec+	BR	
47812	(47657)	**R**	ILRA	xec+	BR	
47813	(47658)	**R**	ILRA	xec+	BR	
47814	(47659)	**O**	ILRA	xec+	BR	
47815	(47660)	**R**	ILRA	xec+	BR	
47816	(47661)	**R**	ILRA	xec+	BR	
47817	(47662)	**R**	ILRA	xec+	BR	
47818	(47663)	**M**	ILRA	xec+	BR	
47819	(47664)	**R**	ILRA	xec+	BR	
47820	(47665)	**R**	ILRA	xe+	BR	
47821	(47607)	**M**	ILRA	xec+	BR	Royal Worcester
47822	(47571)	**R**	ILRA	xec+	BR	
47823	(47610)	**M**	ILRA	xec+	BR	
47824	(47602)	**M**	ILRA	xec+	BR	Glorious Devon
47825	(47590)	**M**	ILRA	xec+	BR	Thomas Telford
47826	(47637)	**M**	ILRA	xec+	BR	
47827	(47589)	**M**	ILRA	xec+	BR	
47828	(47629)	**O**	ILRA	xec+	BR	
47829	(47619)	**M**	ILRA	xec+	BR	
47830	(47649)	**O**	ILRA	xec+	BR	
47831	(47563)	**M**	ICDA	xec+	CD	Bolton Wanderer
47832	(47560)	**M**	ILRA	xec+	BR	Tamar
47833	(47608)	**I**	ILRA	xec+	BR	
47834	(47609)	**I**	ILRA	xec+	BR	
47835	(47620)	**IO**	ICDA	xec+	CD	
47836	(47618)	**I**	ICDA	xec+	CD	
47837	(47611)	**I**	ICDA	xec+	CD	
47838	(47612)	**I**	ICDA	xec+	CD	
47839	(47621)	**I**	ICDA	xec+	CD	
47840	(47613)	**I**	ICDA	xec+	CD	
47841	(47622)	**I**	ICDA	xec+	CD	
47842	(47606)			xec+		
47843	(47)			xec+		
47844	(47)			xec+		
47845	(47638)	**R**	ICDA	xec+	CD	County of Kent
47846	(47647)	**R**	ICDA	xec+	CD	THOR
47847	(47577)	**O**	ICDA	xec+	CD	
47848	(47632)	**R**	ICDA	xec+	CD	
47849	(47570)	**M**	ICDA	xec+	CD	
47850	(47648)			xec+		
47851	(47639)			xec+		

CLASS 47/9 BRUSH TYPE 5 Co–Co

Built: 1964 by BR at Crewe Works. Converted 1980 at Crewe Works. Details as class 47/0 except:
Engine: Ruston Paxman RP12RK3CT of 2455 kW (3300 hp) at 900 rpm.
Main Alternator: Brush BA1101.
Max. Tractive Effort: 255 kN (57325 lbf).
Cont. Tractive Effort: 145 kN (32600 lbf) at 32.3 mph.

Power At Rail: 2089 kW (2808 hp). **Max. Speed**: 75 (95) mph.
Brake Force: 61 t. **Weight**: 114 t.
Formerly 1628 then 47046. RA6.

47901 (47601)	**F**	FAWK	aoc	CF

CLASS 47/4 continued.

47971 (97480)	**R**	DRTC	xec	CD	Robin Hood
47972 (97545)	**R**	DRTC	xec	CD	
47973 (97561)	**0**	DRTC	xec	CD	Midland Counties Railway 150 1839–1989

CLASS 50 ENGLISH ELECTRIC TYPE 4 Co–Co

Built: 1967–68 by English Electric Co. at Vulcan Foundry, Newton le Willows.
Engine: English Electric 16CVST of 2010 kW (2700 hp) at 850 rpm
Main Generator: English Electric 840/4B.
Traction Motors: English Electric 538/5A.
Max. Tractive Effort: 216 kN (48500 lbf).
Cont. Tractive Effort: 147 kN (33000 lbf) at 23.5 mph.
Power At Rail: 1540 kW (2070 hp). **ETH Index**: 61
Brake Force: 59 t. **Length over Buffers**: 20.88 m.
Weight: 117 t. **Wheel Diameter**: 1092 mm.
Max. Speed: 100 or 60 (100†) mph. **RA**: 6.
Multiple Working: Orange Square coupling code. (Within class only).
Non-standard livery: 50007 is in Great Western Railway green livery and carries its number on a cast plate.
Formerly numbered 401–449, 400.

50001	**N**	NSSA	xec	LA	Dreadnought
50002	**N**	NSSA	xec	LA	Superb
50003	**N**	NSSA	xec	LA	Temeraire
50004	**R**	DCWA	xec†	LA	St.Vincent
50005	**N**	DCWA	xec†	LA	Collingwood
50007	**0**	NSSA	xec	LA	SIR EDWARD ELGAR
50008	**R**	DCWA	xe†	LA	Thunderer
50009	**N**	NSSA	xe	LA	Conqueror
50015	**R**	DCWA	xec†	LA	Valiant
50016	**N**	NSSA	xec	LA	Barham
50017	**N**	NSSA	xec	LA	Royal Oak
50018	**N**	NSSA	xec	LA	Resolution
50019		DCWA	xec†	LA	Ramillies
50020	**R**	DCWA	xe†	LA	Revenge
50021	**R**	DCWA	xe†	LA	Rodney
50023	**N**	NWRA	xe	OC	Howe
50024	**N**	NWRA	xec	OC	Vanguard
50026	**N**	NWRA	xec	OC	Indomitable
50027	**N**	NSSA	xec	LA	Lion
50028	**N**	NSSA	xec	LA	Tiger
50029	**N**	NSSA	xec	LA	Renown
50030	**N**	NSSA	xec	LA	Repulse
50031	**R**	NWRA	xec	OC	Hood
50032	**N**	NWRA	xec	OC	Courageous

50033	**R**	NWRA	xec	OC	Glorious
50034	**N**	NWRA	xe	OC	Furious
50035	**N**	NWRA	xec	OC	Ark Royal
50036	**R**	NWRA	xec	OC	Victorious
50037	**N**	DCWA	xec†	LA	Illustrious
50040	**R**	DCWA	xe†	LA (U)	Centurion
50041	**N**	NSSA	xec	LA	Bulwark
50042	**R**	DCWA	xec†	LA	Triumph
50043	**N**	NSSA	xec	LA	Eagle
50044	**N**	NSSA	xec	LA	Exeter
50045	**N**	NSSA	xec	LA	Achilles
50046	**N**	DCWA	xec†	LA	Ajax
50048	**N**	NSSA	xec	LA	Dauntless
50049	**N**	NSSA	xec	LA	Defiance
50050	**N**	NSSA	xec	LA	Fearless

CLASS 56 BRUSH TYPE 5 Co–Co

Built: 1976–84 by Electroputere at Craiova, Romania (as sub contractors for Brush) or BREL at Doncaster or Crewe Works.
Engine: Ruston Paxman 16RK3CT of 2460 kW (3250 hp) at 900 rpm.
Main Alternator: Brush BA1101A.
Traction Motors: Brush TM73-62.
Max. Tractive Effort: 275 kN (61800 lbf).
Cont. Tractive Effort: 240 kN (53950 lbf) at 16.8 mph.
Power At Rail: 1790 kW (2400 hp).
Brake Force: 60 t.
Weight: 125 t.
Max. Speed: 80 mph.
Length over Buffers: 19.36 m.
Wheel Diameter: 1143 mm.
RA: 7.
Multiple Working: Red Diamond coupling code.

† Temporarily derated to 1320 kW (2400 hp).

56001	**F**	FAWK	aosc	CF	Whatley
56002	**FO**	FECN	aosc	TO	
56003	**F**	FECN	aosc	TO	
56004		FECN	aosc	TO	
56005	**FO**	FECN	aosc	TO	
56006	**F**	FECN	aosc	TO	
56007	**FO**	FECN	aosc	TO	
56008		FECN	aosc	TO	
56009	**FO**	FECN	aosc	TO	
56010		FECN	aosc	TO	
56011	**FR**	FECN	aosc	TO	
56012	**F**	FECN	aosc	TO	Maltby Colliery
56013	**F**	FECN	aosc	TO	
56014	**F**	FECN	aosc	TO	
56015	**F**	FECN	aosc	TO	
56016	**F**	FECN	aosc	TO	
56017	**F**	FECN	aosc	TO	
56018		FECN	aosc	TO	
56019	**FO**	FECN	aosc	TO	

56020		FECN	aosc	TO	
56021		FECN	aosc	TO	
56022		FECN	aosc	TO	
56023	**F**	FECN	aosc	TO	
56024	**FO**	FECN	aosc	TO	
56025	**F**	FECN	aosc†	TO	
56026		FECN	aosc	TO	
56027	**F**	FECN	aosc	TO	
56028	**F**	FECN	aosc	TO	West Burton Power Station
56029	**F**	FEDN	aosc	TO	
56030	**F**	FEDN	aosc	TO	Eggborough Power Station
56031	**F**	FAWK	aosc	CF	Merehead
56032	**F**	FAWK	aosc	CF	Sir De Morgannwg/ County of South Glamorgan
56033	**FO**	FAWK	aosc	CF	
56034	**F**	FAWK	aosc	CF	Castell Ogwr/Ogmore Castle
56035	**FO**	FAWK	aosc	CF	
56036	**F**	FAWK	aos	CF	
56037	**F**	FAWK	aosc	CF	Richard Trevithick
56038	**FO**	FAWK	aosc	CF	Western Mail
56039	**F**	FAWK	aosc	CF	
56040	**FO**	FAWK	aosc	CF	Oystermouth
56041	**FO**	FAWK	aosc	CF	
56042	**F**	FAXN	aosc	TO	
56043	**F**	FAWK	aosc	CF	
56044	**FR**	FAWK	aosc	CF	
56045	**FO**	FAWK	aosc	CF	
56046	**F**	FAWK	aosc	CF	
56047	**F**	FEDN	aosc	TO	
56048	**FO**	FAWK	aos	CF	
56049	**FR**	FAWK	aosc	CF	
56050	**F**	FAWK	aos	CF	
56051	**F**	FAWK	aosc	CF	
56052	**F**	FAWK	aosc	CF	
56053	**F**	FAWK	aosc	CF	Sir Morgannwg Ganol/ County of Mid Glamorgan
56054	**F**	FAXN	aosc	TO	
56055	**F**	FAWK	aosc	CF	
56056	**FO**	FAWK	aosc	CF	
56057	**FO**	FAWK	aosc	CF	
56058	**FO**	FAXN	aosc	TO	
56059	**F**	FAXN	aosc	TO	
56060	**F**	FAXN	aosc	TO	
56061	**FO**	FAXN	aosc	TO	
56062	**F**	FAXN	aosc	TO	Mountsorrel
56063	**F**	FAXN	aosc	TO	Bardon Hill
56064	**FO**	FAXN	aosc	TO	
56065	**F**	FAXN	aosc	TO	
56066	**F**	FEDN	aosc	TO	
56067	**FR**	FEDN	aosc	TO	
56068	**FR**	FEDN	aosc	TO	
56069	**F**	FEDN	aosc†	TO	

56070	**F**	FAXN	aosc	TO	
56071	**F**	FEDN	aosc	TO	
56072	**F**	FEDN	aosc	TO	
56073	**F**	FEDN	aosc	TO	
56074	**F**	FEDN	aosc	TO	Kellingley Colliery
56075	**F**	FEDN	aosc	TO	West Yorkshire Enterprise
56076	**FR**	FEDN	aosc	TO	
56077	**FO**	FEDN	aosc	TO	
56078	**F**	FAXN	aosc	TO	
56079	**FO**	FEDN	aosc	TO	
56080	**F**	FEDN	aosc†	TO	Selby Coalfield
56081	**F**	FEDN	aosc	TO	
56082	**F**	FEDN	aosc	TO	
56083	**FO**	FEDN	aosc	TO	
56084	**F**	FEDN	aosc	TO	
56085	**F**	FEDN	aosc	TO	
56086	**F**	FEDN	aosc	TO	
56087	**FO**	FEDN	aosc	TO	
56088	**F**	FEDN	aosc	TO	
56089	**FR**	FEDN	aosc	TO	
56090	**F**	FEDN	aosc†	TO	
56091	**F**	FEDN	aosc	TO	Castle Donington Power Station
56092	**F**	FEDN	aosc	TO	
56093	**F**	FEDN	aosc	TO	The Institution of Mining Engineers
56094	**F**	FEDN	aosc	TO	
56095	**F**	FEDN	aosc	TO	Harworth Colliery
56096	**FO**	FEDN	aosc	TO	
56097	**F**	FEDN	aosc	TO	
56098	**F**	FEDN	aosc	TO	
56099	**F**	FEDN	aosc	TO	Fiddlers Ferry Power Station
56100	**FR**	FEDN	aosc	TO	
56101	**R**	FEDN	aosc	TO	
56102	**F**	FEDN	aosc	TO	
56103	**R**	FAXN	aosc	TO	
56104	**F**	FEDN	aosc	TO	
56105	**R**	FAXN	aosc	TO	
56106	**R**	FEDN	aosc	TO	
56107	**FR**	FEDN	aosc†	TO	
56108	**FR**	FEDN	aosc	TO	
56109	**R**	FEDN	aosc	TO	
56110	**R**	FAXN	aosc	TO	
56111	**R**	FEDN	aosc	TO	
56112	**R**	FEEN	aosc	TO	
56113	**F**	FEEN	aosc	TO	
56114	**R**	FEEN	aosc	TO	
56115	**F**	FEEN	aosc	TO	
56116	**R**	FEEN	aosc	TO	
56117	**R**	FEEN	aosc	TO	
56118	**R**	FEEN	aosc	TO	
56119	**R**	FEEN	aosc	TO	
56120	**R**	FEEN	aosc	TO	
56121	**F**	FEEN	aosc	TO	

56122	**F**	FEEN	aosc	TO	Wilton–Coalpower
56123	**F**	FEDN	aosc	TO	Drax Power Station
56124	**R**	FEEN	aosc	TO	
56125	**F**	FEEN	aosc	TO	
56126	**R**	FEEN	aosc	TO	
56127	**F**	FEEN	aosc	TO	
56128	**F**	FEEN	aosc	TO	
56129	**F**	FEEN	aosc	TO	
56130	**R**	FEEN	aosc	TO	
56131	**F**	FEEN	aosc†	TO	Ellington Colliery
56132	**F**	FEEN	aosc	TO	
56133	**F**	FEEN	aosc	TO	Crewe Locomotive Works
56134	**F**	FEEN	aosc	TO	Blyth Power
56135	**F**	FEEN	aosc	TO	Port of Tyne Authority

CLASS 58 BREL TYPE 5 Co–Co

Built: 1983–87 by BREL at Doncaster Works.
Engine: Ruston Paxman RK3ACT of 2460 kW (3300 hp) at 1000 rpm.
Main Alternator: Brush BA1101B.
Traction Motors: Brush TM73-62.
Max. Tractive Effort: 275 kN (61800 lbf).
Cont. Tractive Effort: 240 kN (53950 lbf) at 17.4 mph.
Power At Rail: 1780 kW (2387 hp).
Brake Force: 62 t.
Weight: 129 t.
Max. Speed: 80 mph.
Length over Buffers: 19.13 m.
Wheel Diameter: 1120 mm.
RA: 7.
Multiple Working: Red Diamond coupling code.

58001	**F**	FEAN	aosc	TO	
58002	**F**	FEBN	aosc	TO	Daw Mill Colliery
58003	**F**	FEAN	aosc	TO	Markham Colliery
58004	**F**	FEAN	aosc	TO	
58005	**FR**	FEAN	aosc	TO	
58006	**F**	FEAN	aosc	TO	
58007	**FR**	FEAN	aosc	TO	
58008	**FR**	FEBN	aosc	TO	
58009	**FR**	FEAN	aosc	TO	
58010	**F**	FEAN	aosc	TO	
58011	**F**	FEAN	aosc	TO	
58012	**F**	FEAN	aosc	TO	
58013	**FR**	FEAN	aosc	TO	
58014	**F**	FEAN	aosc	TO	Didcot Power Station
58015	**F**	FEAN	aosc	TO	
58016	**F**	FEBN	aosc	TO	
58017	**F**	FEAN	aosc	TO	
58018	**F**	FEBN	aosc	TO	High Marnham Power Station
58019	**F**	FEAN	aosc	TO	Shirebrook Colliery
58020	**F**	FEBN	aosc	TO	Doncaster Works
58021	**FR**	FEBN	aosc	TO	
58022	**FR**	FEBN	aosc	TO	

58023	**F**	FEBN	aosc	TO	
58024	**F**	FEBN	aosc	TO	
58025	**F**	FEBN	aosc	TO	
58026	**F**	FEBN	aosc	TO	
58027	**FR**	FEBN	aosc	TO	
58028	**F**	FEBN	aosc	TO	
58029	**F**	FEBN	aosc	TO	
58030	**F**	FEBN	aosc	TO	
58031	**FR**	FEBN	aosc	TO	
58032	**FR**	FEBN	aosc	TO	
58033	**FR**	FEBN	aosc	TO	
58034	**FR**	FEBN	aosc	TO	Bassetlaw
58035	**F**	FEBN	aosc	TO	
58036	**FR**	FEBN	aosc	TO	
58037	**FR**	FEBN	aosc	TO	
58038	**F**	FEBN	aosc	TO	
58039	**FR**	FEBN	aosc	TO	Rugeley Power Station
58040	**FR**	FEBN	aosc	TO	Cottam Power Station
58041	**F**	FEBN	aosc	TO	Ratcliffe Power Station
58042	**FR**	FEBN	aosc	TO	Ironbridge Power Station
58043	**F**	FEBN	aosc	TO	
58044	**F**	FEBN	aosc	TO	
58045	**F**	FEBN	aosc	TO	
58046	**FR**	FEBN	aosc	TO	
58047	**FR**	FEBN	aosc	TO	
58048	**FR**	FEBN	aosc	TO	
58049	**FR**	FEBN	aosc	TO	Littleton Colliery
58050	**F**	FEBN	aosc	TO	Toton Traction Depot

CLASS 59 GENERAL MOTORS TYPE 5 Co–Co

Built: 1985 (59001–4), 1989 (59005), 1990 (59101–4) by General Motors, La Grange, Illinois, U.S.A.
Engine: General Motors 645E3C two stroke of 2460 kW (3300 hp) at 900 rpm.
Main Alternator: General Motors AR11 MLD-D14A.
Traction Motors: General Motors D77B.
Max. Tractive Effort: 506 kN (113550 lbf).
Cont. Tractive Effort: 291 kN (65300 lbf) at 14.3 mph.
Power At Rail: 1889 kW (2533 hp).
Brake Force: 69 t. **Length over Buffers**: 21.35 m.
Weight: 126 t. **Wheel Diameter**: 1067 mm.
Max. Speed: 60 mph. **RA**: 7.
Non-standard Livery: Painted in the liveries of the owners (see below).

Class 59/0. Owned by Foster-Yeoman Ltd. Blue/silver/blue livery with white lettering and cast numberplates.

59001	**0**	ao	FY	YEOMAN ENDEAVOUR
59002	**0**	ao	FY	YEOMAN ENTERPRISE
59003	**0**	ao	FY	YEOMAN HIGHLANDER
59004	**0**	ao	FY	YEOMAN CHALLENGER
59005	**0**	ao	FY	KENNETH J. PAINTER

Class 59/1. Owned by Associated Roadstone Company. Livery details awaited.

59101	**0**	ao
59102	**0**	ao
59103	**0**	ao
59104	**0**	ao

CLASS 60 BRUSH TYPE 5 Co–Co

Built: 1989 onwards by Brush Traction.
Engine: Mirrlees MB275T of 2310 kW (3100 hp) at 1000 rpm.
Main Alternator: Brush .
Traction Motors: Brush separately excited.
Max. Tractive Effort: 500 kN (106500 lbf).
Cont. Tractive Effort: 336 kN (71570 lbf) at 17.4 mph.
Power At Rail: 1000 kW (2415 hp).
Brake Force:
Weight: 126 t.
Max. Speed: 62 mph.
Multiple Working: Within class.
Length over Buffers: 21.34 m.
Wheel Diameter: 1118 mm.
RA: 7.

60001	**F**	FAXN	aoc	TO	Steadfast
60002	**F**	FP	aoc	TO	Capability Brown
60003	**F**		aoc	TO	Christopher Wren
60004	**F**	FE	aoc	TO	Lochnagar
60005	**F**		aoc	TO	Skiddaw
60006	**F**		aoc		Great Gable
60007	**F**		aoc		Robert Adam
60008	**F**		aoc		
60009	**F**		aoc		
60010	**F**		aoc		
60011	**F**		aoc		
60012	**F**		aoc		
60013	**F**		aoc		
60014	**F**		aoc		
60015	**F**		aoc		
60016	**F**		aoc		
60017	**F**		aoc		
60018	**F**		aoc		
60019	**F**		aoc		
60020	**F**		aoc		
60021	**F**		aoc		
60022	**F**		aoc		
60023	**F**		aoc		
60024	**F**		aoc		
60025	**F**		aoc		
60026	**F**		aoc		
60027	**F**		aoc		
60028	**F**		aoc		
60029	**F**		aoc		
60030	**F**		aoc		
60031	**F**		aoc		

60032	**F**	aoc
60033	**F**	aoc
60034	**F**	aoc
60035	**F**	aoc
60036	**F**	aoc
60037	**F**	aoc
60038	**F**	aoc
60039	**F**	aoc
60040	**F**	aoc
60041	**F**	aoc
60042	**F**	aoc
60043	**F**	aoc
60044	**F**	aoc
60045	**F**	aoc
60046	**F**	aoc
60047	**F**	aoc
60048	**F**	aoc
60049	**F**	aoc
60050	**F**	aoc
60051	**F**	aoc
60052	**F**	aoc
60053	**F**	aoc
60054	**F**	aoc
60055	**F**	aoc
60056	**F**	aoc
60057	**F**	aoc
60058	**F**	aoc
60059	**F**	aoc
60060	**F**	aoc
60061	**F**	aoc
60062	**F**	aoc
60063	**F**	aoc
60064	**F**	aoc
60065	**F**	aoc
60066	**F**	aoc
60067	**F**	aoc
60068	**F**	aoc
60069	**F**	aoc
60070	**F**	aoc
60071	**F**	aoc
60072	**F**	aoc
60073	**F**	aoc
60074	**F**	aoc
60075	**F**	aoc
60076	**F**	aoc
60077	**F**	aoc
60078	**F**	aoc
60079	**F**	aoc
60080	**F**	aoc
60081	**F**	aoc

On order to 60100.

1.2. ELECTRIC LOCOMOTIVES

CLASS 73 ELECTRO–DIESEL Bo–Bo

Supply System: 660–750 volts d.c. from third rail.
Engine: English Electric 4SRKT of 447 kW (600 hp) at 850 rpm.
Main Generator: English Electric 824/3D (73/0), 824/5D (73/1).
Traction Motors: English Electric 542A (73/0), 546/1B (73/1).
Brake Force: 31 t. **Length over Buffers**: 16.36 m.
Weight: 76t (73/0), 77t (73/1) **Wheel Diameter**: 1016 mm.
ETH Index (Elec. power): 66 **RA**: 6.
Multiple Working: Within each sub-class and with Class 33/1 and also various SR e.m.u.s.
Max. Speed: 60 (80 or 90) mph or 90* mph.
Couplings: Drop-head buckeye.
Formerly numbered E 6001–26/28–49 (not in order).

Class 73/0. Max Speed: 60 (80) mph.

Built: 1962 by BR at Eastleigh Works.
Max. Tractive Effort: Electric 187 kN (42000 lbf). Diesel 152 kN (34100 lbf).
Continuous Rating: Electric 1060 kW (1420 hp) giving a tractive effort of 43 kN (9600 lbf) at 55.5 mph.
Cont. Tractive Effort: Diesel 72 kN (16100 lbf) at 10 mph.
Maximum Rail Power: Electric 1830 kW (2450 hp) at 37 mph.
Non-standard Livery:

73004 is Network South-East blue with yellow lower cab sides.
73005 is Network South-East blue.

73001		DMEA	xe	SL	
73002	**R**	DMEA	xe	SL	
73003	**R**	DCSA	xe	SL	
73004	**0**	DCSA	xe	SL	The Bluebell Railway
73005	**0**	NSSB	xe	SL	Mid hants WATERCRESS LINE
73006	**R**	NXXA	xe	SL	

CLASS 73/1. (73/2*) Max Speed 60 (90) or *90 mph. Class 73/2 are locos dedicated to InterCity and Network SouthEast services.

Built: 1965–67 by English Electric Co. at Vulcan Foundry, Newton le Willows.
Max. Tractive Effort: Electric 179 kN (40000 lbf). Diesel 160 kN (36000 lbf).
Continuous Rating: Electric 1060 kW (1420 hp) giving a tractive effort of 35 kN (7800 lbf) at 68 mph.
Cont. Tractive Effort: Diesel 60 kN (13600 lbf) at 11.5 mph.
Maximum Rail Power: Electric 2350 kW (3150 hp) at 42 mph.

73101	**IO**	DCSA	xe	SL	Brighton Evening Argus
73103	**IO**	DCSA	xe	SL	
73104	**IO**	NSSB	xe*	SL	
73105	**IO**	DCSA	xe	SL	Quadrant
73106	**IO**	DMEA	xe	SL	
73107	**IO**	DMEA	xe	SL	

73108		**D**	DCSA	xe	SL	
73109		**IO**	NSSB	xe*	SL	
73110		**IO**	NXXA	xe	SL	
73111		**IO**	DCSA	xe	SL	
73112		**IO**	NSSB	xe	SL	
73114		**IO**	DCSA	xe	SL	
73117		**IO**	DCSA	xe	SL	University of Surrey
73118		**IO**	DCSA	xe	SL	The Romney Hythe and Dymchurch Railway
73119		**IO**	DCSA	xe	SL	Kentish Mercury
73126		**IO**	DCSA	xe	SL	
73128		**IO**	DCSA	xe	SL	
73129		**IO**	DCSA	xe	SL	City of Winchester
73130		**M**	NSSB	xe*	SL	City of Portsmouth
73131		**IO**	DCSA	xe	SL	County of Surrey
73132		**IO**	DCSA	xe	SL	
73133		**IO**	NXXA	xe	SL	
73134		**IO**	NSSB	xe*	SL	Woking Homes 1885–1985
73135		**IO**	DMEA	xe	SL	
73136		**D**	DMEA	xe	SL	
73138		**M**	DCSA	xe	SL	Post Haste 150 YEARS OF TRAVELLING POST OFFICES
73139		**IO**	DCSA	xe	SL	
73140		**IO**	FASB	xe	SL	
73141		**IO**	FASB	xe	SL	
73201	(73142)	**M**	IVGA	xe	SL	Broadlands
73202	(73137)	**IO**	IVGA	xe	SL	Royal Observer Corps
73203	(73127)	**IO**	IVGA	xe	SL	
73204	(73125)	**IO**	IVGA	xe	SL	Stewarts Lane 1860–1985
73205	(73124)	**M**	IVGA	xe	SL	London Chamber of Commerce
73206	(73123)	**IO**	IVGA	xe	SL	Gatwick Express
73207	(73122)	**IO**	IVGA	xe	SL	County of East Sussex
73208	(73121)	**IO**	IVGA	xe	SL	Croydon 1883–1983
73209	(73120)	**IO**	IVGA	xe	SL	
73210	(73116)	**IO**	IVGA	xe	SL	Selhurst
73211	(73113)	**IO**	IVGA	xe	SL	County of West Sussex
73212	(73102)	**IO**	IVGA	xe	SL	Airtour Suisse
73251			NSSB	xe		
73252			NSSB	xe		
73253			NSSB	xe		
73254			NSSB	xe		

CLASS 81 BRCW/BTH DESIGN Bo–Bo

Built: 1960–61 by the Birmingham Rly Carriage & Wagon Co. at Smethwick.
Supply System: 25 kV a.c. from overhead equipment.
Traction Motors: AEI 189 frame mounted.
Max. Tractive Effort: 222 kN (50000 lbf).
Continuous Rating: 2390 kW (3200 hp) giving a tractive effort of 76 kN (17000 lbf) at 71 mph.
Maximum Rail Power: 3580 kW (4800 hp) at 44 mph.
Brake Force: 40 t. **Length over Buffers**: 17.22 m.

Weight: 79t
ETH Index: 66
Max. Speed: 80 (100) mph †40 (100) mph.
Formerly numbered E 3003–8/10–16/20/22.
Wheel Diameter: 1219 mm.
RA: 6.

81002	IWCA	xe*	WN
81004	IWCA	xe*	WN
81009	FGKG	xe	GW
81010	FGKG	xe	GW (U)
81012	FGKG	xe	GW
81017	FGKG	xe	GW

CLASS 85 BR DESIGN Bo–Bo

Built: 1961–65 by BR at Doncaster Works.
Supply System: 25 kV a.c. from overhead equipment.
Traction Motors: AEI 189 frame mounted.
Max. Tractive Effort: 222 kN (50000 lbf).
Continuous Rating: 2390 kW (3200 hp) giving a tractive effort of 76 kN (17000 lbf) at 71 mph.
Maximum Rail Power: 3800 kW (5100 hp) at 45 mph.
Brake Force: 41 t.
Weight: 83t
ETH Index: 66
Max. Speed: 100 or 80 (100)* mph.
Electric Brake: Rheostatic.
Formerly numbered in series E 3058–95.
Length over Buffers: 17.22 m.
Wheel Diameter: 1219 mm.
RA: 6.

Class 85/0. Max Speed: 100 or †80 (100) mph.

85003	RXLE	xe	CE
85005	IWCA	xe*	CE (U)
85007	FGHC	xe*	CE
85008	ICCA	xe*	CE
85011	RXLE	xe	CE
85013	FGHC	xe	CE (U)
85015	FGHC	xe*	CE
85018	FGHC	xe*	CE
85020	FGHC	xe*	CE (U)
85023	RXLE	xe	CE
85025	RXLE	xe	CE
85026	FGHC	xe*	CE
85028	FGHC	xe	CE
85030	ICCA	xe*	CE
85031	FGHC	xe*	CE
85034	FGHC	xe*	CE
85037	ICCA	xe*	CE
85040	FGHC	xe*	CE

Class 85/1. Max Speed: 75 mph. Train heating jumpers removed.

85101	(85006)	FGRC	xy	CE
85102	(85009)	FGRC	xy	CE
85103	(85010)	FGRC	xy	CE
85104	(85012)	FGRC	xy	CE
85105	(85016)	FGRC	xy	CE
85106	(85021)	FGRC	xy	CE
85107	(85024)	FGRC	xy	CE (U)
85108	(85032)	FGRC	xy	CE
85109	(85035)	FGRC	xy	CE
85110	(85036)	FGRC	xy	CE
85111	(85004)	FGRC	xy	CE

CLASS 86 BR DESIGN Bo–Bo

Built: 1965–66 by English Electric Co. at Vulcan Foundry, Newton le Willows or BR at Doncaster Works.
Supply System: 25 kV a.c. from overhead equipment.
Brake Force: 40 t.
Weight: 83–86t
ETH Index: 74
Length over Buffers: 17.83 m.
Wheel Diameter: 1156 mm.
RA: 6.

Electric Brake: Rheostatic.

All locos are equipped with driver–guard telephone.
Formerly numbered E 3101–3200 not in order.

Class 86/1. Rebuilt with class 87 type bogies and motors.

Traction Motors: GEC G412AZ frame mounted.
Max. Tractive Effort: 258 kN (58000 lbf)
Continuous Rating: 3730 kW (5000 hp) giving a tractive effort of 95 kN (21300 lbf) at 87 mph.
Maximum Rail Power: 5860 kW (7860 hp) at ?? mph.
Weight: 85 t.
Max. Speed: 110 mph.

86101 (86201)	**M**	IWCA	xec	WN	Sir William A Stanier FRS
86102 (86202)	**IO**	IWCA	xec	WN	Robert A Riddles
86103 (86203)	**M**	IWCA	xec	WN	André Chapelon

Class 86/2.

Traction Motors: AEI 282BZ.
Max. Tractive Effort: 207 kN (46500 lbf).
Continuous Rating: 3010 kW (4040 hp) giving a tractive effort of 85 kN (19200 lbf) at 77.5 mph.
Maximum Rail Power: 4550 kW (6100 hp) at 49.5 mph.
Weight: 85–86 t.
Max. Speed: 100 (§110) mph.
Multiple Working: With other tdm fitted vehicles (p fitted locos only).

86204	**IO**	IEBA	xecp	WN	City of Carlisle
86205 (86503)	**I**	IWCA	xec	WN	City of Lancaster
86206	**M**	ICCA	xec	WN	City of Stoke on Trent
86207	**IO**	ICCA	xecp	WN	City of Lichfield
86208	**M**	IEBA	xecp	WN	City of Chester
86209	**I**	IEBA	xecp§	WN	City of Coventry
86210	**M**	IEBA	xecp	WN	City of Edinburgh
86212	**IO**	ICHA	xecp	WN	Preston Guild
86213	**IO**	IEBA	xecp	WN	Lancashire Witch
86214	**M**	IANA	xe	IL	Sans Pareil
86215	**IO**	IANA	xec	IL	Joseph Chamberlain
86216	**I**	IANA	xe	IL	Meteor
86217 (86504)	**IO**	IWCA	xecp	WN	Halley's Comet
86218	**IO**	IANA	xe	IL	Planet
86219	**M**	IEBA	xecp	WN	Phoenix
86220	**M**	IANA	xe	IL	The Round Tabler
86221	**M**	IANA	xe	IL	B.B.C. Look East
86222 (86502)	**I**	ICCA	xec	WN	LLOYD'S LIST 250th ANNIVERSARY
86223	**IO**	IANA	xe	IL	Norwich Union
86224	**IO**	IEBA	xecp§	WN	Caledonian
86225	**IO**	IEBA	xecp§	WN	Hardwicke
86226	**M**	IWCA	xep	WN	Royal Mail Midlands
86227	**IO**	IANA	xe	IL	Sir Henry Johnson
86228	**M**	ICCA	xecp	WN	Vulcan Heritage

86229		**M**	RXLE	xec	WN	Sir John Betjeman
86230		**M**	IANA	xep	IL	The Duke of Wellington
86231		**M**	ICCA	xecp§	WN	Starlight Express
86232		**M**	IANA	xec	IL	Harold Macmillan
86233	(86506)	**M**	RXLE	xecp	WN	Laurence Olivier
86234		**M**	RXLE	xecp	WN	J B Priestley OM
86235		**IO**	IANA	xe	IL	Novelty
86236		**M**	IEBA	xecp	WN	Josiah Wedgwood MASTER POTTER 1736–1795
86237		**M**	IWCA	xecp	WN	Sir Charles Hallé
86238		**M**	IANA	xe	IL	European Community
86239	(86507)	**M**	RXLE	xecp	WN	L S Lowry
86240		**M**	ICCA	xecp	WN	Bishop Eric Treacy
86241	(86508)	**M**	RXLE	xecp	WN	Glenfiddich
86242		**M**	IEBA	xecp	WN	James Kennedy GC
86243		**IO**	RXLE	xecp	WN	The Boys' Brigade
86244		**IO**	IANA	xec	IL	The Royal British Legion
86245		**IO**	RXLE	xecp	WN	Dudley Castle
86246	(86505)	**M**	IANA	xec	IL	Royal Anglian Regiment
86247		**IO**	IEBA	xecp	WN	Abraham Darby
86248		**M**	IWCA	xecp	WN	Sir Clwyd/County of Clwyd
86249		**M**	IEBA	xecp	WN	County of Merseyside
86250		**M**	ICCA	xecp	WN	The Glasgow Herald
86251		**IO**	IWCA	xec	WN	The Birmingham Post
86252		**IO**	IWCA	xecp	WN	The Liverpool Daily Post
86253	(86044)	**M**	IEBA	xecp	WN	The Manchester Guardian
86254	(86047)	**M**	RXLE	xec	WN	William Webb Ellis
86255	(86042)	**M**	ICCA	xecp	WN	Penrith Beacon
86256	(86040)	**I**	ICHA	xecp	WN	Pebble Mill
86257	(86043)	**M**	IEBA	xecp	WN	Snowdon
86258	(86501)	**IO**	ICCA	xec	WN	Talyllyn The First Preserved Railway
86259	(86045)	**M**	IANA	xec	IL	Peter Pan
86260	(86048)	**M**	ICCA	xep	WN	Driver Wallace Oakes G.C.
86261	(86041)	**M**	RXLE	xep	WN	Driver John Axon G.C.

Class 86/4.

Traction Motors: AEI 282AZ.
Max. Tractive Effort: 258 kN (58000 lbf).
Continuous Rating: 2680 kW (3600 hp) giving a tractive effort of 89 kN (20000 lbf) at 67 mph.
Maximum Rail Power: 4400 kW (5900 hp) at 38 mph.
Weight: 83–84 t.
Max. Speed: 100 mph.
Multiple Working: With other tdm fitted vehicles (p fitted locos). All class 86/4 and 86/6 locos are also fitted with an older system of multiple working and can work in multiple with each other and with class 87/0.

86401	(86001)	**N**	RXLE	xec	CE	Northampton Town
86403	(86003)	**M**	FGLC	xec	CE	
86404	(86004)	**M**	FGLC	xec	CE	
86405	(86005)	**M**	FGLC	xec	CE	

86406	(86006)	**M**	FGLC	xec	CE	
86410	(86010)	**M**	FGLC	xec	CE	
86411	(86311)	**IO**	FGLC	xec	CE	Airey Neave
86412	(86312)	**M**	FGLC	xec	CE	Elizabeth Garrett Anderson
86414	(86314)	**M**	FGLC	xe	CE	Frank Hornby
86415	(86315)	**M**	FGLC	xec	CE	Rotary International
86416	(86316)	**M**	ICCA	xec	CE	Wigan Pier
86417	(86317)	**IO**	RXLE	xec	CE	The Kingsman
86418	(86318)	**IO**	FGLC	xec	CE	
86419	(86319)	**M**	ICCA	xec	CE	
86422	(86322)	**M**	FGLC	xec	CE	
86423	(86323)	**M**	FGLC	xec	CE	
86424	(86324)	**M**	RXLE	xecp	CE	
86425	(86325)	**M**	IWCA	xecp	CE	
86426	(86326)	**M**	RXLE	xecp	CE	
86428	(86328)	**M**	FGLC	xec	CE	Aldaniti
86430	(86030)	**M**	ICCA	xe	CE	Scottish National Orchestra
86431	(86031)	**M**	FGLC	xep	CE	
86434	(86034)	**M**	FGLC	xec	CE	University of London
86437	(86037)	**M**	FGLC	xec	CE	
86438	(86038)	**M**	FGLC	xec	CE	
86439	(86039)	**M**	FGLC	xec	CE	

Class 86/6. Being renumbered from 86/4. Dedicated for Railfreight use.

Details as for class 86/4 except:

Max Speed: 75 (100) mph.

86601	(86401)					
86602	(86402)	**F**	FGPC	xyc	CE	
86603	(86403)					
86604	(86404)					
86605	(86405)					
86606	(86406)					
86607	(86407)	**M**	FGPC	xyc	CE	The Institution of Electrical Engineers
86608	(86408)	**M**	FGPC	xy	CE	St. John Ambulance
86609	(86409)	**M**	FGPC	xyc	CE	
86610	(86410)					
86611	(86411)					
86612	(86412)					
86613	(86413)	**F**	FGPC	xyc	CE	County of Lancashire
86614	(86414)					
86615	(86415)					
86616	(86416)					
86617	(86417)					
86618	(86418)					
86619	(86419)					
86620	(86420)	**M**	FGPC	xy	CE	
86621	(86421)	**M**	FGPC	xyc	CE	London School of Economics
86622	(86422)					
86623	(86423)					
86624	(86424)					

86625	(86425)					
86626	(86426)					
86627	(86427)	**F**	FGPC	xec	CE	The Industrial Society
86628	(86428)					
86630	(86430)					
86631	(86431)					
86632	(86432)	**M**	FGPC	xyc	CE	Brookside
86633	(86433)	**M**	FGPC	xyc	CE	Wulfruna
86634	(86434)					
86635	(86435)	**M**	FGPC	xyc	CE	
86636	(86436)	**M**	FGPC	xyc	CE	
86637	(86437)					
86638	(86438)					
86639	(86439)					

CLASS 87 BR DESIGN Bo–Bo

Built: 1973–75 by BREL at Crewe Works. Class 87/1 has thyristor control instead of HT tap changing.
Supply System: 25 kV a.c. from overhead equipment.
Traction Motors: GEC G412AZ frame mounted (87/0), G412BZ (87/1).
Max. Tractive Effort: 258 kN (58000 lbf).
Continuous Rating: 3730 kW (5000 hp) giving a tractive effort of 95 kN (21300 lbf) at 87 mph (Class 87/0), 3620 kW (4850 hp) giving a tractive effort of 96 kN (21600 lbf) at 84 mph (Class 87/1).
Maximum Rail Power: 5860 kW (7860 hp) at ?? mph.
Brake Force: 40 t. **Length over Buffers**: 17.83 m.
Weight: 83t **Wheel Diameter**: 1150 mm.
ETH Index: 74 **RA**: 6.
Electric Brake: Rheostatic.
Multiple Working: With other tdm fitted vehicles (p fitted locos). All class 86/4 and 86/6 locos are also fitted with an older system of multiple e working and can work in multiple with each other and with class 87/0.
Max. Speed: 110 mph. (87/0), 100 mph (87/1).
All locos are equipped with driver–guard telephone.

Class 87/0. Standard Design.

87001	**I0**	IWCA	aec	WN	Royal Scot
87002	**I0**	IWCA	aec	WN	Royal Sovereign
87003	**M**	!WCA	aecp	WN	Patriot
87004	**M**	IWCA	aecp	WN	Britannia
87005	**M**	IWCA	aec	WN	City of London
87006	**I0**	IWCA	aec	WN	City of Glasgow
87007	**M**	IWCA	aecp	WN	City of Manchester
87008	**M**	IWCA	aec	WN	City of Liverpool
87009	**I**	ICCA	aecp	WN	City of Birmingham
87010	**M**	ICCA	aecp	WN	King Arthur
87011	**M**	IWCA	aec	WN	The Black Prince
87012	**M**	IWCA	aecp	WN	The Royal Bank of Scotland
87013	**I0**	IWCA	aecp	WN	John O' Gaunt
87014	**I0**	IWCA	aec	WN	Knight of the Thistle
87015	**I0**	IWCA	aecp	WN	Howard of Effingham

87016	**M**	IWCA	aec	WN	
87017	**IO**	IWCA	aec	WN	Iron Duke
87018	**M**	IWCA	aec	WN	Lord Nelson
87019	**IO**	IWCA	aec	WN	Sir Winston Churchill
87020	**IO**	IWCA	aec	WN	North Briton
87021	**IO**	IWCA	aecp	WN	Robert the Bruce
87022	**M**	IWCA	aec	WN	Cock o' the North
87023	**M**	IWCA	aecp	WN	Velocity
87024	**IO**	IWCA	aec	WN	Lord of the Isles
87025	**IO**	IWCA	aec	WN	County of Cheshire
87026	**IO**	IWCA	aecp	WN	Sir Richard Arkwright
87027	**IO**	IWCA	aec	WN	Wolf of Badenoch
87028	**M**	IWCA	aec	WN	Lord President
87029	**IO**	IWCA	aecp	WN	Earl Marischal
87030	**IO**	IWCA	aec	WN	Black Douglas
87031	**M**	ICCA	aec	WN	Hal o' the Wynd
87032	**IO**	ICCA	aec	WN	Kenilworth
87033	**M**	IWCA	aecp	WN	Thane of Fife
87034	**M**	IWCA	aec	WN	William Shakespeare
87035	**M**	IWCA	aec	WN	Robert Burns

Class 87/1. Thyristor Control. Push & pull fitted.

87101		FGHC	ae	CE	STEPHENSON

CLASS 89 BRUSH DESIGN Co–Co

Built: 1986 by BREL at Crewe Works (under sub-contract for Brush Traction, Loughborough).
Supply System: 25 kV a.c. from overhead equipment.
Traction Motors: Brush TM2201A.
Max. Tractive Effort: 205 kN (46100 lbf).
Continuous Rating: 4350 kW (5850 hp) giving a tractive effort of 105 kN (23540 lbf) at 92 mph.
Maximum Rail Power: 5860 kW (7860 hp) at ?? mph.
Brake Force: 50 t.
Weight: 104 t.
ETH Index: 95.
Electric Brake: Rheostatic.
Length over Buffers: 17.83 m.
Wheel Diameter: 1150 mm.
RA: 6.
Max. Speed: 125 mph.

Equipped with driver–guard telephone.

89001	**I**	IECA	aecp	BN	Avocet

CLASS 90 GEC DESIGN Bo–Bo

Built: 1987–90 by BREL at Crewe Works. Thyristor control.
Supply System: 25 kV a.c. from overhead equipment.
Traction Motors: GEC G412BZ separately excited frame mounted.
Max. Tractive Effort: 192 kN (43150 lbf).
Continuous Rating: 3730 kW (5000 hp) giving a tractive effort of 95 kN (21300 lbf) at 87 mph.
Maximum Rail Power: 5860 kW (7860 hp) at ?? mph.
Brake Force: 40 t.
Length over Buffers: 18.90 m.

Weight: 83 t. **Wheel Diameter**: 1150 mm.
ETH Index: 95. **RA**: 7.
Electric Brake: Rheostatic. **Max. Speed**: 110 mph.
Multiple Working: With other tdm fitted vehicles.

All locos are equipped with driver–guard telephone.

90001	**I**	IPSA	aecp	WN	BBC Midlands Today
90002	**I**	IWCA	aec	WN	
90003	**I**	IWCA	aecp	WN	
90004	**I**	IWCA	aec	WN	
90005	**I**	IWCA	aecp	WN	Financial Times
90006	**I**	IWCA	aec	WN	
90007	**I**	IWCA	aec	WN	
90008	**I**	IPSA	aecp	WN	
90009	**I**	IPSA	aecp	WN	Royal Show
90010	**I**	IWCA	aecp	WN	275 Railway Squadron (Volunteers)
90011	**I**	IWCA	aec	WN	The Chartered Institute of Transport
90012	**I**	IPSA	aecp	WN	
90013	**I**	IWCA	aec	WN	
90014	**I**	IWCA	aecp	WN	
90015	**I**	IPSA	aecp	WN	BBC North West
90016	**I**	FGQW	aecp	WN	
90017	**I**	FGQW	aec	WN	
90018	**I**	FGQW	aecp	WN	
90019	**I**	RXLE	aecp	WN	
90020	**I**	RXLE	aecp	WN	
90021	**I**	FGQW	aecp	WN	
90022	**I**	FGQW	aecp	WN	
90023	**I**	FGQW	aecp	WN	
90024	**I**	FGQW	aecp	WN	
90025	**I**	FGQW	aecp	WN	
90026	**M**	FGQW	aecp	WN	
90027	**M**	FGQW	aecp	WN	
90028	**M**	FGQW	aecp	WN	
90029	**M**	FGQW	aecp	WN	
90030	**M**	FGQW	aecp	WN	
90031	**M**	FGQW	aecp	WN	
90032	**M**	FGQW	aecp	WN	
90033	**M**	FGQW	aecp	WN	
90034	**M**	FGQW	aecp	WN	
90035	**M**	FGQW	aecp	WN	
90036	**M**	FGYX	aec		
90037	**F**	FGYX	aec		
90038	**F**	FGYX	aec		
90039	**F**	FGYX	aec		
90040	**F**	FGYX	aec		
90041		FGYX	aec		
90042		FGYX	aec		
90043		FGYX	aec		
90044		FGYX	aec		
90045		FGYX	aec		
90046		FGYX	aec		

90047	FGYX	aec
90048	FGYX	aec
90049	RXXA	aec
90050	RXXA	aec

CLASS 91 GEC DESIGN Bo–Bo

Built: 1988 onwards by BREL/GEC at Crewe Works. Thyristor control.
Supply System: 25 kV a.c. from overhead equipment.
Traction Motors: GEC G426AZ.
Max. Tractive Effort: .
Continuous Rating: 4540 kW (6090 hp).
Maximum Rail Power: 4700 kW (6300 hp).
Brake Force: 45 t. **Length over Buffers**: 19.40 m.
Weight: 82 t. **Wheel Diameter**: 1000 mm.
ETH Index: 95. **RA**: 7.
Electric Brake: Rheostatic.
Multiple Working: With other tdm fitted vehicles.
Max. Speed: 125 (140) mph.
All locos are equipped with driver–guard telephone.

91001 **I**	IECA	aecp	BN	Swallow
91002 **I**	IECA	aecp	BN	
91003 **I**	IECA	aecp	BN	
91004 **I**	IECA	aecp	BN	The Red Arrows
91005 **I**	IECA	aecp	BN	
91006 **I**	IECA	aecp	BN	
91007 **I**	IECA	aecp	BN	
91008 **I**	IECA	aecp	BN	
91009 **I**	IECA	aecp	BN	
91010 **I**	IECA	aecp	BN	
91011		aecp		
91012		aecp		
91013		aecp		
91014		aecp		
91015		aecp		
91016		aecp		
91017		aecp		
91018		aecp		
91019		aecp		
91020		aecp		
91021		aecp		
91022		aecp		
91023		aecp		
91024		aecp		
91025		aecp		
91026		aecp		
91027		aecp		
91028		aecp		
91029		aecp		
91030		aecp		
91031		aecp		

1.3. DEPARTMENTAL LOCOMOTIVES

CLASS 97/2 BR TYPE 2 Bo–Bo

Formerly class 25. Now used as electric heating units.
Built: 1966 by Beyer–Peacock (97251) or BR Derby Loco. (97252).
Engine: Sulzer 6LDA28-B of 930 kW (1250 hp) at 750 rpm.
Main Generator: AEI RTB15656.
Traction Motors: AEI 253AY (isolated).
Max. Tractive Effort: 200 kN (45000 lbf).
Cont. Tractive Effort: 93 kN (20800 lbf) at 17.1 mph.
Power At Rail: 708 kW (949 hp).
Brake Force: 38 t.
Weight: 72 t.
Max. Speed: 90 mph.
Length over Buffers: 15.39 m.
Wheel Diameter: 1143 mm.
RA: 5.
Multiple Working: Blue Star coupling code.

97251 (25305) **M** ICHA xe CL
97252 (25314) **M** ICHA xe OC

CLASS 97/4 BR TYPE 4 1 Co–Co 1

Built: 1962 by BR at Derby Loco Works.
Engine: Sulzer 12LDA28-B of 1860 kW (2500 hp) at 750 rpm.
Main Generator: Brush TG160-60.
Traction Motors: Brush TM73-68 MkIII.
Max. Tractive Effort: 245 kN (55000 lbf).
Cont. Tractive Effort: 141 kN (31600 lbf) at 23.3 mph.
Power At Rail: 1460 kW (1960 hp).
Brake Force: 63 t.
Weight: 141 t.
Max. Speed: 90 mph.
Length over Buffers: 20.70 m.
Wheel Diameters: 1143 mm (914 mm pony).
RA: 7.
Multiple Working: Blue Star coupling code. (97403 only).
Non-standard Livery: RTC blue/red.

97403 (46035) **0** DRTC xo RTC Ixion
97404 (46045) DRTC xo RTC (U)

CLASS 97/6 RUSTON SHUNTER 0–6–0

Built: 1959 by Ruston & Hornsby at Lincoln.
Engine: Ruston 6VPH of 123 kW (165 hp).
Main Generator: British Thomson Houston RTB6034.
Traction Motor: One British Thomson Houston RTA5041.
Max. Tractive Effort: 75 kN (17000 lbf).
Brake Force: 16 t.
Weight: 31 t.
Max. Speed: 20 mph.
Length over Buffers: 7.62 m.
Wheel Diameter: 978 mm.
RA: 1.
Non-Standard Livery: Departmental Yellow.

97651 (PWM 651)	**0**	DCWC	vo	CF
97653 (PWM 653)	**0**	DCWC	vo	CF
97654 (PWM 654)	**0**	DCWC	vo	RG

CLASS 97/7 BATTERY LOCOS Bo–Bo

Built: 1973–80 by BREL at Doncaster and Wolverton Works. Converted from class 501 e.m.u. cars.
Supply System: 750 volts dc third rail or 320 volts dc batteries.
Traction Motors: GEC WT344A.
Max. Tractive Effort: 73 kN (16400 lbf).
Brake Force: 45 t.
Weight: 59 t.
Max. Speed: 25 mph.
Length over Buffers: 18.44 m.
Wheel Diameter: 1071 mm.
RA: 4.
Multiple Working: Work in pairs.
Non-Standard Livery: Bright blue with yellow stripe.

97701	(61136)	**0**	DMEC	vo	BD
97702	(61139)	**0**	DMEC	vo	BD
97703	(61182)		DMEC	vo	HE
97704	(61185)		DMEC	vo	HE
97705	(61184)		DMEC	vo	HE
97706	(61189)		DMEC	vo	HE
97707	(61166)	**N**	DMEC	vo	HE
97708	(61173)	**N**	DMEC	vo	HE
97709	(61172)		DMEC	vo	HE (S)
97710	(61175)		DMEC	vo	HE (S)

CLASS 97/8 SHUNTER 0–6–0

For details see class 09. Severn Tunnel emergency train locomotive).

Non-Standard Livery: Grey cab.

97806 (09017) **0** DOPA xo CF

DB 968xxx SERIES

This number series was introduced in 1969 and is for former capital stock locomotives which no longer operate under their own power.

ADB 968021 (84009)	x	CE	Mobile load bank.
ADB 968026 (25908)	x	TO	Training loco.
TDB 968030 (33018)	x		Moreton-in-Marsh Training loco.

OTHER DEPARTMENTAL LOCOS.

These locos are in departmental service, but are not allocated departmental numbers.

D2991 Stationary generator at Eastleigh Major Depot.
20001 TO. Training loco.
D 7672(25912) TAMWORTH CASTLE. Holbeck Training loco.
47403 Ministry of Defence. Moreton-on-Lugg.
45111 RTC. Dead load.
45125 RTC. Dead load.
45126 RTC. Dead load.
45148 RTC. Dead load.

1.4. LOCOMOTIVES AWAITING DISPOSAL

Codes used in this section:

(1) Closed depots and depots without allocation (official codes).

CW	Cricklewood
HM	Healey Mills
MG	Margam Yard
MR	March

(2) Unofficial codes.

Ba	Basford Hall Yard, Crewe
KM	Kingmoor Yard, Carlisle
LN	Holmes Yard, Lincoln
RFS	RFS Industries Ltd, Doncaster
ST	Stanton Gate

03073 CH
03084 MR
03158 MR
03189 MR
08141 TI
08200 TE
08202 CF
08222 BN
08224 DR
08239 NL
08242 LN
08250 SR
08254 GD
08285 DR
08295 TE
08305 KY
08334 TE
08354 CF
08367 DR
08385 DR
08386 LN
08406 MR
08408 SR
08412 MR
08420 DR
08427 MR
08438 MR
08439 IM
08458 SR
08461 SR
08463 RFS
08468 SP
08477 AN
08478 IM
08486 CF
08487 SR
08488 CF
08491 CF
08494 SR
08501 SR
08504 TE
08518 MR
08549 MR
08579 HM
08608 GD
08626 AN
08640 OC
08671 GD
08763 HA
08769 MG
08779 SR
08785 CF
08852 ZH
08943 ZC
20002 ZF
20005 ST
20008 TE
20009 TE
20011 ZE
20022 ZF
20044 TO
20054 ST
20068 IM
20077 ZF
20086 ST
20089 IM
20097 ST
20100 ST
20126 TI
20134 TO
20147 ST
20150 ZF
20158 ST
20161 ZF
20171 TO
20178 TI
20179 ST
20180 ZF
20181 BS
20192 TO
20193 ST
20201 ST
20202 TO
20203 ML
20205 IM
20217 ST
20218 TO
20220 KM
20221 KM
20223 KM
20226 TI
25080 Ba
25194 BS
25205 BS
25206 BS
25209 TO
25211 BS
25235 BS
25259 BS
26029 ZH
26031 ED
26034 ED
31131 TE
31152 SR
31173 SR
31222 TE
31225 ZF
31226 SF
31259 ZF
31280 SR
31281 TE
31283 TE
31284 ZF
31320 TE
31323 SR
31443 ZF
33003 EH
33005 SA
33011 EH
33013 ZG
33015 EH
33016 EH
33017 EH
33022 AF
33031 ZG
33034 EH
33035 SL
33037 EH
33038 SR
33039 ZG
33043 EH
33044 SA
33045 SA
33055 AF
33059 ZG
33062 SA

33105 SA
33107 ZG
33112 ZG
33115 RFS
33119 EH
33209 ZG
33210 EH
33212 EH
37062 ZF
37260 ZF
45007 TI
45012 TI
45013 MR
45015 TO
45017 TI
45033 TI
45037 TI
45041 TE
45046 TI
45052 TI
45058 MR
45062 MR
45076 MR
45103 TI
45104 TI
45105 MR
45106 TI
45107 TI
45108 MR
45110 TI
45112 MR
45113 TI
45114 MR
45115 TI
45118 MR
45119 MR
45120 TI
45121 DY
45122 MR
45124 TI
45127 MR
45128 TI
45130 TI
45132 MR
45133 MR
45134 TI
45136 DY
45137 MR
45138 MR
45139 MR
45140 TI
45142 MR
45143 MR
45145 TI
45149 CW
45150 TI
46010 DR
46023 Ba
47001 Ba
47008 SR
47011 Ba
47012 TI
47015 Ba
47093 TI
47101 CD
47104 TI
47110 TE
47113 SR
47124 TI
47130 TI
47137 ED
47143 DR
47159 TE
47189 ZF
47191 SP
47198 CF
47199 CL
47202 BR
47203 ZF
47235 ZF
47342 ZF
47404 MR
47410 MR
47411 IM
47412 TI
47415 MR
47419 MR
47428 CD
47542 SR
47713 ZF
50010 LA
50011 ZC
50039 OC
81003 CE
81005 CE
81006 CE
81007 CE
81008 CE
81011 CE
81013 CE
81014 CE
81015 CE
81018 CE
81019 CE
81020 CE
81021 CE
82003 CE
82005 CE
82008 CE
83009 CE
83012 CE
83015 CE
85002 CE
85014 CE
85017 CE
85019 CE
85022 CE
85029 CE
97250 OC
97409 TI
97410 TI
97411 TI
97412 TI
97413 TI
97650 RG
97652 LA
968000 ME
968028 ED

Notes:

Locos shown as MR may be in Whitemoor Yard.
Locos shown as TI may be in Tinsley Yard.

1.5. PRESERVED LOCOMOTIVES

The following is a list of preserved locomotives (all steam) which may be used on BR main lines.

Old No	*BR No.*	*Name*	*Wheels*	*Location*
3717	3440	CITY OF TRURO	4–4–0	National Railway Museum
3822	3822		2–8–0	Severn Valley Railway
5080	5080	DEFIANT	4–6–0	Birmingham Rly. Museum
6000	6000	KING GEORGE V	4–6–0	Bulmers R.C., Hereford
6024	6024	KING EDWARD I	4–6–0	Buckinghamshire Railway Centre
6960	6960	RAVENINGHAM HALL	4–6–0	Severn Valley Railway
6998	6998	BURTON AGNES HALL	4–6–0	Didcot Railway Centre
7029	7029	CLUN CASTLE	4–6–0	Birmingham Rly. Museum
7754	7752		0–6–0PT	Birmingham Rly. Museum
7819	7819	HINTON MANOR	2–6–0	Severn Valley Railway
777	30777	SIR LAMIEL	4–6–0	HLPG, Hull Dairycoates
850	30850	LORD NELSON	4–6–0	Steamtown, Carnforth
34092	34092	CITY OF WELLS	4–6–2	Keighley & Worth Valley Rly.
34027	34027	TAW VALLEY	4–6–2	Severn Valley Railway
35028	35028	CLAN LINE	4–6–2	Bulmers R.C., Hereford
43106	43106		4–6–0	Severn Valley Railway
4767	44767		4–6–0	North Yorks Moors Rly.
44871	44871		4–6–0	Steamtown, Carnforth
44932	44932		4–6–0	Midland Railway Centre
5305	45305		4–6–0	HLPG, Hull Dairycoates
5407	45407		4–6–0	Steamtown, Carnforth
45596	45596	BAHAMAS	4–6–0	Dinting Rly. Centre
6201	46201	PRINCESS ELIZABETH	4–6–2	Bulmers R.C., Hereford
6203	46203	PRINCESS MARGARET ROSE	4–6–2	Midland Railway Centre
46229	46229	DUCHESS OF HAMILTON	4–6–2	National Railway Museum
48151	48151		2–8–0	Midland Railway Centre
53809	53809		2–8–0	Midland Railway Centre
4498	60007	SIR NIGEL GRESLEY	4–6–2	Steamtown, Carnforth
60009	60009	OSPREY	4–6–2	
4468	60022	MALLARD	4–6–2	National Railway Museum
4472	60103	FLYING SCOTSMAN	4–6–2	Steamtown, Carnforth
60532	60532	BLUE PETER	4–6–2	North Yorkshire Moors
4771	60800	GREEN ARROW	2–6–2	National Railway Museum
3442	61994	THE GREAT MARQUESS	2–6–2	Severn Valley Railway
62005	62005		2–6–0	North Yorks Moors Rly.
673	65243	MAUDE	0–6–0	SRPS Falkirk
70000	70000	BRITANNIA	4–6–2	Nene Valley Railway
71000	71000	DUKE OF GLOUCESTER	4–6–2	Midland Railway Centre
75069	75069		4–6–0	Severn Valley Railway
80080	80080		2–6–4T	Midland Railway Centre
92220	92220	EVENING STAR	2–10–0	National Railway Museum

Note: The above list is only a guide, as locos are being added to and taken off the BR approved list, and certain locos are still under renovation.

2. LOCO-HAULED PASSENGER STOCK

INTRODUCTION

This section contains full details and differences of all BR locomotive-hauled coaching stock. Coaches are listed in batches, according to their class, with lot number information for the various batches being shown above the listings. Where a coach has been renumbered, the former number is shown in parentheses. If the coach has been renumbered more than once, the original number is shown in parentheses, with the intermediate numbers being given in the text. Where the old number of a coach due to be converted or renumbered e.g. 3401, is known and the conversion or renumbering has not yet taken place, the coach is listed both under its old number with its depot allocation, and under its new number without an allocation. Thanks are due to all who have helped by contributing information for this book especially the staff of the BR Rolling Stock Library.

NUMBERING SYSTEMS

Six different numbering systems were in use on BR. These were the BR series, the four pre-nationalisation companies' series' and the Pullman Car Company's series. Only BR number series loco-hauled coaches now exist.

All coaching stock vehicles have now been given depot allocations, regional prefixes being phased out.

DETAILED INFORMATION

After the heading, the following details are shown:

(1) Diagram number. This consists of the first three characters of the TOPS code followed by two numbers which relate to the particular design of vehicle.

(2) 'Mark' of coach (see below).

(3) Number of first class seats , second class seats and lavatory compartments shown as nF nS nL respectively.

(4) Bogie type (see below).

(5) Brake type. Codes are:

a Air braked.
b Air braked with through vacuum pipe.
v Vacuum braked.
x Dual braked (air and vacuum).
w Vacuum braked with through air pipe.

(6) Heating type. Codes are:

d Dual heated (steam & electric).
e electric heated
o No heating apparatus.
p Steam piped.
q Steam piped & electric heated.
r Steam piped & electric wired.
s Steam heated.
u UIC/BR electric heat.
y Electric wired.

(7) Additional features. Codes are:

f Facelifted or fluorescent lighting provided.
k Composition brake blocks (instead of cast iron).
n Fitted with day/night lighting for use on "Nightrider services"
pa Public address speakers installed.
pg Public address transmission and driver–guard communication.
pt Public address transmission and speakers.
r Refurbished vehicle.
to Public address transmission only.
t Fitted with public telephone.
w Fitted with wheelchair space & disabled persons' toilet.
z Fitted with disabled persons' toilet.

(8) ETH Index. The sum of ETH indices in a train must not be more than that of the locomotive. Suffix 'S' denotes SR 750 V heaters, the usual voltage on BR being 1000, and suffix 'X' denotes 600 amp wiring instead of 400 amp. Trains whose ETH index comes to more than 66 must be formed completely with 600 amp wired stock.

TOPS CODES

TOPS (Total operations processing system) codes have now been allocated to all coaching stock. For passenger stock the code consists of:

(1) Two letters denoting the layout of the vehicle as follows:

AA Gangwayed Corridor
AB Gangwayed Corridor Brake
AC Gangwayed Open (2+2 seating)
AD Gangwayed Open (2+1 seating)
AE Gangwayed Open Brake
AF Gangwayed Driving Open Brake
AG Micro-Buffet
AH Brake Micro-Buffet
AI As 'AC' but with end equipment.
AJ Restaurant Buffet with Kitchen
AL As 'AC' but with disabled toilet
AN Miniature Buffet
AO Private Owner
AP Pullman Kitchen
AQ Pullman
AR Pullman Brake
AS Sleeping Car
AT Royal Train Coach
AU Sleeping Car with Pantry

(2) A digit for the class of passenger accommodation (1 First, 2 Second, 3 Composite, 4 Unclassified, 5 None).

(3) A suffix relating to the build of coach.

1 Mark 1	A Mark 2A	C Mark 2C	E Mark 2E	G Mark 3 or	H Mark 3B
Z Mark 2	B Mark 2B	D Mark 2D	F Mark 2F	Mark 3A	J Mark 4

For non-passenger carrying coaching stock, the suffix denotes the brake type:

A Air braked
B Air braked with through vacuum pipe
V Vacuum braked
W Vacuum braked with through air pipe
X Dual braked

OPERATOR CODES

The normal operator codes are given in brackets after the TOPS codes. These are as follows:

F First
S Second
C Composite

B Brake
O Open
K Side corridor with lavatory

Various other letters are in use and the meaning of these can be ascertained by referring to the titles at the head of each class.

BOGIE TYPES

BR Mk 1 (BR1). Standard double bolster leaf spring bogie. Generally 90 m.p.h. but certain vehicles were allowed to run at 100 m.p.h. with special maintenance. Weight: 6.1t.
BR Mk 1 (heavy duty). Similar to above. Weight 6.5t.
BR Mk 2 (BR2). Later variant of BR Mark 1 used on certain vans. Single bolster. Leaf spring bogie fitted to one batch of SKs. Weight: 5.3t.
COMMONWEALTH (C). Heavy, cast steel soil spring bogie. 100 m.p.h. Weight: 6.75t.
B4. Coil spring fabricated bogie for 100 m.p.h. Certain BGs (classified NHA) are allowed to run at 110 m.p.h. with special maintenance. Weight: 5.2t.
B5. Heavy duty version of B4. 100 m.p.h. Weight: 5.3t.
BT10. A fabricated bogie designed for 125 m.p.h. Air suspension.
BT15. Modification of BT10, but with rubber primary suspension instead of coil springs.
BT7. Similar to BX1 e.m.u. bogie.
T4. The latest 125 m.p.h. bogie from BREL.
The new Mark 4 vehicles are fitted with a bogie from the Swiss firm of SIG.

THE DEVELOPMENT OF BR STANDARD COACHES

The standard BR coach built from 1951 to 1963 is the mark 1. This has a separate underframe and body. The underframe is normally 64′6″ long, but certain vehicles were built on short (57′) frames. Tungsten lighting is standard and until 1961, BR mark 1 bogies were generally provided. In 1959 TSOs to lot No. 30525 appeared with fluorescent lighting and melamine interior panels and from 1961 onwards Commonwealth bogies were fitted in an attempt to improve the quality of ride which became very poor when the tyre profiles on the wheels of the Mark 1 bogies became worn. The further batches of TSOs and BSOs retained the features of lot 30525, but the BSKs, SKs, BCKs and CKs, whilst utilising melamine panelling in second class, still retained tungsten lighting. Wooden interior finish was retained in first class compartments. The FOs had fluorescent lighting with wooden panelling except for lot No. 30648 which had tungsten lighting. In later years many mark 1s had their mark 1 bogies replaced by B4s.

In 1964, a new train was introduced. Known as "XP64", it featured new seat designs, pressure ventilation, aluminium compartment doors and corridor partitions, foot pedal operated toilets, and B4 bogies. The vehicles were on standard mark 1 underframes. Folding doors were fitted but these proved troublesome and were later replaced with hinged doors. All XP64 coaches have now been withdrawn, but some have been preserved.

The prototype mark 2 vehicle (E 13252) was produced in 1963. This was an FK of semi-integral construction and was pressure ventilated. Tungsten lighting as provided and B4 bogies. This vehicle has been preserved by the National Railway Museum. The production build was similar, but wider windows were used. The second class open vehicles used the new seat design similar to that in the XP64 and fluorescent lighting was provided. Interior finish reverted to wood. MK 2s were built from 1964–66.

The mark 2As, built 1967–68, incorporated the rest of the novel features first used in the XP64 set, i.e. foot pedal operated toilets (except BSOs), new first class seat design, aluminium compartment doors and partitions together with fluorescent lighting in first class compartments. Folding gangway doors (lime green coloured) were used instead of the traditional variety. The following list summarises the changes made in the later Mk 2 variants:

Mk 2B: Wide wrap round doors, no centre doors, slightly longer body. In second class, one toilet at each end instead of two at one end as previously . Red gangway doors.
Mk 2C: Lowered ceiling with twin strips of fluorescent lighting, ducting for air conditioning, but no air conditioning.
Mk 2D: Air conditioning. No opening lights in windows.
Mk 2E: Smaller toilets with luggage racks opposite. Fawn gangway doors.
Mk 2F: Plastic interior panels. Inter-City 70 seats. Modified air conditioning system.

The Mark 3 coach has BT10 bogies, is 75'long and is of fully integral construction. Inter-City 70 seats are fitted. Gangway doors are yellow (red in RFB). Loco-hauled coaches are classified Mark 3A, Mark 3 being reserved for HST trailers. A new batch of FOs and BFOs classified Mark 3B was built in 1985. These vehicles have APT style seating and revised lighting. A batch of driving brake vehicles is under construction for the West Coast Main Line.

The Mark 4 coach is being built by Metro-Cammell for the East Coast Main Line electrification scheme and features a body profile suitable for tilting trains, although tilt is not fitted, and is not intended to be. They are suitable for 140 mph running, although initially they will be restricted to 125 mph pending the installation of automatic train protection (ATP) on the East Coast Main Line.

2.1. BR HAULED PASSENGER STOCK

AP1Z (PK) PULLMAN FIRST WITH KITCHEN

Dia. AP101. Mark 2. Pressure ventilated (ac locos only). 18F 1L. B5 bogies. xe. ETH 6.

Lot No. 30755 Derby 1966. 40 t.

504 I	ICHV	EN	Ullswater
506 I	ICHV	EN	Windermere

AQ1Z (PC) PULLMAN PARLOUR FIRST

Dia. AQ101. Mark 2. Pressure ventilated (ac locos only). 36F 2L. B4 bogies. xe. ETH 5.

Lot No. 30754 Derby 1966. 35 t.

546 I	ICHV	EN	Coniston Water
548 I	ICHV	EN	Grasmere
549 I	ICHV	EN	Bassenthwaite
550 I	ICHV	EN	Rydal Water
551 I	ICHV	EN	Buttermere
552 I	ICHV	EN	Ennerdale Water
553 I	ICHV	EN	Crummock Water

AR1Z (PK) PULLMAN BRAKE FIRST

Dia. AR101. Mark 2. Pressure ventilated (ac locos only). 30F 1L. B4 bogies. xe. ETH 4.

Lot No. 30753 Derby 1966. 35 t.

586 I	ICHV	EN	Derwent Water

AJ1F(RFB) BUFFET OPEN FIRST

Dia. AJ104. Mark 2F. Air conditioned. Converted 1988 onwards from Mark 2F FOs. 1200/1/3/6/11/14–17/20/21 have Stones equipment, others have Temperature Ltd. 26F 1L. Public telephone. B4 bogies. ETH 6X.

1200/3/6/11/14/16/20. Lot No. 30845 Derby 1973. 33 t.
1201/4/5/7/8/10/12/13/15/17–9/21. Lot No. 30859 Derby 1973–4. 33 t.
1202/9. Lot No. 30873 Derby 1974–5. 33 t.

Note: 1200–2/5/7/9/10/2/4/7/20 were also numbered 6459/45/56/38/22/57/62/53/33/44/32 when declassified prior to conversion.

1200 (3287)	I	ICCX		EC
1201 (3361)	I	ICCX		EC
1202 (3436)	I	ICCX		EC
1203 (3291)	I	ICCX		EC
1204 (3401)	I	ICCX		EC
1205 (3329)	I	ICCX		PC
1206 (3319)	I	ICCX		PC
1207 (3328)	I	ICCX		PC
1208 (3393)	I	ICCX		PC
1209 (3437)	I	ICCX		PC
1210 (3405)	I	ICCX		MA
1211 (3305)	I	ICCX		MA
1212 (3427)	I	ICCX		MA
1213 (3419)	I	ICCX		EC
1214 (3317)	I	ICCX		EC
1215 (3377)	I	ICCX		EC
1216 (3302)	I	IWRX		LA
1217 (3357)	I	IWRX		LA
1218 (3332)	I	IWRX		LA
1219 (3418)	I	ICCX		OY
1220 (3315)	I	ICCX		OY
1221 (3371)	I	ICCX	n	OY

AJ41 (RBR) RESTAURANT BUFFET

Dia. AJ403. Mark 1. Gas cooking. Built with 23 loose chairs (dia. AJ402). All remaining vehicles refurbished with 23 fixed polypropylene chairs and fluorescent lighting. ETH 2 (2X*).

Lot No. 30628 Pressed Steel 1960–61. d. 39 t.

1644 I	ICHV	a to	BN
1645 I	ICHV	a pt	BN
1646 I	ICHV	a pt	OM
1647 I	IANX	a pt	NC
1648 I	ICCX	a pt	MA
1649 I	ICCX	a pt	MA
1650	ICCX	a pt	MA
1651 I	ICCX	a pa	MA
1652 I	ICCX	a pt	MA
1653 I	ICCX	a pt	EC
1654 I	ICCX	a pa	MA
1655 I	IANL	a pt	NC
1656 I	IXXH	a to	CL
1657 I	ICCX	a pt	MA
1658 I	ICCX	a pa	EC
1659 I	ICHV	a pt	BN
1663 I	IANL	x*to	NC
1665	IXXH	x*	CL
1666 I	ICHV	x*	OM
1667 I	ICHV	x	BN
1670	ICCX	x*	EC
1671 I	IANX	x*to	NC
1672 I	ICHV	x*	BN
1673	ICCX	a pt	MA
1674 I	ICHV	a pa	BN
1675 I	ICHV	x*	BN
1678 I	ICHV	x*	OM
1679 I	ICHV	a pt	BN
1680	ICCX	x*to	MA
1681 I	IANX	a pt	NC
1683 I	IANX	a pt	NC
1684	ICCX	x*pt	EC
1686 I	IANX	a pt	NC
1687	IXXH	a to	CL
1688	ICCX	a	MA
1689 I	IANX	a	NC
1691 I	IANX	a	NC
1692 I	IANX	a pt	NC
1693 I	ICHV	x* p	BN
1694	ICCX	a	MA
1695	ICCX	a	MA
1696 I	IANX	a pt	NC
1697 I	IANX	a pt	NC
1698 I	ICHV	a	BN
1699 I	IANX	a pt	NC

AN21 (RMB) MINIATURE BUFFET CAR

Dia. AN203. Mark 1. Gas cooking. 44S 2L. These vehicles are basically an open second with two full window spaces removed to accommodate a buffet counter, and four seats removed to for a stock cupboard. All remaining vehicles now have fluorescent lighting. All vehicles have Commonwealth bogies except 1805 (B4) and 1850 (B5). d. ETH 3 (3X*).

1805. Lot No. 30485 York 1957–8. 34.5 t.
1813–1833. Lot No. 30520 Wolverton 1960. 38 t.
1838–1850. Lot No. 30507 Wolverton 1960. 37t (1850 is 36t).
1853–1864. Lot No. 30670 ton 1961–2. 38 t.
1865–1882. Lot No. 30702 Wolverton 1962. 38 t.

r Refurbished vehicle. Fitted with microwave oven. Dia. AN208.

1805 I	IWCX	a*pa	PC
1813	IWRX	x*	OM
1816	IWRX	x*	OM
1832 I	IWRX	x	OM
1833 I	IWCX	x*	MA
1838	IXXZ	x	OM
1842	NNEX	xrpt	CA
1845	ICHS	x	OY
1848 I	IWRX	x	OM
1850 **N**	NNEX	v p	CA

1853 **I**	IWRX	x	OM
1854 **I**	IWRX	x	OM
1857	ICHL	x	BN
1860 **I**	IWCX	x	WB
1863	IWCX	x	WB
1864	IWCX	x	WB
1865 **N**	NNEX	v	CA
1867 **I**	IWRX	a pa	OM
1869 **I**	IANX	a pa	NC
1870 **I**	IANX	a pa	NC
1871 **N**	NNEX	x p	CA
1875 **I**	IWRX	a pa	OM
1876 **I**	IWCX	a	WB
1878 **I**	IWRX	a pa	OM
1879 **I**	IWRX	a pa	OM
1880	IWCX	a	WB
1882 **I**	IWRX	a pa	OM

AJ41 (RBR) RESTAURANT BUFFET

Dia. AJ414. Mark 1. Gas cooking. These vehicles were built as unclassified restaurant (RU). All remaining vehicles were rebuilt with buffet counter and 23 fixed polypropylene chairs (RBS). They were then further refurbished by fitting fluorescent lighting and reclassified RBR. d. ETH 2 (2X*).

1923/4. Lot No. 30476 Swindon 1958. B4/B5 bogies. 36.5 t.
1944–1956. Lot No. 30575 Ashford/Eastleigh 1960. B4/B5 bogies. 36.5 t.
1959–1984. Lot No. 30632 Ashford/Eastleigh 1960–61. Commonwealth bogies. 39 t.

1923 **I**	ICCX	x*	MA
1924	IXXZ	x*	ZN
1944	ICCX	a pa	MA
1945	ICCX	a pt	MA
1946	ICCX	a pt	MA
1947	ICCX	x*	MA
1948 **I**	ICCX	x*	MA
1953	ICHL	x*	OM
1954	ICCX	a pt	MA
1956	ICCX	a pa	EC
1959 **I**	ICCX	a pt	EC
1961 **I**	ICCX	a pt	EC
1962	ICCX	a pt	EC
1966 **I**	ICCX	a pt	MA
1969 **I**	ICCX	a pt	MA
1970	ICCX	a pt	EC
1971 **I**	ICCX	a pt	EC
1972	ICCX	a pt	MA
1973 **I**	ICCX	a pt	EC
1981 **I**	ICCX	a pa	EC
1984	ICCX	a pt	MA

AU51 CHARTER TRAIN STAFF COACHES

Dia. AU501. Mark 1. Converted from BCKs. ETH 2.

Lot No. 30732 Derby 1964. Commonwealth bogies. 37 t.

2833 (21270)	ICHV	BN
2834 (21267) **I**	ICHV	BN

AT5 ROYAL SALOONS

Non-standard livery: All Royal vehicles are in Royal purple.

AT51. Royal Saloon (Royal family or household). Dia AT501. Mark 1. This vehicle is a side corridor with a lounge, four bedrooms and a bathroom. Air conditioned. Converted 1977 from vs to ae and B5 bogies.

Lot No. 30130 Wolverton 1955. ETH 5X. 42 t.

2900 **0** ICHX ZN

AT51. Royal Saloon (Private secretary and Royal household). Dia AT503. Mark 1. This vehicle is a side corridor with an attendant's compartment, three bedrooms two bathrooms and a lounge/office compartment. Air conditioned. Converted 1977 from vs to ae and B5 bogies.

Lot No. 30131 Wolverton 1957. ETH 5X. 42 t.

2901 **0** ICHX ZN

AT51. Royal Dining Car. Dia AT502. Mark 1. This vehicle has a kitchen, a service compartment and a dining saloon. Air conditioned. Renumbered and converted 1977 from vs to ae and B5 bogies.

Lot No. 30129 Wolverton 1956. ETH 3X. 43 t.

2902 (499) **0** ICHX ZN

AT5G. The Queen's Saloon. Dia. AT525. Mark 3. Converted from a mark 3 FO built 1972. Consists of a lounge, bedroom and bathroom for the Queen, and a combined bedroom and bathroom for the Queen's dresser. One entrance vestibule has double doors. Air conditioned. ae. BT10 bogies.

Lot No. 30886 Wolverton 1977. ETH 9X. 36 t.

2903 (11001) **0** ICHX ZN

AT5G. The Duke of Edinburgh's Saloon. Dia. AT526. Mark 3. Converted from a mark 3 TSO built 1972. Consists of a combined lounge/dining room, a bedroom and a shower room for the Duke, a kitchen and a valet's bedroom and bathroom. Air conditioned. ae. BT10 bogies.

Lot No. 30887 Wolverton 1977. ETH 15X. 36 t.

2904 (12001) **0** ICHX ZN

AT5B. Staff Couchette/Power Brake. Dia. AT527. Mark 2B. Converted from a Mk. 2B BFK built 1969. Consists of luggage accommodation, guard's compartment, 350 kW diesel generator and staff sleeping accommodation. Pressure ventilated. ae. B5 bogies.

Lot No. 30888 Wolverton 1977. ETH 5X. 46 t.

2905 (14105) **0** ICHX ZN

AT5B. Staff Couchette. Dia. AT528. Mark 2B. Converted from a Mk. 2B BFK built 1969. Pressure ventilated. ae. B5 bogies.

Lot No. 30889 Wolverton 1977. ETH 4X. 35.5 t.

2906 (14112) **0** ICHX ZN

AT51 (RF). First Class Restaurant. Dia. AT517. Mark 1. Gas cooking. 24F. ae. B5 bogies. ETH2X.
Lot No. 30633 Swindon 1961. 41 t. Part of Royal Train.

2907 (325) **0** ICHX ZN

AT50. SPECIAL SALOON. Dia. AT530 (LMS dia. 12). Built by LNWR as Chairman's saloon. Now part of Royal train fleet. B4 bogies. ae.

Lot No. ? Wolverton 1920. 35.5 t.

2911 (45000, LNWR 5000) **0** ICHX ZN

AT5G. ROYAL TRAIN STAFF SLEEPING CARS. Dia. AT531. Mark 3A Details as for 10646–732 except that controlled emission toilets are not fitted.

Lot No. 31002 Derby/Wolverton 1985. 42.5 t.

2914	**0** ICHX	ZN	2915	**0** ICHX	ZN

AT5G NEW ROYAL VEHICLES. Full details not available.

Lot No. 31059 (2916), 30884 (others) Wolverton 1986. Dia. AT532.

2916	(40512)	**0**	ICHX	ZN	2918	(40515)	**0**	ICHX	ZN
2917	(40514)	**0**	ICHX	ZN	2919	(40518)	**0**	ICHX	ZN

AT5B. ROYAL STAFF VEHICLE. Full details not available.

Lot No. 31034 Wolverton 1986. Dia. AT536.

2920 (17109) **0** ICHX ZN

AT5G. NEW ROYAL VEHICLES. No details available.

Lot No. 31035 Derby/Wolverton 1987. Dia. AT534.

2922 **0** ICHX ZN

Lot No. 31036 Derby/Wolverton 1987. Dia. AT535.

2923 **0** ICHX ZN

AD11 (FO) OPEN FIRST

Dia. AD103. Mark 1. 42F 2L. ETH 3. d. Now fitted with table lights for use in first class charter trains.

3045. Lot No. 30091 Doncaster 1954. B4 bogies. 33 t.
3097–3100. Lot No. 30576 BRCW 1959. B4 bogies. 33 t.

3045	**I**	ICHV	vk	BN	3098	**I**	ICHV	v	BN
3096	**I**	ICHV	v	BN	3100	**I**	ICHV	v	BN
3097	**I**	ICHV	a	BN					

Later design with fluorescent lighting, aluminium window frames and Commonwealth bogies.

3107–3127. Lot No. 30697 Swindon 1962–3. 36 t.
3131–3150. Lot No. 30717 Swindon 1963. 36 t.

Note: 3136/41/3/4/6/7/8 were renumbered 1060/3/5/6/8/9/70, then 3605/8/9/2/6/4/10, but have now regained their original numbers.

3107	**I**	ICHV	x	BN	3131	**I**	ICHV	x	BN
3111	**I**	ICHV	x	BN	3132	**I**	ICHV	x pa	BN
3114	**I**	ICHV	x	BN	3133	**I**	ICHV	x pa	BN
3115	**I**	ICHV	x	BN	3134	**I**	ICHV	x pa	BN
3118	**I**	ICHV	x	BN	3136	**I**	ICHV	a	BN
3119	**I**	ICHV	x	BN	3140	**I**	ICHV	x	BN
3120	**I**	ICHV	x	BN	3141	**I**	ICHV	a pa	BN
3121	**I**	ICHV	a	BN	3143	**I**	ICHV	a pa	BN
3123	**I**	ICHV	a	BN	3144	**I**	ICHV	a pa	BN
3124	**I**	ICHV	a	BN	3146	**I**	ICHV	a pa	BN
3127	**I**	ICHV	a	BN	3147	**I**	ICHV	a	BN

3148 I	ICHV	a	BN
3149 I	ICHV	a	BN
3150 I	ICHV	a	BN

AD1D (FO) OPEN FIRST

Dia. AD105. Mark 2D. Air conditioned. 3170–90/3211–6 have Stones equipment and 3191–3210 have Temperature Ltd. 3189–3216 require at least 800 V train heating supply. 42F 2L (41F 1L†). B4 bogies. ae. pa. ETH 5.

Lot No. 30821 Derby 1971–2. 32.5 t.

3172 I	IANX	NC
3174 I	IANX	NC
3178 I	ICHV	OM
3181 I	ICHV	BN
3182 I	IANX	NC
3186 I	IANX	NC
3187 I	ICHV	OM
3188 I	IANX	NC
3192 I	ICHM	MA
3202 I	ICHM	MA

AD1E (FO) OPEN FIRST

Dia. AD106. Mark 2E. Air conditioned. Stones equipment. Require at least 800 V train heating supply. 42F 2L. B4 bogies. ae. pa. ETH 5.

Lot No. 30843 Derby 1973–4. 32.5 t.

3221 I	IANX	w	NC
3222	ICHM		MA
3223 I	ICHV	n	OM
3224 I	ICCX		MA
3225 I	ICHM		MA
3226 I	ICHV		OM
3227 I	ICHV		OM
3228 I	ICCX		MA
3229 I	ICCX		MA
3230 I	ICHV		OM
3231 I	ICCX		MA
3232 I	IANX	w	NC
3233	ICHM		MA
3234 I	IANX	w	NC
3235 I	ICCX		MA
3236 I	ICCX		MA
3237 I	ICHM		MA
3238 I	IMLX		DY
3239 I	ICCX		MA
3240 I	ICCX		MA
3241 I	ICCX		MA
3242 I	IANX	w	NC
3244 I	IANX	w	NC
3245 **PS**	PECX		EC
3246 I	IANX	w	NC
3247 **PS**	PECX		EC
3248 **PS**	PECX		EC
3249 I	IANX	w	NC
3250 I	IANX	w	NC
3251 I	IANX	w	NC
3252 I	IANX	w	NC
3253	IMLX		DY
3254 I	ICHV		OM
3255 I	ICCX	n	MA
3256 I	IANX	w	NC
3257 I	IANX	w	NC
3258 I	IANX	n	NC
3259 I	IANX	w	NC
3260 I	ICHV		OM
3261 I	IANX	w	NC
3262 I	IANX		NC
3263 I	ICHV		OM
3264 I	ICHV		OM
3265 **PS**	PECX		EC
3266 I	IANX		NC
3267 I	IMLX		DY
3268 I	ICHM		MA
3269	ICCX		MA
3270	ICHM		MA
3271 I	ICCX		MA
3272	ICHM		OM
3273 I	ICCX		MA
3274 I	ICHV		OM
3275	ICHM		MA

AD1F (FO) OPEN FIRST

Dia. AD107. Mark 2F. Air conditioned. 3277–3318/58–81 have Stones equipment, others have Temperature Ltd. 42F 2L. B4 bogies. ae. pa. ETH 5X.

3277–3319. Lot No. 30845 Derby 1973. 33 t.
3322–3428. Lot No. 30859 Derby 1973–4. 33 t.
3429–3439. Lot No. 30873 Derby 1974–5. 33 t.

f–Facelifted with new panels and seat trim.
r–Refurbished. Power-operated vestibule doors, new panels and seat trim.
Note: Not all vehicles fitted were known at time of going to press.

3277	**I**	IWCX		PC
3278	**I**	IWCX		MA
3279	**I**	IWCX	f	OY
3280		IWCX		WB
3284	**PS**	PECX		EC
3285	**I**	IWCX	f	OY
3290		IWCX		WB
3292	**I**	IWCX	r	OY
3293	**I**	IWCX		MA
3295	**I**	IWCX	r	OY
3296		IWCX	n	WB
3299	**I**	IWCX	r	OY
3300	**I**	IWCX	f	OY
3303		IWCX		WB
3304	**I**	IWCX	r	OY
3309		IWCX		PC
3312	**I**	IWCX		MA
3313	**I**	IWCX	r	OY
3314	**I**	IWCX		WB
3318	**I**	IWCX	n	PC
3322		IWCX		PC
3325	**I**	IWCX	r	OY
3326	**I**	IWCX	f	OY
3330	**I**	IWCX	r	OY
3331	**I**	IWCX		PC
3333	**I**	IWCX	r	OY
3334	**I**	IWCX	f	OY
3336	**I**	IWCX		PC
3337	**I**	IWCX	r	OY
3338		IWCX		PC
3340	**I**	IWCX	r	OY
3344	**I**	IWCX		PC
3345	**I**	IWCX	r	OY
3348	**I**	IWCX	r	OY
3350	**I**	IWCX	r	OY
3351	**I**	IWCX	f	OY
3352	**I**	IWCX		MA
3353	**I**	IWCX	f	OY
3354	**I**	IWCX	rp	OY
3356	**I**	IWCX	f	MA
3358	**I**	IWCX	r	OY
3359	**I**	IWCX	r	OY
3360	**I**	IWCX	f	OY
3362	**I**	IWCX	f	OY
3363	**I**	IWCX	r	OY
3364	**I**	IWCX	f	MA
3366	**I**	IWCX	r	OY
3368	**I**	IWCX	f	OY
3369	**I**	IWCX	r	OY
3372		IWCX		WB
3373	**I**	IWCX	r	OY
3374	**I**	IWCX	f	OY
3375	**I**	IWCX	r	OY
3378		IWCX		WB
3379	**I**	IWCX	r	OY
3381	**I**	IWCX	r	OY
3383		IWCX		WB
3384	**I**	IWCX	r	OY
3385	**I**	IWCX	r	OY
3386	**I**	IWCX	r	OY
3387	**I**	IWCX	r	OY
3388	**I**	IWCX	rn	OY
3389	**I**	IWCX	r	OY
3390	**I**	IWCX	f	OY
3391		IWCX		PC
3392	**I**	IWCX	r	OY
3395	**I**	IWCX	r	OY
3397	**I**	IWCX	r	OY
3399	**I**	IWCX	r	OY
3400	**I**	IWCX	r	OY
3402	**I**	IWCX	r	OY
3403	**I**	IWCX	r	OY
3408	**I**	IWCX		PC
3411	**I**	IWCX	f	OY
3414	**I**	IWCX	f	OY
3416	**I**	IWCX	f	OY
3417	**I**	IWCX	f	OY
3424	**I**	IWCX	r	OY
3425	**I**	IWCX	r	OY
3426	**I**	IWCX	f	OY

3428 I	IWCX	rn	OY
3429 I	IWCX	rn	OY
3431 I	IWCX	r	OY
3432 I	IWCX		WB
3433 I	IWCX	r	OY
3434 I	IWCX	r	OY
3438 I	IWCX	r	OY
3439	IWCX		WB

AC21/AD21 (TSO/SO) OPEN SECOND

Mark 1. Most vehicles have 2+2 seating (dia. AC204, 64S 2L) and are classified TSO ('Tourist second open'–a former LNER designation). TSOs numbered below 4098 have narrower seats than later vehicles. These narrow seats are also fitted to the later batch adjacent to the vestibule end doors. Other vehicles were built with 2+1 seating (48S 2L), and were often used as second class dining cars when new. These are classified SO (dia. AD201). All vehicles are TSO and have BR1 bogies except where stated otherwise. ETH 4 (4S•).

3749. Lot No. 30043 Doncaster 1953. 32.5 t.
3754–3771. Lot No. 30079 York 1953. 32.5 t.
3923–3961. Lot No. 30086 Ashford/Eastleigh 1954–5. 33t (35t C).
3991–3993. Lot No. 30090 York 1954. 34.5 t.
4058–4076. Lot No. 30149 Ashford/Eastleigh 1956–7. 33 t.

3749	PDYX	ae•C	DY
3754	PNLX	vd	NL
3766 **E**	ICHS	vd	BN
3767 **E**	ICHS	vd	BN
3769	PNLX	vd	NL
3771	PNLX	vd	NL
3923	ICHH	aeC	CL
3950	PNLX	vd	NL
3958	PNLX	vd	NL
3961	PNLX	vd	NL
3991	PDYX	ae•C	DY
3993	ICHH	ae•C	CL
4058	PDYX	aeC	DY
4076	ICHH	ae•C	CL

4366 (SO). Lot No. 30121 Eastleigh 1955. 32 t.

4366	PHTX	vd	HT

4376–4393. Lot No. 30219 Ashford/Eastleigh 1957. 33t (34.5t C).
4419–4435. Lot No. 30226 BRCW 1956–7. 33 t.

4376	ICHH	aeC	CL
4393	PDYX	aeC	DY
4419 **E**	ICHS	vd	BN
4420	PNLX	vd	NL
4435 **E**	ICHS	vd	BN

4478–4485 (SO). Lot No. 30227 BRCW 1957. 33 t.

4478	PLLX	vd	LL
4479	PLLX	vd	LL
4483	PPCX	vd	PC
4485	PLLX	vd	LL

4634. Lot No. 30375 York 1957. 33 t. Reinstated for steam loco crew training on the West Highland Line.

4634	ICHS	vs	FW

4779–4809 (SO). Lot No. 30376 York 1957. 33 t.

4779	PLLX	vd	LL
4785	PHTX	vd	HT
4795	PLLX	vd	LL
4798	PLLX	vd	LL
4809	PPCX	vd	PC

AC21 (TSO) OPEN SECOND

Dia. AC201. Mark 1. Development of dia. AC204 with fluorescent lighting and modified design of seat headrest. Built with BR1 bogies but most rebuilt with B4 bogies. 64S 2L. ETH 4.

Lot No. 30506 Wolverton 1959. BR1 bogies. xdk. 35 t.

4830 **N** NNWX EN
4831 PISX IS
4834 **N** NNWX EN
4836 PNLX NL

Lot No. 30525 Wolverton 1959–60. d. B4 bogies except where otherwise stated. 33 t. (35 t BR1).

4842 ICHM x MA
4843 PNLX v NL
4845 ICHC v OY
4846 **N** NNWX v pa EN
4848 **N** NNWX v pa EN
4849 **N** NNWX v EN
4852 **N** NNWX v pa EN
4853 **N** NNWX v pa EN
4854 **N** NNWX v pa EN
4855 ICHC v OY
4856 ICHM x MA
4857 ICHS v OY
4858 ICHM x MA
4860 ICHM x MA
4861 ICHC v OY
4862 **N** NNWX v EN
4864 PPCX vBR1 PC
4866 **N** NNWX v pa EN
4867 **N** NNWX v pa EN
4869 ICHM x MA
4873 **N** NNWX v EN
4875 **N** NNWX v EN
4876 **N** NNWX v EN
4880 **N** NNWX v EN
4884 **N** NNWX v EN
4885 PNLX v NL
4886 ICHC v OY
4888 **N** NNWX v EN
4891 **N** NNWX v EN
4894 ICHS v OY
4895 **N** NNWX v EN
4899 **N** NWRX v OM

Lot No. 30646 Wolverton 1961. Built with Commonwealth bogies, but BR1 bogies substituted by the SR. Some now re-rebogied. BR1 bogies except where stated otherwise. d. 35 t. (34 t B4, 36 t C).

4900 **E** ICHS v pa BN
4901 **N** NWRX v C OM
4902 **I** ICHX x B4 OY
4903 PNLX v NL
4904 PNLX v NL
4905 **N** NNWX v C EN
4906 PNLX v NL
4907 PNLX v NL
4908 **N** NNWX v C EN
4909 **I** ICHX x B4 OY
4910 **N** NNWX v C EN
4911 **E** ICHS v pa BN
4912 **E** ICHS v pa BN
4913 **N** NNWX v C EN
4915 **I** ICHX x B4 OY
4916 ICHX x B4 OY
4917 NNWX vCpa EN

Lot No. 30690 Wolverton 1961–2. Commonwealth bogies and aluminium window frames. d. 37 t.

f–Facelifted with new laminate, new diffusers etc.

4918 ICHC v OY
4919 **I** ICHB vf OY
4920 **N** NNWX v pa EN
4922 ICHS v OY
4923 NNWX v EN
4924 ICHC v OY
4925 **I** ICHD a OY
4926 **N** NNWX v pa EN
4927 **I** ICHB vf OY
4928 ICHC v OY

4929 **N**	NNWX	v	pa	EN
4930 **I**	ICHB	vf		OY
4931 **N**	NNWX	v		EN
4932 **N**	ICHS	v		BN
4933 **N**	NNWX	v		EN
4934	ICHC	v		OY
4935 **N**	NNWX	v	pa	EN
4936 **N**	NNWX	v	pa	EN
4937	ICHC	v		OY
4938 **I**	IWCX	a		WB
4939 **I**	IWCX	a		WB
4940 **N**	NNWX	v		EN
4943	NNWX	v		EN
4944	ICHC	v		OY
4945 **N**	NNWX	v	pa	EN
4946 **I**	ICHB	vf		OY
4947	ICHS	v		BN
4948	ICHC	v		OY
4949	IWCX	a		WB
4950 **N**	NWRX	v		OM
4951 **N**	NNWX	v	pa	EN
4952	ICHS	v		OY
4953 **N**	NNWX	v	pa	EN
4954	ICHS	v		OY
4955 **N**	NNWX	v	pa	EN
4956 **I**	ICHD	a		OY
4958	ICHC	v		OY
4959 **I**	IWCX	a		WB
4960 **N**	ICHS	v		BN
4961 **I**	ICHB	vf		OY
4962	ICHC	v		OY
4963 **I**	IWCX	a		WB
4964	ICHS	v		OY
4965	ICHC	v		OY
4966 **I**	ICHB	vf		OY
4973 **N**	NNWX	v	pa	EN
4974	ICHS	v		OY
4975 **N**	NNWX	v	pa	EN
4976	ICHS	v		OY
4977 **I**	IWCX	a		WB
4978	ICHC	v		OY
4979 **I**	ICHB	vf		OY
4980 **N**	NNWX	v	pa	EN
4981	ICHX	v		OY
4982	ICHS	v		OY
4983	ICHS	v		OY
4984 **N**	ICHS	v		BN
4985 **N**	NNWX	v	pa	EN
4986 **I**	ICHD	a		OY
4987	ICHS	v		BN
4988	ICHS	v		BN
4989 **N**	NNWX	v	pa	EN
4990 **N**	NNWX	v	pa	EN
4991 **I**	IWCX	v		WB
4992 **N**	NNWX	v	pa	EN
4993 **I**	IWCX	a		WB
4994 **N**	NNWX	v	pa	EN
4995	ICHC	v		OY
4996 **I**	ICHB	vf		OY
4997	ICHC	v		OY
4998	IWCX	a		WB
4999	ICHC	v		OY
5000 **N**	NNWX	v	pa	EN
5001 **I**	ICHB	vf		OY
5002	IWCX	a		WB
5003 **N**	NNWX	v	pa	EN
5004	ICHC	v		OY
5005 **I**	ICHD	a		OY
5006 **N**	NNWX	v	pa	EN
5007 **I**	ICHD	a		OY
5008 **I**	ICHB	vf		OY
5009 **I**	ICHD	a		OY
5010 **I**	ICHD	a		OY
5023 **I**	ICHD	a		OY
5024 **N**	NNWX	v		EN
5025 **I**	ICHD	a		OY
5027 **I**	ICHD	a		OY
5028	ICHM	x		MA
5029 **I**	IWCX	a		WB
5030 **I**	IWCX	a		WB
5031	ICHC	v		OY
5032	ICHM	x		MA
5033 **I**	IWCX	a		WB
5035	ICHM	x		MA
5036	ICHC	v		OY
5037 **I**	ICHD	a		OY
5038 **I**	ICHM	x		MA
5039 **N**	NNWX	v	pa	EN
5040	ICHM	x		MA
5041 **I**	IWCX	a		WB
5042	ICHM	x		MA
5043 **N**	NNWX	v	pa	EN
5044	IWCX	a		WB

AC2Z (TSO) OPEN SECOND

Dia. AC205. Mark 2. Pressure ventilated. 64S 2L. B4 bogies. vd. ETH 4.

*–Air braked by means of disc brakes.
Note: The vehicles shown allocated to 'MO' are IS vehicles outstationed at Muir of Ord until the Ness Viaduct is repaired.

Lot No. 30751 Derby 1965–7. 32 t.

No.				
5073	**N**	NWRX		OM
5080	**N**	NWRX		OM
5085	**N**	NNEX		CA
5086	**N**	NWRX		OM
5087	**N**	NWRX		OM
5090	**N**	NNEX		CA
5097		PNLX	pa	NL
5104	**N**	NWRX		OM
5105	**N**	NNEX		CA
5112	**N**	NWRX		OM
5113	**N**	NWRX		OM
5118	**N**	NWRX		OM
5124		PISX	pa	IS
5128		PNLX	pa	NL
5131		PNLX		NL
5132		PISX	pa	IS
5133		PISX		IS
5135		PISX		IS
5136	**N**	NNEX		CA
5138	**N**	NWRX		OM
5139		PISX	pa	IS
5140	**N**	NNEX		CA
5141	**N**	NNEX		CA
5142		ICCX	*	PC
5144		PCLX	*	CL
5145		PISX	pa	IS
5146	**PS**	PECX	pa	EC
5147	**P**	PHTX	*	HT
5148		PISX	pa	IS
5149		PISX	pa	IS
5150	**N**	NWRX		OM
5151		PISX	pa	IS
5152	**PS**	PECX	pa	EC
5153	**PS**	PECX	pa	EC
5154	**E**	PMOX	pa	MO
5155		PISX	pa	IS
5156		PISX	pa	IS
5157	**N**	NWRX		OM
5158	**N**	NWRX		OM
5159	**N**	NWRX		OM
5160	**P**	PHTX	*	HT
5161	**N**	NWRX		OM
5162	**N**	NNEX		CA
5163		PISX		IS
5166	**E**	PMOX	pa	MO
5167	**N**	NWRX		OM
5170		PISX	pa	IS
5171	**N**	NNEX		CA
5172	**N**	NNEX		CA
5173	**N**	NWRX		OM
5174		PISX	pa	IS
5175	**N**	NNEX		CA
5176	**PS**	PECX	pa	EC
5177	**N**	NWRX		OM
5178	**N**	NWRX		OM
5179	**N**	NWRX		OM
5180	**N**	NWRX		OM
5181	**N**	NNEX		CA
5182		PCLX	*	CL
5183	**N**	NWRX		OM
5184	**N**	NWRX		OM
5186	**N**	NWRX		OM
5187		ICCX	*	PC
5189		PCLX	*	CL
5191	**E**	PMOX	pa	MO
5192	**P**	PHTX	*	HT
5193	**E**	PMOX	pa	MO
5194	**N**	NWRX		OM
5196		PISX	pa	IS
5197	**PS**	PECX	pa	EC
5198	**N**	NWRX		OM
5199	**N**	NNEX		CA
5200	**N**	NNEX		CA
5204	**N**	NNEX		CA
5207		PISX	pa	IS
5208	**N**	NNEX		CA
5209	**N**	NWRX		OM
5210	**N**	NWRX		OM
5211	**N**	NNEX		CA
5212	**E**	PMOX	pa	MO
5213	**N**	NWRX		OM
5214	**N**	NWRX		OM
5215	**N**	NNEX		CA
5216	**N**	NNEX		CA
5217		PISX	pa	IS
5219	**N**	NNEX		CA
5220	**N**	NNEX		CA
5221		PISX	pa	IS
5222	**N**	NNEX		CA
5224	**N**	NNEX		CA
5225	**N**	NWRX		OM
5226	**N**	NWRX		OM
5228	**N**	NWRX		OM

Named vehicles:

5154	CLAN FRAZER	5193	CLAN MACLEOD
5166	CLAN MACKENZIE	5212	CAPERKAILZIE
5191	CLAN DONALD	5230	CORRIEMOILLIE

AD2Z (SO) OPEN SECOND

Dia. AD203. Mark 2. Pressure ventilated. 48S 2L. B4 bogies. Originally used as restaurant cars. d. ETH 4. (5251–6 are pa fitted).

Lot No. 30752 Derby 1966.

5230 **E**	PMOX	v	MO
5231	PISX	v	IS
5232	PISX	v	IS
5233	PISX	v	IS
5234	PISX	v	IS
5235	PNLX	v	NL
5236	PNLX	v	NL
5237	PNLX	v	NL
5239	PNLX	v	NL
5240	PNLX	v	NL
5241	PNLX	v	NL
5243	PNLX	v	NL
5244	PNLX	v	NL
5246	PLLX	a	LL
5247	PLLX	a	LL
5248 **N**	PDYX	a	DY
5249	PNLX	v	NL
5251	PLLX	a	LL
5253	PLLX	a	LL
5254	PLLX	a	LL
5255	PDYX	a	DY
5256 **N**	PDYX	a	DY

AC2A (TSO) OPEN SECOND

Dia. AC206. Mark 2A. Pressure ventilated. 64S 2L. B4 bogies. ad. ETH 4.

5257–5345. Lot No. 30776 Derby 1967–8. 32 t.
5346–5433. Lot No. 30787 Derby 1968. 32 t.

5257	PCLX		CL
5258	PDYX		DY
5259	IWCX		WB
5260	PCLX		CL
5261 **N**	NSSX	pa	LA
5262 **N**	PCLX		CL
5263 **N**	NSSX	pa	LA
5264 **N**	NWRX		OM
5265 **N**	NSSX	pa	LA
5266	PDYX		DY
5267	PCLX		CL
5268	PLLX		LL
5269	PISX	pa	IS
5270	PISX	pa	IS
5271	PISX	pa	IS
5272	PISX	pa	IS
5273	PISX	pa	IS
5274	PISX	pa	IS
5275 **N**	NWRX		OM
5276 **N**	NSSX	pa	LA
5277	PISX	pa	IS
5278 **N**	NSSX	pa	LA
5279	PISX	pa	IS
5281 **I**	ICCX		EC
5282	PISX	pa	IS
5283 **P**	PHTX		HT
5284	PLLX		LL
5286	IWCX		WB
5287 **I**	ICCX		EC
5288	PCLX		CL
5290 **N**	NSSX	pa	LA
5291 **P**	PHTX		HT
5292	PISX	pa	IS
5293 **N**	NWRX		OM
5294	PCLX		CL
5295	PLLX		LL
5296	PCLX		CL
5298 **P**	PHTX		HT
5299	PCLX		CL
5300	PISX	pa	IS
5301 **P**	PHTX		HT
5303 **N**	NSSX	pa	LA
5304	IMLX		DY
5306	PLLX		LL
5307	PISX	pa	IS
5308 **N**	NSSX	pa	LA
5309 **N**	NWRX		OM
5310	PCLX		CL
5311 **P**	PHTX		HT
5312 **N**	NSSX	pa	LA

5314	IMLX		DY
5315	PCLX		CL
5316 **P**	PHTX		HT
5317	PCLX		CL
5318	PISX	pa	IS
5321 **P**	PHTX		HT
5322 **N**	NSSX		EH
5323 **P**	PHTX		HT
5325	PISX	pa	IS
5326 **N**	NWRX		OM
5327	PISX	pa	IS
5329 **P**	PHTX		HT
5330	PISX	pa	IS
5331 **N**	NSSX	pa	LA
5333	PISX	pa	IS
5335 **N**	NSSX		EH
5336 **N**	NSSX	pa	LA
5337	PLLX		LL
5338	IMLX		DY
5339	IMLX		DY
5340 **P**	PHTX		HT
5341 **N**	NWRX		OM
5342	PLLX		LL
5343 **P**	PHTX		HT
5345 **N**	NWRX	pa	LA
5346 **I**	ICCX		EC
5347 **P**	PHTX		HT
5349	PISX		IS
5350 **N**	NSSX	pa	LA
5351 **P**	PHTX		HT
5352	PCLX	pa	CL
5353	PISX	pa	IS
5354 **N**	NWRX		OM
5355	NSSX	pa	LA
5356	ICCX		PC
5357	IWCX		WB
5359	PLLX	pa	LL
5360 **N**	NWRX		OM
5361 **I**	ICCX		PC
5362 **N**	NSSX		LA
5363	PCLX	pa	CL
5364	IMLX		DY
5365 **P**	PHTX		HT
5366 **N**	NWRX		OM
5367	ICCX		PC
5368	IWCX		WB
5369	IWCX		WB
5370	ICCX		MA
5371 **I**	ICCX		PC
5372	IWCX		WB
5373 **N**	NSSX		LA
5374	PCLX		CL
5375	PDYX		DY
5376 **N**	NWRX		OM
5378 **N**	NWRX		OM
5379 **P**	PHTX		HT
5381 **N**	NSSX	pa	LA
5382 **N**	ICCX		PC
5383	PCLX		CL
5384 **N**	NSSX	pa	LA
5385	PLLX		LL
5386 **N**	NWRX		OM
5387 **P**	PHTX		HT
5388	PCLX		CL
5389	NSSX	pa	LA
5391 **P**	PHTX		HT
5392	PISX	pa	IS
5393	PISX	pa	IS
5394	PISX	pa	IS
5395 **N**	IMLX		DY
5396 **P**	PHTX		HT
5397	PLLX		LL
5398 **N**	NSSX	pa	LA
5399	PCLX		CL
5400	IWCX		WB
5401	PISX	pa	IS
5402	PCLX		CL
5403	IMLX		DY
5404 **N**	NSSX	pa	LA
5406 **P**	PHTX		HT
5408 **N**	NSSX	pa	EH
5409 **P**	PHTX		HT
5410 **N**	NSSX		EH
5412 **N**	NWRX		OM
5413	PCLX		CL
5414 **P**	PHTX		HT
5415	ICCX		MA
5416	PDYX		DY
5417 **I**	ICCX		PC
5418 **N**	NSSX		EH
5419	NSSX	pa	EH
5420 **N**	NSSX	pa	LA
5421	PLLX		LL
5422	IMLX		DY
5423	ICCX		PC
5424	PDYX		DY
5425 **P**	PHTX		HT
5426	PCLX		CL
5427	PISX		IS
5429	PLLX		LL
5430 **I**	ICCX		PC
5431 **P**	PHTX		HT
5432 **P**	PHTX		HT
5433 **N**	NSSX	pa	LA

AC2B (TSO) OPEN SECOND

Dia. AC207. Mark 2B. Pressure ventilated. 62S 2L. B4 bogies. ad. ETH 4.

Lot No. 30791 Derby 1969. 32 t.

5434	PDYX		DY
5435 **N**	NSSX	pa	LA
5436 **N**	NSSX	pa	LA
5437	PISX		IS
5438	ICCX		MA
5439 **N**	NSSX		EH
5440	PCLX		CL
5442 **N**	NWRX		OM
5443 **N**	NSSX		EH
5444 **N**	NSSX	pa	LA
5445	PECX		EC
5446 **N**	NSSX	pa	LA
5447 **N**	NSSX	pa	LA
5448 **N**	NSSX	pa	LA
5449 **N**	NSSX	pa	LA
5450 **N**	NSSX	pa	LA
5451 **N**	NSSX	pa	LA
5452 **N**	NWRX		OM
5453 **P**	PDYX		DY
5454 **N**	NSSX	pa	LA
5455 **N**	NSSX		LA
5456 **N**	NSSX	pa	LA
5457 **N**	NWRX		OM
5458 **N**	NSSX		EH
5459 **I**	IWCX		WB
5461 **I**	IWCX		WB
5462 **N**	NSSX	pa	LA
5463 **P**	PHTX		HT
5464 **N**	NSSX	pa	LA
5465 **N**	NSSX	pa	LA
5466 **I**	IMLX		DY
5467 **I**	ICCX		PC
5468 **N**	NSSX	pa	LA
5470 **N**	NSSX	pa	LA
5471 **N**	NSSX		EH
5472 **N**	NSSX	pa	LA
5473	PLLX		LL
5474 **N**	NSSX	pa	LA
5475 **N**	NSSX	pa	LA
5476 **I**	ICCX		PC
5477 **I**	IWCX		WB
5478	IWCX		WB
5479 **P**	PHTX		HT
5480	NSSX	pa	LA
5481 **N**	NSSX	pa	LA
5482 **N**	NSSX	pa	LA
5483	PCLX		CL
5484 **N**	NSSX		EH
5486	IWCX		WB
5487	IMLX		DY
5488	PECX		EC
5489	PCLX		CL
5490 **N**	NSSX	pa	EH
5491 **N**	NWRX		OM
5492 **N**	NSSX	pa	LA
5493 **N**	NWRX		OM
5494 **N**	NSSX	pa	LA
5495 **N**	NSSX	pa	LA
5496	IWCX		WB
5497 **N**	NSSX	pa	LA

AC2C (TSO) OPEN SECOND

Dia. AC208. Mark 2C. Pressure Ventilated. 62S 2L. B4 bogies. ad. pa (except 5552 which is not fitted. ETH 5.

Lot No. 30795 Derby 1969–70. 32.4 t.

5500	IWCX	WB
5501	IWCX	WB
5502	IWCX	WB
5505	ICCX	PC
5508	IWCX	WB
5509	ICCX	PC
5512	IWCX	WB
5515	IWCX	WB
5517	IWCX	WB
5519	ICCX	MA
5520 **P**	PHTX	HT
5522	ICCX	PC
5523 **N**	NSSX	LA
5524	ICCX	WB
5525 **N**	NSSX	LA
5526	IWCX	WB
5527	ICCX	PC
5529	ICCX	MA
5530	IWCX	WB
5532	ICCX	PC
5533	IWCX	WB
5536	ICCX	MA
5539	ICCX	PC
5540	ICCX	MA
5541	ICCX	PC
5542	IWCX	WB
5546	ICCX	MA

5548	ICCX	PC
5549	IWCX	WB
5550	ICCX	MA
5552	IWCX	WB
5554	ICCX	PC
5557	ICCX	PC
5560	ICCX	PC
5561	ICCX	PC
5565	ICCX	WB
5566	ICCX	PC
5569	IWCX	WB
5572	IWCX	WB
5574	ICCX	MA
5576	ICCX	MA
5581	ICCX	MA
5582	IWCX	WB
5584	IWCX	WB
5585	IWCX	WB
5586	ICCX	PC
5589	ICCX	MA
5590	IWCX	WB
5591	IWCX	WB
5594	ICCX	MA
5595 **I**	ICCX	WB
5596	ICCX	PC
5597	ICCX	MA
5600	ICCX	WB
5601	IWCX	WB
5604	ICCX	MA
5605	IWCX	WB
5607	IWCX	WB
5609	ICCX	MA
5610	ICCX	WB
5613	ICCX	MA
5614 **P**	PHTX	HT
5615	ICCX	WB

AC2D (TSO) OPEN SECOND

Dia. AC209. Mark 2D. Air conditioned. Stones (5653 has Temperature Ltd.) equipment. 62S 2L. B4 bogies. ae. pa. ETH 5.

Lot No. 30822 Derby 1971. 33 t.

5616 **I**	ICCX		MA
5617 **I**	ICCX		MA
5618	IANR		NC
5619 **I**	IANR		NC
5620 **I**	IANR		NC
5621 **I**	IANX		NC
5623 **PS**	PECX		EC
5624 **I**	IANR		NC
5625 **I**	IANX		NC
5626 **I**	IANR		NC
5628 **I**	IANR		NC
5629 **I**	IANR		NC
5630 **I**	ICCX		MA
5631 **I**	IANR		NC
5632 **I**	IANR		NC
5633 **I**	ICCX		MA
5634 **I**	IANR		NC
5636 **I**	IANR		NC
5637	ICCX		MA
5638 **I**	IANR		NC
5639 **I**	ICCX		MA
5640 **I**	ICCX		OY
5642	ICCX		OY
5643 **I**	IANX		NC
5645 **I**	ICCX		MA
5646	IANR		NC
5647	ICCX		MA
5648 **I**	ICCX		MA
5650 **I**	IANR		NC
5651 **I**	IANR		NC
5652 **I**	IANR		NC
5653 **PS**	PECX	*	EC
5654 **I**	IANR		NC
5657 **I**	IANR		NC
5658 **I**	IANR		NC
5659 **PS**	PECX		EC
5660 **I**	ICCX		OY
5661 **I**	IANX		NC
5662 **PS**	PECX		EC
5663 **PS**	PECX		EC
5665 **I**	IANR		NC
5666 **I**	ICCX		MA
5667 **I**	ICCX		MA
5668 **I**	IANX		NC
5669 **I**	IANR		NC
5670 **I**	ICCX		MA
5671 **PS**	PECX		EC
5672 **I**	IANX		NC
5673 **PS**	PECX		EC
5674 **I**	IANX		NC
5675	ICCX		MA
5676 **I**	IANR		NC
5677 **I**	ICCX		MA
5678 **I**	ICCX		MA
5679 **I**	IANR		NC
5680 **I**	ICCX		MA
5681 **I**	ICCX		MA
5682 **I**	IANR		NC
5684	ICCX		MA
5685 **I**	IANR		NC
5686 **I**	IANR		NC
5687	ICCX		MA
5689 **I**	ICCX		MA
5690 **I**	IANR		NC
5692 **I**	IANR		NC
5693 **I**	IANR		NC

5694 **PS**	PECX	EC
5695 **I**	IANR	NC
5699	ICCX	MA
5700 **I**	IANR	NC
5701 **I**	ICCX	MA
5703 **I**	IANR	NC
5704	ICCX	MA
5705 **I**	IANX	NC
5706 **I**	ICCX	MA
5707 **I**	IANX	NC
5708 **I**	ICCX	MA
5709	ICCX	MA
5710 **I**	IANR	NC
5711 **PS**	PECX	EC
5712 **I**	IANX	NC
5713 **I**	IANX	NC
5714 **I**	IANX	NC
5715 **I**	IANR	NC
5716 **PS**	PECX	EC
5717 **I**	ICCX	MA
5718 **I**	IANX	NC
5719 **I**	ICCX	OY
5722 **I**	ICCX	OY
5723 **I**	IANR	NC
5724 **I**	IANR	NC
5726 **PS**	PECX	EC
5727 **I**	ICCX	MA
5728 **I**	IANR	NC
5729 **I**	IANX	NC
5730 **I**	IANR	NC
5731 **I**	IANX	NC
5732 **I**	IANX	NC
5734 **I**	ICCX	MA
5735	IANR	NC
5737 **I**	IANR	NC
5738	ICCX	MA
5739 **I**	IANX	NC
5740 **PS**	PECX	EC
5742 **I**	IANX	NC
5743 **I**	IANR	NC

AC2E (TSO) — OPEN SECOND

Dia. AC210. Mark 2E. Air conditioned. Stones equipment. 64S 2L. B4 bogies. Require at least 800 V train heat supply. ae. pa. ETH 5.

5744–5804. Lot No. 30837 Derby 1972. 33.5 t.
5809–5907. Lot No. 30844 Derby 1972–3. 33.5 t.

5744 **I**	ICCX	MA
5745 **I**	ICCX	PC
5746 **I**	ICCR	PC
5747 **I**	ICCX	MA
5748 **I**	ICCR	PC
5749 **I**	ICCX	MA
5750 **I**	IWRX	LA
5751 **I**	ICCR	PC
5752 **I**	ICCX	MA
5753 **I**	ICCX	PC
5754 **I**	ICCR	PC
5755 **I**	ICCX	PC
5756 **I**	IWRX	LA
5757 **I**	ICCX	MA
5758 **I**	ICCX	MA
5759 **I**	ICCX	PC
5760 **I**	ICCR	PC
5761 **I**	ICCX	MA
5762 **I**	ICCX	PC
5763 **I**	ICCX	MA
5764 **I**	ICCX	MA
5765 **I**	ICCX	MA
5766 **I**	ICCX	PC
5767 **I**	ICCX	MA
5768 **I**	ICCX	OY
5769 **I**	ICCR	PC
5770 **I**	ICCX	MA
5771 **I**	ICCX	MA
5772 **I**	ICCR	PC
5773 **I**	ICCX	MA
5774 **I**	ICCX	MA
5775 **I**	ICCX	PC
5776 **I**	ICCR	PC
5777 **I**	ICCX	MA
5778 **I**	ICCR	PC
5779 **I**	ICCX	PC
5780 **I**	ICCR	PC
5781 **I**	ICCX	PC
5782 **I**	ICCX	MA
5783 **I**	ICCX	MA
5784 **I**	ICCR	PC
5785 **I**	ICCX	MA
5786 **I**	ICCX	MA
5787 **I**	ICCX	MA
5788 **I**	ICCX	MA
5789 **I**	ICCR	PC
5790 **I**	ICCX	PC
5791 **I**	ICCR	PC
5792 **I**	ICCR	PC
5793 **I**	ICCR	PC
5794 **I**	ICCR	PC
5795 **I**	ICCX	PC
5796 **I**	ICCR	PC
5797 **I**	ICCR	PC
5798 **I**	ICCX	PC
5799 **I**	IWRX	LA
5800 **I**	ICCX	MA
5801 **I**	ICCR	PC
5802 **I**	ICCX	PC
5803 **I**	ICCX	MA
5804 **I**	ICCX	OY
5809 **I**	ICCX	PC
5810 **I**	ICCX	PC
5811 **I**	ICCX	MA
5812 **I**	ICCR	PC
5813 **I**	ICCX	MA
5814 **I**	ICCR	PC
5815 **I**	IWRX	LA
5816 **I**	ICCR	PC
5817 **I**	ICCX	MA
5818 **I**	ICCX	MA
5819 **I**	ICCX	MA

5820 I ICCX PC
5821 I ICCR PC
5822 I ICCX PC
5823 I ICCX MA
5824 I ICCR PC
5825 I ICCR PC
5826 I ICCR PC
5827 I ICCX MA
5828 I ICCX PC
5829 I ICCX PC
5830 I ICCX MA
5831 I ICCX MA
5832 I ICCX MA
5833 I ICCX MA
5834 I ICCX PC
5835 I ICCX PC
5836 I ICCX MA
5837 I ICCX OY
5838 I ICCX MA
5839 I ICCX MA
5840 I ICCX PC
5841 I ICCX MA
5842 I ICCX OY
5843 I ICCR PC
5844 I ICCX PC
5845 I ICCX MA
5846 I ICCX OY
5847 I ICCR PC
5848 I ICCX MA
5849 I ICCX PC
5850 I ICCX MA
5851 I ICCX MA
5852 I ICCX MA
5853 I ICCX MA
5854 I ICCX MA
5855 I IWRX LA
5856 I ICCX MA
5857 I ICCX MA
5858 I ICCX MA
5859 I ICCX PC
5860 I ICCX PC
5861 I ICCX PC
5862 I ICCX PC
5863 I ICCX MA
5864 I ICCX MA
5865 I ICCR PC
5866 I ICCX MA
5867 I ICCX PC
5868 I ICCR PC
5869 I ICCR PC
5870 I ICCX MA
5871 I ICCX MA
5872 I IWRX LA
5873 I ICCX OY
5874 I ICCX MA
5875 I ICCX MA
5876 I ICCR PC
5877 I ICCX PC
5878 I ICCX PC
5879 I IWRX LA
5880 I ICCX PC
5881 I IWRX LA
5882 I ICCX MA
5883 I ICCX MA
5884 I ICCX PC
5885 I ICCX MA
5886 I ICCX MA
5887 I ICCR PC
5888 I ICCR PC
5889 I ICCR PC
5890 I ICCX MA
5891 I ICCX MA
5892 I ICCR PC
5893 I ICCR PC
5894 I ICCX OY
5895 I ICCX OY
5896 I ICCR PC
5897 I ICCR PC
5898 I ICCX MA
5899 I ICCR PC
5900 I ICCR PC
5901 I ICCX PC
5902 I ICCX MA
5903 I ICCX MA
5904 I ICCX PC
5905 I ICCR PC
5906 I ICCR PC
5907 I ICCX MA

AC2F (TSO) — OPEN SECOND

Dia. AC211. Mark 2F. Air conditioned. Temperature Ltd. equipment. 64S 2L. B4 bogies. ae. pa. ETH 5X. Inter-City 70 seats.

*–Early Mark 2 style seats.
f–Facelifted with new panels & red seat trim.
r–Refurbished. Power-operated vestibule doors, new panels and seat trim.
Note: Not all vehicles fitted were known at time of going to press.

5908–5958. Lot No. 30846 Derby 1973. 33 t.
5959–6170. Lot No. 30860 Derby 1973–4. 33 t.
6171–6184. Lot No. 30874 Derby 1974–5. 33 t.

5908 I IWCX r OY
5910 I IWCX rw OY
5911 I IWCX r OY
5912 I IWCX r OY
5913 I IWCX r OY
5914 I IWCX r OY
5915 I IWCX r OY
5916 I IWCX rw PC
5917 I IWCX PC
5918 I IWCX rw OY
5919 I IWCX r OY
5920 I IWCX r OY
5921 I IWCX PC
5922 I IWCX PC
5924 IWCX n WB
5925 I IWCX f WB

No.		Code		
5926	I	IWCX	f	MA
5927	I	IWCX		PC
5928	I	IWCX		PC
5929	I	PECX		EC
5930	I	IWCX	r	OY
5931	I	IWCX	rw	MA
5932	I	IWCX		MA
5933	I	IWCX	r	OY
5934	I	IWCX		MA
5935		IWCX		WB
5936	I	IWCX		PC
5937	I	IWCX	f	OY
5939	I	IWCX	r	OY
5940	I	IWCX	rw	MA
5941	I	IWCX	r	OY
5943	I	IWCX	rw	OY
5944	I	IWCX	rw	OY
5945	I	IWCX	f	MA
5946	I	IWCX	r	OY
5947		IWCX		PC
5948	I	IWCX		OY
5949	I	IWCX	rw	OY
5950		IWCX		PC
5951		IWCX	n	WB
5952		IWCX		WB
5953	I	IWCX		MA
5954	I	IWCX	f	WB
5955	I	IWCX	r	OY
5956	I	IWCX	f	WB
5957	I	IWCX		MA
5958	I	IWCX	r	OY
5959	I	PECX	n	EC
5960	I	IWCX	f	MA
5961	I	IWCX	f	MA
5962	I	IWCX	r	OY
5963	I	IWCX	f	OY
5964	I	PECX	n	EC
5965		IWCX		PC
5966		IWCX		MA
5967	I	IWCX		PC
5968	I	PECX		EC
5969	I	IWCX	rw	OY
5971		IWCX		WB
5973	I	IWCX		MA
5975	I	IWCX	*n	WB
5976	I	IWCX	f	OY
5977	I	IWCX	r	OY
5978	I	IWCX	*n	WB
5980	I	IWCX	r	OY
5981	I	IWCX	r	OY
5983	I	IWCX	f	OY
5984	I	IWCX	*r	OY
5985	I	IWCX	r	OY
5986	I	IWCX	r	OY
5987	I	IWCX	*r	OY
5988	I	IWCX	rw	OY
5989		IWCX		PC
5991	I	IWCX	f	OY
5993	I	ICCX	*	WB
5994	I	IWCX	*	MA
5995		IWCX		WB
5996	I	IWCX	r	OY
5997	I	IWCX	r	OY
5998	I	IWCX	r	EC
5999	I	IWCX		PC
6000	I	IWCX	r	OY
6001	I	IWCX	r	OY
6002	I	IWCX	r	OY
6005	I	IWCX	*f	OY
6006	I	IWCX	r	OY
6008	I	IWCX	r	OY
6009	I	IWCX	r	OY
6010		IWCX	n	WB
6011	I	IWCX	f	WB
6012	I	IWCX	*r	OY
6013	I	PECX	*f	EC
6014	I	IWCX	f	WB
6015	I	IWCX	f	OY
6016	I	IWCX	r	OY
6018	I	IWCX	*	PC
6021	I	IWCX	f	MA
6022	I	IWCX	rw	OY
6024	I	IWCX	f	OY
6025	I	IWCX	*	MA
6026	I	IWCX	*f	OY
6027	I	IWCX	rw	OY
6028	I	IWCX	r	OY
6029	I	IWCX	r	OY
6030	I	IWCX	f	OY
6031	I	IWCX	r	OY
6034	I	IWCX	r	OY
6035	I	IWCX	rw	OY
6036	I	IWCX	*	PC
6037	I	IWCX	f	OY
6038	I	IWCX	r	OY
6041	I	IWCX	r	OY
6042	I	IWCX	r	OY
6043	I	IWCX	f	MA
6045	I	IWCX	rw	OY
6046	I	IWCX	r	OY
6047	I	IWCX	*n	WB
6049	I	IWCX		OY
6050	I	IWCX	f	WB
6051	I	IWCX	*r	OY

6052 I	IWCX	rw	OY
6053	IWCX	*	PC
6054 I	IWCX	r	OY
6055 I	IWCX	r	OY
6056 I	IWCX	f	OY
6057 I	IWCX	r	OY
6059 I	IWCX	f	WB
6060 I	IWCX	*r	OY
6061 I	IWCX	*f	WB
6062 I	IWCX	*r	OY
6063 I	IWCX	rw	OY
6064 I	IWCX	r	OY
6065 I	IWCX	f	OY
6066 I	IWCX	r	OY
6067 I	IWCX	r	OY
6073	IWCX	n	WB
6100 I	IWCX	*	MA
6101 I	IWCX	r	OY
6102 I	IWCX	f	MA
6103 I	PECX	f	EC
6104 I	IWCX	r	OY
6105 I	IWCX	r	OY
6106 I	IWCX	f	MA
6107 I	IWCX	f	OY
6110 I	IWCX	rw	WB
6111 I	IWCX	f	OY
6112	IWCX	n	WB
6113 I	IWCX		PC
6115 I	IWCX	r	OY
6116 I	IWCX	r	OY
6117 I	IWCX	r	OY
6119 I	IWCX	r	OY
6120 I	IWCX	r	OY
6121 I	IWCX	r	OY
6122 I	IWCX	r	OY
6123	IWCX		PC
6124	IWCX	r	OY
6134 I	IWCX	r	OY
6135 I	IWCX	r	OY
6136 I	IWCX	f	MA
6137 I	IWCX	f	OY
6138 I	IWCX	f	WB
6139 I	PECX	*n	EC
6141 I	IWCX	rw	OY
6142 I	IWCX	*n	WB
6144 I	IWCX	*	PC
6145 I	IWCX	*f	OY
6146 I	IWCX	*r	OY
6147 I	IWCX	*r	OY
6148 I	IWCX	*	WB
6149 I	IWCX	*rw	OY
6150 I	IWCX	*	PC
6151 I	IWCX	*	PC
6152 I	IWCX	*	WB
6153 I	IWCX	*	PC
6154 I	IWCX	*	PC
6155 I	IWCX	*	PC
6157 I	IWCX	*	PC
6158 I	IWCX	*f	OY
6159 I	IWCX	*n	WB
6160 I	IWCX	*	EC
6161 I	IWCX	*r	OY
6162 I	IWCX	f	OY
6163 I	IWCX	f	EC
6164 I	IWCX	r	OY
6165 I	IWCX	r	OY
6166 I	IWCX	f	WB
6167 I	PECX	f	EC
6168 I	IWCX	r	OY
6170 I	IWCX	r	OY
6171 I	IWCX	r	OY
6172 I	IWCX	r	OY
6173 I	IWCX	r	OY
6174 I	IWCX	f	MA
6175 I	IWCX	r	OY
6176 I	IWCX	rw	OY
6177 I	IWCX	r	OY
6178 I	IWCX	rw	OY
6179 I	IWCX	r	OY
6180 I	IWCX	rw	PC
6181 I	IWCX	f	WB
6182 I	IWCX	r	OY
6183 I	IWCX	r	OY
6184 I	IWCX	*	WB

AC2D (TSO) OPEN SECOND

Dia. AC217. Mark 2D. Air conditioned. Stones. 58S 2L. B4 bogies. ae. pa. ETH 5X. Rebuilt from FO with new style 2+2 seats.

Lot No. 30821 Derby 1971–2. 32.5 t.

6200 (3198)	I	IWRX	OM
6201 (3210)	I	IWRX	OM
6202 (3191)	I	IWRX	OM
6203 (3180)	I	IWRX	OM
6204 (3216)	I	IWRX	OM
6205 (3193)	I	IWRX	OM

6206 (3183)	I	IWRX		OM
6207 (3204)	I	ICHX		OM
6208 (3205)	I	ICHX		OM
6209 (3177)	I	ICHX		OM
6210 (3196)	I	ICHX		OM
6211 (3215)	I	IWRX		OM
6212 (3176)	I	IWRX		OM
6213 (3208)	I	IWRX	n	OM
6214 (3211)	I	IWRX		OM
6215 (3170)	I	IWRX	n	OM
6216 (3179)	I	IWRX		OM
6217 (3184)	I	IWRX		OM
6218 (3209)	I	IWRX		OM
6219 (3213)	I	IWRX		OM
6220 (3175)	I	ICHX		OM
6221 (3173)	I	ICHX		OM
6222 (3171)	I	IWRX		OM
6223 (3194)	I	IWRX		OM
6224 (3195)	I	IWRX		OM
6225 (3200)	I	IWRX		OM
6226 (3203)	I	IWRX		OM
6227 (3197)	I	IWRX		OM
6228 (3201)	I	IWRX		OM
6229 (3212)	I	IWRX		OM
6230 (3185)	I	IWRX	n	OM
6231 (3189)	I	IWRX		OM
6232 (3199)	I	IWRX		OM
6233 (3206)	I	IWRX		OM
6234 (3207)	I	IWRX		OM
6235 (3190)	I	IWRX		OM

AD4Z (OC) OBSERVATION CAR

Dia.AD401. Converted 1987 from d.m.u. DTCL. D.m.u. bogies. 42U 1L. v. pa.

Lot No. 30468 Metro-Cammell 1958. 25.5 t.

Note: This vehicle is an IS vehicle outstationed at Muir of Ord until the Ness Viaduct is repaired.

6300 (54346) **E** PISX IS I

Name: 6300 HEBRIDEAN

GX51 GENERATOR VAN

Dia. GX501. Renumbered 1989 from BR departmental series. Three-phase supply generator van for use with HST trailers. Often used at times of low availability of HST power cars. Rebuilt from NDA 81448. B5 bogies.

Lot No. 30400 Pressed Steel 1958. t.

6310 (ADB 975325) **I** IECX BN

AZ5Z SPECIAL SALOON

Dia. AZ501. Renumbered 1989 from LMR departmental series. Formerly the LMR General Manager's saloon. Rebuilt from LMS period 1 BFK M 5033 M to dia. 1654 and mounted on the underframe of BR suburban BS M 43232. B5 bogies. This vehicle has a maximum speed of 100 mph, but is restricted to 60 mph when carrying passengers with screw coupling operative.

LMS Lot No. 326 Derby 1927. t.

6320 (TDM 395707) **I** ICHV BN

GF50 (HSBV) BARRIER VEHICLE

Dia. GF501 (502*, 503†, 504§, 505x, 506y). Renumbered from departmental stock. Various original types and characteristics. ae. B4 bogies (Commonwealth bogies xy). Only 6331/39/41 had been renumbered by 01/01/90.

6330/9. Lot No. 30786 Derby 1968. 32 t.
6331/6. Lot No. 30008 BRCW 1954. 33 t.
6332/3/7. Lot No. 30229 Metro-Cammell 1956–7. 34 t.
6334/5/8. Lot No. 30775 Derby 1967–8. 32 t.
6340. Lot No. 30669 Swindon 1962. 36 t.
6341/2. Lot No. 30632 Ashford/Swindon 1961. 38 t.
6343. Lot No. 30091 Doncaster 1954. 33 t.

6330	(14084, ADB 975629)	IWRX	†	OM
6331	(3016, ADB 975650)	IWRX	*	PM
6332	(34964, ADB 975651)	IWRX		PM
6333	(34988, ADB 975652)	IWRX		PM
6334	(14067, ADB 975654)	IECX	†	NL
6335	(14065, ADB 975655)	IWRX	†	OM
6336	(3014, ADB 975658)	IECX	*	BN
6337	(35006, ADB 975660)	IECX		BN
6338	(14071, ADB 975665)	IECX	†	NL
6339	(14078, ADB 975666)	IECX	†	NL
6340	(21251, ADB 975678)	ICCX	x	LA
6341	(1967, ADB 975680)	IWRX	y	PM
6342	(1983, ADB 975681)	ICCX	y	LA
6343	(3054, ADB 975699)	IWRX	*	NL

AD2C (SO) OPEN SECOND

Dia. AD205. Renumbered 1982–4 from FO. Headrests removed from seats. Mark 2C. Pressure ventilated. B4 bogies. 42S 2L. ad. pa. ETH 4.

Lot No. 30810 Derby 1970. 33 t.

6400 (3167)	PDYX		DY
6405 (3154)	IWCD	k	WB
6406 (3156)	IWCD	k	WB
6414 (3161)	PDYX		DY
6415 (3155)	PDYX	k	DY

AD2F (SO) OPEN SECOND

Dia. AD206. Renumbered 1985–6 from FO. Headrests removed from seats. Mark 2F. Air conditioned. B4 bogies. 42S 2L. ae. pa. ETH 5X.

Note: All these vehicles should have been reseated to 76S 2L by May 1990, for use on commuter services.

6420. 6423–34. 6458–61. Lot No. 30845 Derby 1973. 33 t.
6435–52. 6463–65. Lot No. 30859 Derby 1973–4. 33 t.
6454–6455. Lot No. 30873 Derby 1974–5. 33.5 t.

6420 (3282)	I	IECX	BN
6423 (3288)		IXXH	WB
6424 (3289)		IXXH	WB
6425 (3294)		IXXH	WB
6427 (3298)	I	IECX	BN
6428 (3306)	I	IECX	BN
6429 (3307)		IXXH	WB
6431 (3311)	I	IECX	BN
6434 (3320)	I	IECX	BN
6435 (3323)	I	IECX	BN
6436 (3324)	I	IECX	BN
6437 (3327)		IXXH	WB
6439 (3339)	I	IECX	BN
6440 (3342)		IXXH	WB
6442 (3349)		IECX	BN
6443 (3355)	I	IECX	BN

6446 (3365)	I	IECX	BN
6447 (3376)		IXXH	WB
6448 (3394)		IXXH	WB
6449 (3396)	I	IECX	BN
6451 (3404)		IXXH	WB
6452 (3423)		IXXH	WB
6454 (3430)		IXXH	PC
6455 (3435)		IXXH	WB
6458 (3281)		IXXH	WB
6460 (3301)		IXXH	WB
6461 (3316)	I	IECX	BN
6463 (3410)		IXXH	WB
6464 (3380)	I	IECX	BN
6465 (3422)		IXXH	WB

AG2C (TSOT) OPEN SECOND (MICRO-BUFFET)

Dia. AG201. Mark 2C. Converted from TSO by removal of one seating bay and replacing this by a counter with a space for a trolley. Adjacent toilet removed and converted to steward's washing area/store. Pressure ventilated. 54S 1L. B4 bogies. ad. pa. ETH 4.

Lot No. 30795 Derby 1969–70. 32.5 t.

6500 (5603)	N	NSSX	pt	LA
6501 (5547)	N	NSSX	pt	LA
6502 (5602)	N	NSSX	pt	LA
6503 (5510)	N	NSSX	pt	LA
6505 (5553)		IXXZ		LA
6506 (5608)		ICCX		MA
6508 (5511)		ICCX		MA
6509 (5513)		IMLX		DY
6510 (5518)		PISX	pt	IS
6511 (5587)		IMLX		DY
6512 (5534)		NXXX		LA
6513 (5538)	N	NSSX	pt	LA
6514 (5528)		PISX	pt	IS
6515 (5556)		PNLX		NL
6516 (5551)		IWCX		WB
6517 (5499)	N	NSSX	pt	LA
6518 (5503)		PISX	pt	IS
6519 (5544)		PISX	pt	IS
6520 (5598)		IMLX		DY
6521 (5558)	N	NSSX		LA
6522 (5611)	N	NSSX	pt	LA
6523 (5568)		ICCX		MA
6524 (5564)		ICCX		MA
6526 (5537)		PDYX		DY
6527 (5563)	N	NSSX	pt	LA
6528 (5592)		PDYX		DY
6529 (5579)		PDYX		DY

AG2D (TSOT) OPEN SECOND (MICRO-BUFFET)

Dia. AG202. Mark 2D. Converted from TSO by removal of one seating bay and replacing this by a counter with a space for a trolley. Adjacent toilet removed and converted to steward's washing area/store. Air conditioned. Stones equipment. 54S 1L. B4 bogies. ae. pa. ETH 5.

Lot No. 30822 Derby 1971. 33 t.

6600 (5688)	I	ICCX	EC
6601 (5697)	PS	PECX	EC
6603 (5691)		PISX	IS
6604 (5656)	PS	PECX	EC
6605 (5741)	PS	PECX	EC
6607 (5625)		PECX	EC
6608 (5696)	I	ICCX	EC
6609 (5698)	I	IANX	NC
6613 (5702)	PS	PECX	EC
6614 (5725)	PS	PECX	EC
6616 (5720)	I	IANX	NC
6617 (5733)	I	IANX	NC
6618 (5644)	I	ICCX	EC
6619 (5655)	PS	PECX	EC

AN2D (RMBT) MINIATURE BUFFET CAR

Dia. AN207. Mark 2D. Converted from TSOT by the removal of another seating bay and fitting a proper buffet counter with boiler and microwave oven. Air conditioned. Stones equipment. 46S 1L. B4 bogies. ae. pa. ETH 5.

Lot No. 30822 Derby 1971. 33 t.

Note: Original numbers shown in parentheses. These vehicles carried 6602/10–2/5 when they were TSOTs.

6652 (5622)	**I**	IWCX	WB
6660 (5627)	**I**	IWRX	LA
6661 (5736)	**I**	IWRX	LA
6662 (5641)	**I**	IWRX	LA
6665 (5721)	**I**	IWCX	WB

AN1F (RLO) SLEEPER RECEPTION CAR

Dia. AN101 (AN102*). Mark 2F. Converted from FO, these vehicles consist of pantry, microwave cooking facilities, seating area for passengers, telephone booth and staff toilet. 6703–8 also have a bar. Converted at RTC, Derby (6700), Ilford (6701–5) and Derby (6706–8). 26F 1L.

6701–2/4/8. Lot No. 30859 Derby 1973–4. 33.5 t.
6703/5–7. Lot No. 30845 Derby 1973. 33.5 t.

Note: 6705–7 were also numbered 6430/21/18 when declassified prior to conversion.

6700 (3347)	**I**	IWCX	EC
6701 (3346)	**I**	IWCX	EC
6702 (3421)	**I**	IWCX	EC
6703 (3308)	**I**	IWCX	EC
6704 (3341)	**I**	IWCX	EC
6705 (3310)	**I**	IWCX	EC
6706 (3283)	**I**	IWCX	EC
6707 (3276)	**I**	IWCX	EC
6708 (3370)	**I**	IWCX	EC

AA31 (CK) CORRIDOR COMPOSITE

Mark 1. There are two variants: those built for the Eastern, London Midland, North Eastern or Scottish regions which have 6-seater second class compartments with armrests (24F 18S 2L. Dia. AA301) and those built for the Southern and Western regions which have 8-seater second class compartments without armrests (24F 24S 2L. dia AA302). All vehicles are assumed to seat 24F 18S, unless marked * (24F 24S). vd. All remaining vehicles were renumbered from the 16xxx series, with the 16 being replaced by a 7 and were built with Commonwealth bogies, metal window frames and melamine finish in the second class sections. ETH 4.

7153–7195. Lot No. 30665 Derby 1961. 37 t.
7212–7219. Lot No. 30666 Derby 1961. 37 t.
7232–7237. Lot No. 30729 Derby 1963. 37 t.

f–Facelifted with fluorescent lighting, wide mirrors, blinds removed.

7153	**N**	NNWX		EN
7156	**N**	NWRX		OM
7158	**N**	NWRX		OM
7165	**N**	NNWX	pa	EN
7166	**N**	NNWX		EN
7167	**N**	NNWX		EN
7168	**N**	NNWX		EN
7172	**N**	NNWX	pa	EN
7188	**N**	NNWX		EN
7190	**N**	NNWX		EN
7195	**N**	NNWX		EN
7212	**N**	NNWX	*f	EN
7213	**N**	NNWX	*f	EN
7219	**N**	NNWX	*f	EN
7232	**N**	NNWX	*fpa	EN
7235		NNWX		EN
7237	**N**	NNWX		EN

AA3C (CK) CORRIDOR COMPOSITE

Dia. AA304. Mark 2C. Converted 1985 from FK. Pressure ventilated. 18F 24S 2L. B4 bogies. ad. pa. ETH 4.

Lot No. 30797 Derby 1969–70. 33.5 t.

7550 (13550)	PISX	IS	7558 (13558)	PISX	IS
7551 (13551)	PISX	IS	7561 (13561)	PISX	IS
7553 (13553)	PISX	IS			

AH21 (BSOT) OPEN BRAKE SECOND (MICRO-BUFFET)

Dia. AH202. Mark 1. Converted from BSO by removal of one seating bay and replacing this by a counter with a space for a trolley. Adjacent toilet removed and converted to a steward's washing area/store. 31S 0L. ETH 3.

9007. Lot No. 30698 Wolverton 1963. Commonwealth bogies. Fluorescent lighting. and metal window frames. Reinstated for steam locomotive driver training at Fort William.
9016. Lot No. 30170 Doncaster 1955–6. BR1 bogies. xdk. 34 t.

9007 (9374)	ICHS		FW	9016 (9237)	PCLX	*	CL

AH2Z (BSOT) OPEN BRAKE SECOND (MICRO-BUFFET)

Dia. AH203. Mark 2. Converted from BSO by removal of one seating bay and replacing this by a counter with a space for a trolley. Adjacent toilet removed and converted to a steward's washing area/store. 23S 0L (27S 0L*). vd. ETH 4.

Lot No. 30757 Derby 1966.

9100 (9405)	PISX		IS	9104 (9401)	PCLX	pt	CL
9101 (9398)	PISX		IS	9105 (9404)	PPCX	pt	PC
9102 (9383)	PNLX	pt	NL	9106 (9406)	PCLX	pt	CL
9103 (9389)	PCLX	pt	CL	9107 (9407)	PCLX	pt	CL

AE2Z (BSO) OPEN BRAKE SECOND

Dia. AE203. Mark 2. These vehicles use the same body shell as the mark 2 BFK and have first class seat spacing and wider tables. Pressure ventilated. 31S 1L. B4 bogies. vd. ETH 4.
*–Air braked by means of disc brakes.
Note: The vehicles shown allocated to 'MO' are IS vehicles outstationed at Muir of Ord until the Ness Viaduct is repaired.

Lot No. 30757 Derby 1966. 31.5t (33.5t*)

9382	**N**	NWRX	pt	OM	9391	**N**	NNEX	to	CA
9384	**N**	NNEX	to	CA	9392		PNLX		NL
9385		PNLX		NL	9393		PISX		IS
9386		PNLX		NL	9394		PNLX		NL
9388	**E**	PMOX		MO	9396		NNEX	to	CA
9390	**I**	NNEX	to	CA	9397		PISX	* pt	IS

9400	ICHL	*	IS
9409 **N**	NNEX	to	CA
9410	PNLX		NL
9412	PISX	* pt	IS
9413	ICHS	pt	OY
9414 **E**	PMOX	pt	MO

Named vehicles:

9388 BAILECHAUL

9414 BRAHAN SEER

AE2A (BSO) OPEN BRAKE SECOND

Dia. AE204. Mark 2A. These vehicles use the same body shell as the mark 2A BFK and have first class seat spacing and wider tables. Pressure ventilated. 31S 1L. B4 bogies. ad. ETH 4.

9417–9425. Lot No. 30777 Derby 1967. 31.5 t.
9426–9438. Lot No. 30788 Derby 1968. 31.5 t.

9417	PDYX		DY
9418	PECX	pt	EC
9419	PLLX		LL
9420	PDYX		DY
9421	IMLX		DY
9422 **N**	PLLX		LL
9423	PISX	pt	IS
9424	PISX	pt	IS
9425	PDYX	pt	DY
9426	PLLX		LL
9427	IWRX		OM
9428	PDYX	pt	DY
9429	IWCD		WB
9431	PISX	pt	IS
9432	IWCD		MA
9434	PISX	pt	IS
9435 **N**	PLLX		LL
9436	PCLX	pt	CL
9437	PCLX	pt	CL
9438 **I**	IWRX		OM

AE2C (BSO) OPEN BRAKE SECOND

Dia. AE205. Mark 2C. Pressure ventilated. 31S 1L. B4 bogies. ad. ETH 4.

9439–9448. Lot No. 30798 Derby 1970. 32 t.
9449–9478. Lot No. 30820 Derby 1970. 32 t.

9439	IWCX		WB
9440	PCLX	pt	CL
9441	ICHM	pt	MA
9442	PLLX		LL
9443	IMLX	pt	DY
9444	PISX	pt	IS
9445	PISX	pt	IS
9446	PISX	pt	IS
9447	IMLX	pt	DY
9448	PCLX		CL
9449	PECX		EC
9450	IWCX	pt	WB
9451	PISX	pt	IS
9452	PLLX		LL
9453	IWCX	pt	WB
9454	ICCX	pt	MA
9455	IXXH	pt	DY
9456	ICHM	pt	MA
9457	ICHM	pt	MA
9458	PCLX		CL
9459 **I**	IWRX		OM
9460	PISX	pt	IS
9461	IMLX		DY
9462	PCLX		CL
9463	PISX	pt	IS
9464	ICCX		MA
9465	IWRX		OM
9466	IMLX	pt	DY
9467	ICCX		MA
9468	IWCX		WB
9469	ICHM		MA
9471	IWCX	pt	WB
9472	ICHM	pt	MA
9473	IWCX	pt	WB
9474	ICCX		MA
9475	IWCX	pt	WB
9476	ICHM	pt	MA
9477	IWCD	pt	MA
9478	ICCX		MA

AE2D (BSO) OPEN BRAKE SECOND

Dia. AE206. Mark 2D. Air conditioned (Stones). 31S 1L. B4 bogies. ae. pg. ETH 5.

Lot No. 30824 Derby 1971. 33 t.

No.	Pool	Depot
9479 I	IANX	NC
9480 I	IWRX	OM
9481 I	IWRX	OM
9482 I	IWRX	OM
9483 I	IANX	NC
9484	IWRX	OM
9485 I	IANX	NC
9486 I	IANX	NC
9487 I	IANX	NC
9488 I	IANX	NC
9489 I	IANX	NC
9490 I	IWRX	OM
9492 I	IANX	NC
9493 I	IANX	NC
9494 I	IANX	NC
9495 I	IANX	NC

AE2E (BSO) OPEN BRAKE SECOND

Dia. AE207. Mark 2E. Air conditioned (Stones). 32S 1L. B4 bogies. ae. pg. ETH 5.

Lot No. 30838 Derby 1972. 33 t.

No.	Pool	Depot
9496 I	IWCX	WB
9497 I	IWCX	WB
9498 I	IWRX	LA
9499	IWRX	LA
9500 I	IWCX	OY
9501 I	IWRX	OM
9502 I	IWCX	WB
9503 I	IWRX	OM
9504 I	ICCX	MA
9505 I	IWCX	OY
9506 I	IWRX	LA
9507 I	IWCX	OY
9508 I	IWRX	OM
9509	ICCX	MA

AE2F (BSO) OPEN BRAKE SECOND

Dia. AE208. Mark 2F. Air conditioned (Temperature Ltd.). 32S 1L. B4 bogies. ae. pg. ETH 5X.

f–Facelifted with new panels & red seat trim.
r–Refurbished. Power-operated vestibule doors, new panels and seat trim.
Note: not all vehicles fitted were known at time of going to press.

Lot No. 30861 Derby 1974. 34 t.

No.	Pool	Notes	Depot
9513 I	IWCX	f	WB
9516 I	IWCX	r	OY
9520 I	IWCX	rn	WB
9521 I	IWCX	r	OY
9522 I	IWCX		WB
9523 I	IWCX	f	WB
9524 I	IWCX	f	WB
9525 I	IWCX	r	OY
9526 I	IWCX	fn	WB
9527 I	IWCX	r	WB
9529 I	IWCX	r	WB
9531 I	IWCX	f	WB
9533 I	IWCX		WB
9537 I	IWCX	fn	WB
9538 I	IWCX		WB
9539 I	IWCX	r	OY

AF2F (DBSO) DRIVING OPEN BRAKE SECOND

Dia. AF201. Mark 2F. Air conditioned (Temperature Ltd.). Push & pull (two wire RHC cable system) vehicles with half-cab at brake end. Converted from BSO. Fitted with cowcatchers. 32S 0L. B4 bogies. ae. pg. ETH 5X.

9701–9710. Lot No. 30861 Derby 1974. Converted 1979. Disc brakes. 34 t.
9711–9713. Lot No. 30861 Derby 1974. Converted Glasgow 1985. 34 t.

9714. Lot No. 30861 Derby 1974. Converted Glasgow 1986. Disc brakes. 34 t.

9701	(9528)	**PS**	PECX	EC
9702	(9510)	**PS**	PECX	EC
9703	(9517)	**PS**	PECX	EC
9704	(9512)	**PS**	PECX	EC
9705	(9519)	**PS**	PECX	EC
9707	(9511)	**PS**	PECX	EC
9708	(9530)	**PS**	PECX	EC
9709	(9515)	**PS**	PECX	EC
9710	(9518)	**PS**	PECX	EC
9711	(9532)	**PS**	PECX	EC
9712	(9534)	**PS**	PECX	EC
9713	(9535)	**PS**	PECX	EC
9714	(9536)	**I**	IANX	NC

AJ2G (RSM) RESTAURANT BUFFET FIRST (MODULAR)

Dia. AJ201. Mark 3A. Air conditioned. Converted from HST TRFKs. 24F. BT10 bogies. ae. pt. ETH 14X.

Lot No. 30884 Derby 1977.

10200	(40519)	I	IWCX	WB
10201	(40520)	I	IWCX	WB

AJ1G (RFM) RESTAURANT BUFFET FIRST (MODULAR)

Dia. AJ103. Mark 3A. Air conditioned. Under conversion from HST TRFKs, RFBs and FOs. 24F. BT10 bogies. ae. pt. ETH 14X.

10202–10211. Lot No. 30884 Derby 1977.
10212–10229. Lot No. 30878 Derby 1975–6.
10230–10260. Lot No. 30890 Derby 1979.

10202	(40504)	I	IWCX		MA
10203	(40506)	I	IWCX		WB
10204	(40502)	I	IWCX		OY
10205	(40503)	I	IWCX		OY
10206	(40507)	I	IWCX		OY
10207	(40516)	I	IWCX		OY
10208	(40517)	I	IWCX		WB
10209	(40508)	I	IWCX	p	OY
10210	(40509)	I	IWCX		WB
10211	(40510)	I	IWCX		WB
10212	(11049)	I	IWCX		WB
10213	(11050)	I	IWCX		WB
10214	(11034)	I	IWCX		OY
10215	(11032)	I	IWCX		WB
10216	(11041)	I	IWCX		OY
10217	(11051)	I	IWCX		MA
10218	(11053)	I	IWCX		OY
10219	(11047)	I	IWCX		OY
10220	(11056)	I	IWCX		OY
10221	(11012)	I	IWCX		WB
10222	(11063)	I	IWCX	p	OY
10223	(11043)	I	IWCX		WB
10224	(11062)	I	IWCX	p	WB
10225	(11014)	I	IWCX		OY
10226	(11015)	I	IWCX		WB
10227	(11057)	I	IWCX		WB
10228	(11035)	I	IWCX		OY
10229	(11059)	I	IWCX		MA
10230	(10021)	I	IWCX		WB
10231	(10016)	I	IWCX		WB
10232	(10027)	I	IWCX		OY
10233	(10013)	I	IWCX	p	WB
10234	(10004)	I	IWCX		OY
10235	(10015)	I	IWCX		WB
10236	(10018)	I	IWCX		WB
10237	(10022)	I	IWCX		WB
10238	(10017)	I	IWCX		WB
10240	(10003)	I	IWCX	p	OY
10241	(10009)	I	IWCX		WB
10242	(10002)	I	IWCX		WB
10245	(10019)	I	IWCX		PC
10246	(10014)	I	IWCX		PC
10247	(10011)	I	IWCX		PC
10248	(10005)	I	IWCX	p	WB
10249	(10012)	I	IWCX		OY
10250	(10020)	I	IWCX		OY
10251	(10024)	I	IWCX		PC
10252	(10008)	I	IWCX		WB
10253	(10026)	I	IWCX		WB
10254	(10006)	I	IWCX	p	WB
10255	(10010)	I	IWCX		OY
10256	(10028)	I	IWCX		WB

10257 (10007) I	IWCX		WB	10259 (10025) I	IWCX	p	WB
10258 (10023) I	IWCX		WB	10260 (10001) I	IWCX	p	WB

AJ1J (RFM) RESTAURANT BUFFET FIRST (MODULAR)

Dia. AJ105. Mark 4. Air conditioned. 20F 1L. SIG bogies (BT41). ae. pt. ETH X. Stencilled 'SV' (service vehicle).

Lot No. 31045 Metro-Cammell 1989 onwards. 45.5 t.

10300 I IECX BN
10301 I IECX BN
10302 I IECX BN
10303 I IECX BN
10304 I IECX BN
10305 I IECX BN
10306
10307
10308
10309
10310
10311
10312
10313
10314
10315
10316
10317
10318
10319
10320
10321
10322
10323
10324
10325
10326
10327
10328
10329
10330
10331
10332
10333

AU4G (SLEP) SLEEPING CAR WITH PANTRY

Dia. AU401. Mark 3A. Air conditioned. 12 compartments with a fixed lower berth and a hinged upper berth, plus an attendants compartment with 2L (controlled emission). BT10 bogies. ae. ETH 7X.

† Fitted with short swing link bogies. Able to work over Southern Region.
(S) Stored at MoD Bicester.

Lot No. 30960 Derby 1981–3.

10500	I	ICHV		BN
10501	I	IWCX		EC
10502	I	IWCX		EC
10503	I	ICHV		BN
10504		IWCX		EC
10506	I	IWCX		WB
10507	I	IWCX		WB
10508	I	IWCX		EC
10509		IXXH		S
10510	I	IWCX		EC
10511		IXXH		S
10512		IWCX		EC
10513	I	IWCX		EC
10514	I	ICHV		BN
10515		IWCX		EC
10516	I	IWCX		EC
10517		IXXH		S
10518		IXXH		S
10519	I	IWCX		EC
10520	I	IWCX	†	EC
10521		IXXH		S
10522	I	IWCX		EC
10523	I	IWCX		EC
10525		IXXH		S
10526	I	IWCX	†	EC
10527	I	IWCX		EC
10529	I	IWCX		EC
10530	I	IWCX		EC
10531	I	IWCX		EC
10532	I	ICCX	†	EC
10533		IWCX		EC
10534	I	ICCX		EC
10535		ICCX		EC
10536	I	IWCX		EC
10537		ICCX		EC
10538		IWCX		EC
10539	I	IWCX	†	EC
10540		IWCX		EC
10541	I	IWCX		EC
10542	I	IWCX		EC

10543		IWCX		EC	10582	I	IWCX		EC
10544		IWCX		EC	10583	I	IWCX		EC
10545	I	ICCX	†	EC	10584	I	IWCX		EC
10546	I	IWCX		EC	10585		IWRX		LA
10547	I	IWCX		EC	10586	I	IWRX		LA
10548	I	IWCX		EC	10587	I	IWRX		LA
10549	I	IWCX		EC	10588		IWRX		LA
10550	I	IWCX		EC	10589		IWCX		EC
10551	I	IWCX		EC	10590	I	IWCX		EC
10552		IXXH		ZN	10591		IWCX	†	WB
10553	I	IWRX		LA	10592	I	ICCX	†	EC
10554		ICCX	†	EC	10593	I	IWCX	†	EC
10555	I	IWCX	†	EC	10594	I	IWRX		LA
10556	I	IWRX		LA	10595		PECX	†	EC
10557	I	IWCX		EC	10596	I	IWCX		EC
10558	I	IWRX		LA	10597	I	IWCX	†	EC
10559	I	IWCX		EC	10598	I	IWCX		EC
10560		IWCX		EC	10599		IWCX		WB
10561	I	IWCX		EC	10600		IWCX		EC
10562		IWCX		EC	10601		IWCX		WB
10563	I	IWCX		EC	10602		IWCX		EC
10564	I	IWRX		LA	10603	I	IWRX		LA
10565	I	IWCX		EC	10604	I	IWCX		EC
10566	I	IWCX		EC	10605	I	IWCX		EC
10567		IWCX		WB	10606	I	IWCX		EC
10568		IWCX		EC	10607	I	IWCX		EC
10569		IWCX	†	WB	10608		IXXH		ZN
10570	I	IWRX		LA	10609		PECX	†	EC
10571		IWCX		WB	10610	I	IWCX	†	EC
10572		IWCX		WB	10611		IXXH	†	S
10573	I	IWCX		EC	10612		IWCX	†	EC
10574	I	ICHV		BN	10613	I	IWCX	†	EC
10575	I	ICHV		BN	10614	I	IWCX	†	EC
10576		PECX	†	EC	10615		IWCD	†	WB
10577		PECX	†	EC	10616		ICCX	†	EC
10578	I	IWCX	†	EC	10617		ICCX	†	EC
10579		PECX	†	EC	10618		IXXH		S
10580	I	IWCX		WB	10619		IXXH		S
10581		PECX	†	EC					

AS4G (SLE) SLEEPING CAR

Dia. AS403. Mark 3A. Air conditioned. 13 compartments with a fixed lower berth and a hinged upper berth. 2L (controlled emission). BT10 bogies. ae. ETH 6X.

† Fitted with short swing link bogies. Able to work over Southern Region.
(S) Stored at MoD Bicester.

Lot No. 30961 Derby 1980–4.

10646	I	ICHV	BN	10648	I	IWCX	EC
10647	I	IWCX	EC	10649	I	IWCX	EC

10650	I	IWCX		EC
10651	I	IWCX		EC
10653	I	IWCX		EC
10654		IWCX		EC
10655	I	IWCX		EC
10656	I	IWCX		EC
10657	I	IWCX		EC
10658	I	IWCX		EC
10660	I	IWCX		EC
10661	I	ICHV		BN
10662		ICHV		BN
10663	I	ICCX	†	EC
10665	I	IWCX		EC
10666	I	ICCX	†	EC
10668	I	IWCX		WB
10670	I	IWCX		EC
10671	I	ICCX		EC
10672		ICCX		EC
10673		IWCX		EC
10674		ICCX		EC
10675	I	IWCX		EC
10678		IWCX		EC
10679		IWCX		EC
10680	I	IWCX	†	EC
10682	I	IWCX		EC
10683	I	IWCX		EC
10684		IWCX	†	EC
10685	I	IWCX	†	EC
10686	I	IWCX		WB
10687		IWCX		EC
10688	I	ICCX		EC
10689		IWCX		EC
10690	I	IWCX		EC
10691		IWCX		EC
10692	I	IWCX		EC
10693	I	IWCX	†	EC
10696	I	IWCX	†	EC
10697	I	IWCX	†	EC
10699		IWCX	†	EC
10700		IWCX		EC
10701		IWCX		WB
10702	I	IWCX		EC
10703	I	IWCX		EC
10704	I	IWCX	†	EC
10705	I	IWCX		EC
10706	I	IWCX		EC
10707	I	IWCX		EC
10708	I	IWCX	†	EC
10709	I	IWCX		EC
10710		IWCX		EC
10711	I	IWCX		EC
10712	I	IWCX	†	EC
10713	I	IWCX		EC
10714	I	IWCX		EC
10715	I	IWCX		EC
10716	I	IWCX	†	EC
10717	I	IWCX	†	EC
10718	I	IWCX	†	EC
10719		IWCX	†	EC
10720	I	IWCX		EC
10722	I	IWCX		EC
10723	I	IWCX		EC
10724	I	ICHV		BN
10725		ICHV		BN
10726	I	ICHV		BN
10727	I	ICHV		BN
10728		ICHV		BN
10729		ICHV		BN
10730	I	IWCX		EC
10731	I	IWCX		EC
10732		IWCX		EC

AD1G (FO) OPEN FIRST

Dia. AD108. Mark 3A. Air conditioned. 48F 2L. BT10 bogies (BT15*). ae. pa. ETH 6X. All now facelifted with new upholstery, carpets etc. 11005–7 are being converted from 11905–7.

r–Refurbished. Inter-City 80 seats (APT style).

Lot No. 30878 Derby 1975–6. 33 t.

11005				
11006				
11007				
11011	I	IWCX	z	WB
11013	I	IWCX		WB
11016	I	IWCX		WB
11017	I	IWCX		WB
11018	I	IWCX		WB
11019	I	IWCX		WB
11020	I	IWCX		WB

11022			
11023	I	IWCX	WB
11024	I	IWCX	WB
11026	I	IWCX	WB
11027	I	IWCX	WB
11028	I	IWCX	WB
11029	I	IWCX	WB
11030	I	IWCX	WB
11031	I	IWCX	WB
11033	I	IWCX	WB
11036	I	IWCX	WB
11037	I	IWCX	WB
11038	I	IWCX	WB
11039	I	IWCX	WB
11040	I	IWCX	WB
11042	I	IWCX	WB
11044	I	IWCX	WB
11045	I	IWCX	WB
11046	I	IWCX	WB
11048	I	IWCX	WB
11052	I	IWCX	WB
11054	I	IWCX	WB
11055	I	IWCX	WB
11058	I	IWCX	WB
11060	I	IWCX	WB

AD1H (FO) OPEN FIRST

Dia. AD109. Mark 3B. Air conditioned. 48F 2L. BT10 bogies. ae. pa. ETH 6X. Inter-City 80 seats. P–"Pullman".

Lot No. 30982 Derby 1985. 33 t.

11064	I	IWCX	WB
11065	I	IWCX	WB
11066	I	IWCX	WB
11067	I	IWCX	WB
11068	I	IWCX	WB
11069	I	IWCX	WB
11070	I	IWCX	WB
11071	I	IWCX	WB
11072	I	IWCX	WB

11073	I	IWCX	P	WB	
11074	I	IWCX	P	WB	Thomas Brassey
11075	I	IWCX	P	WB	
11076	I	IWCX	P	WB	
11077	I	IWCX	P	WB	Sir Richard Arkwright
11078	I	IWCX	P	WB	John Owens
11079	I	IWCX	P	WB	Francis Egerton
11080	I	IWCX	P	WB	
11081	I	IWCX	P	WB	
11082	I	IWCX	P	WB	James Joule
11083	I	IWCX	Pt	WB	
11084	I	IWCX	Pt	WB	William Roscoe
11085	I	IWCX	Pt	WB	Sir John Barbirolli
11086	I	IWCX	Pt	WB	
11087	I	IWCX	Pt	WB	Sir William Fairbairn
11088	I	IWCX	Pt	WB	Sir John Brunner
11089	I	IWCX	Pt	WB	George Stubbs
11090	I	IWCX	Pt	WB	
11091	I	IWCX	Pt	WB	Sir Stanley Matthews
11092	I	IWCX	Pt	WB	Ernest Rutherford
11093	I	IWCX	Pt	WB	L. S. Lowry
11094	I	IWCX	Pt	WB	

11095	I	IWCX	p	WB
11096	I	IWCX	p	WB
11097	I	IWCX	p	WB
11098	I	IWCX	p	WB
11099	I	IWCX	p	WB
11100	I	IWCX	p	WB
11101	I	IWCX	p	WB

MARK 1 STOCK. RBR 1973 at Wigan NW on 19th October 1988 in InterCity livery with Commonwealth bogies. *John Augustson*

FO 3147 at Gloucester on 22nd May 1988. This coach is part of the InterCity charter fleet and has been fitted with table lamps. The latest variant of InterCity livery has been applied, as has white roof paint. *Norman Preedy*

SK 18602, seen at Oxford on 8th August 1989 sports Network SouthEast livery and B4 bogies. *John Augustson*

BCK 21269 is kept as spare at Carstairs for attaching to portions of trains from the south which might have no brake vehicle. It was seen on 12th August 1989. *A.O. Wynn*

MARK 2 STOCK. The original Mark 2 stock was built with vacuum brakes and dual heating. This view shows BSO 9388 'BAILECHAUL' at Kyle of Lochalsh on 19th July 1989 in LNER green and cream livery. *N.J. Joynson*

MARK 2A STOCK. The Mark 2As were built with air brakes, the first class vehicles having a new compartment design. All Mark 2 variants have B4 bogies. BFK 17093 is seen at Hull on 23rd July 1989. *John Augustson*

MARK 2B STOCK. This was the first stock to have large wrap-round slam doors with the centre door removed. TSO 5463 at Durham on 30th September 1988 is in Provincial livery with "Trans-Pennine" branding. *John Augustson*

MARK 2C STOCK. These are similar to Mark 2B stock but fitted with twin rows of ceiling lights and ducting for possible future air conditioning (never actually carried out). TSOT 6501 in Network SouthEast livery at Exeter St. Davids was on an Exeter–Waterloo train on 24th June 1989. *John Augustson*

MARK 2D STOCK. The Mark 2D was the first ordinary air-conditioned variant and the FKs were BR's last build of loco-hauled compartment vehicles. FK 13601 is seen at Cheltenham on 17th October 1989. *B.J. Nicolle*

MARK 2E STOCK. The Mark 2Es featured smaller toilets with luggage racks opposite them. FO 3265 in Provincial ScotRail livery is seen at Dundee on 23rd August 1989. *John Augustson*

MARK 2F STOCK. These have InterCity 70 seats and a modified air-conditioning system. Although no catering vehicles were a feature of the entire Mark 2 range, a number of FOs have at last been converted to RFBs. 1203 is seen at Carlisle on 1st April 1989.
John Augustson

Another Mark 2F conversion was that of a number of BSOs to D3SOs, initially for the Glasgow–Edinburgh service but later also for Aberdeen trains. 9714 is seen at Dyce on 7th August 1989 having arrived with the 06.25 ex Perth.

Duncan Street

MARK 3 STOCK. Mark 3 coaches are approximately 23 m in length and are characterised by their ribbed roofs and BT 10 bogies. FO 11021 is seen at Crewe on 5th July 1988. *John Augustson*

40513, seen at Leeds on 20th September 1989 is a first class lounge car, converted from a TRUK and can be incorporated in any HST set. *John Augustson*

MARK 4 STOCK. New Mark 4 stock first class service vehicle (first class with kitchen & bar) 10302 at Leeds on 20th September 1989. *John Augustson*

FO 11205 is seen at York Holgate Sidings on 17th September 1989. *B.J. Nicolle*

An interior view of a Mark 4 FO. *Peter Fox*

SPECIAL SERVICE VEHICLES. Royal Saloon 2923 at Carlisle on 15th June 1989. Like all other Royal vehicles, 2923 is painted in the special Royal livery. *John Augustson*

6300 'HEBRIDEAN' is an observation car converted from a Class 101 DMU car. It is used on the summer extra train from Inverness to Kyle of Lochalsh and was seen at Dingwall on 22nd September 1989. *John Augustson*

DVTs. Mark 3 DVT 82119 arrives at Euston at the head of a service from Wolverhampton on 7th August 1989. *John Augustson*

Mark 4 DVT 82200 at Holgate Sidings, York shortly after arrival from Peterborough on 17th September 1989. *B.J. Nicolle*

Post Office vehicles 80351 and 80439 in Post Office Red livery at Derby on 26th October 1989. *John Eggleshaw*

NEX 92315 in Network SouthEast livery at Oxford on 8th August 1989. *John Augustson*

Newspaper Van 95329 was newly modified with Commonwealth bogies when this photograph was taken on 23rd March 1988. With the demise of the carriage of newspapers by rail these vehicles are now in common use. *John Augustson*

Motorail GUV 96169 is seen at Inverness on 22nd August 1989. *John Augustson*

AD1J (FO) OPEN FIRST

Dia. AD111. Mark 4. Air conditioned. 46F 1L. SIG bogies (BT41). ae. pa. ETH 6. Stencilled 'PO' (Pullman open).

Lot No. 31046 Metro-Cammell 1989 onwards. 39.7 t.

11200 **I** IECX BN
11201 **I** IECX BN
11202 **I** IECX BN
11203 **I** IECX BN
11204 **I** IECX BN
11205 **I** IECX BN
11206 **I** IECX BN
11207 **I** IECX BN
11208 **I** IECX BN
11209 **I** IECX BN
11210
11211
11212
11213
11214
11215
11216
11217
11218
11219
11220
11221
11222
11223
11224
11225
11226
11227
11228
11229
11230
11231
11232
11233
11234
11235
11236
11237
11238
11239
11240
11241
11242
11243
11244
11245
11246
11247
11248
11249
11250
11251
11252
11253
11254
11255
11256
11257
11258
11259
11260
11261
11262
11263
11264
11265
11266
11267
11268
11269
11270
11271

AD3G (CO) OPEN COMPOSITE

Dia. AD301. Mark 3A. Air conditioned. 23F 38S 2L. Converted from FO. BT10 bogies. ae. pa. ETH 6X. Note: These vehicles are being converted to FO.

Lot No. 30878 Derby 1975. 33 t.

11905 (11005) **PS** PECX EC
11906 (11006) **PS** PECX EC
11907 (11007) **PS** PECX EC
11908 (11008) **PS** PECX EC
11909 (11009) **PS** PECX EC
11910 (11010) **PS** PECX EC
11922 (11022) **PS** PECX EC

AC2G (TSO) OPEN SECOND

Dia. AC212 (AC216 w, AC220 zw). Mark 3A. Air conditioned. 72S 2L. BT10 bogies. ae. pa. ETH 6X.

f Facelifted with new upholstery, carpets etc.
r Refurbished with modified seat backs in addition to above. 76S 2L.
z Refurbished and with toilet for disabled.
§ Fitted with SIG bogies.

Lot No. 30877 Derby 1975–7.

12004	**PS**	IWCX	r	WB
12005	**I**	IWCX	r	WB
12007	**I**	IWCX	r	WB
12008	**I**	IWCX		WB
12009	**I**	IWCX	r	WB
12010	**I**	PECX		EC
12011	**I**	IWCX	r	WB
12012	**PS**	IWCX		WB
12013	**I**	IWCX	r	WB
12014	**I**	IWCX	r	WB
12015	**PS**	IWCX		WB
12016	**PS**	PECX		EC
12017	**I**	IWCX	r	WB
12019	**PS**	PECX		EC
12020	**PS**	PECX		EC
12021	**PS**	IWCX		WB
12022	**I**	IWCX	r	WB
12023	**I**	IWCX		WB
12024	**I**	IWCX	r	WB
12025	**I**	IWCX	r	WB
12026	**PS**	PECX		EC
12027	**I**	IWCX	r	WB
12028	**I**	IWCX	r	WB
12029	**PS**	PECX		EC
12030	**PS**	IWCX		WB
12031	**I**	PECX	r	EC
12032	**I**	IWCX	f	WB
12033	**I**	IWCX		WB
12034	**I**	IWCX	f	WB
12035	**I**	IWCX	f	WB
12036	**I**	IWCX	rw	WB
12037	**I**	IWCX	f	WB
12038	**I**	IWCX	r	WB
12040	**I**	IWCX	r	WB
12041	**I**	IWCX	f	WB
12042	**I**	IWCX	rw	WB
12043	**I**	PECX	r	EC
12044	**I**	IWCX	f	WB
12045	**I**	PECX	f	EC
12046	**I**	IWCX	r	WB
12047	**I**	IWCX	rw	WB
12048	**I**	IWCX		WB
12049	**I**	IWCX	f	WB
12050	**I**	IWCX	f	WB
12051	**I**	IWCX	r	WB
12052	**I**	IWCX	f	WB
12053	**I**	PECX	f	EC
12054	**I**	IWCX	rw	WB
12055	**I**	IWCX		WB
12056	**I**	IWCX	f†	WB
12057	**I**	IWCX	r	WB
12058	**I**	IWCX	r	WB
12059	**I**	IWCX	rw	WB
12060	**I**	IWCX	f	WB
12061	**I**	IWCX	f	WB
12062	**I**	IWCX	*	WB
12063	**I**	IWCX		WB
12064	**I**	IWCX	f	WB
12065	**I**	IWCX	f	WB
12066	**I**	PECX	f	EC
12067	**I**	PECX	f	EC
12068	**I**	IWCX	f	WB
12069	**I**	PECX	f	EC
12070	**I**	IWCX	r	WB
12071	**I**	IWCX	f	WB
12072	**I**	IWCX	f	WB
12073	**I**	IWCX	f	WB
12075	**I**	IWCX	f	WB
12076	**I**	IWCX	f	WB
12077	**I**	IWCX		WB
12078	**I**	PECX	f	EC
12079	**I**	IWCX	f	WB
12080	**I**	IWCX		WB
12081	**I**	IWCX	r	WB
12082	**I**	IWCX	r	WB
12083	**I**	IWCX		WB
12084	**I**	IWCX		WB
12085	**I**	IWCX	rw	WB
12086	**I**	IWCX	rw	WB
12087	**I**	IWCX	rw	WB
12088	**I**	IWCX	zw	WB
12089	**I**	IWCX	r	WB
12090	**I**	IWCX	f	WB
12091	**I**	IWCX	f	WB
12092	**I**	IWCX	r	WB
12093	**I**	IWCX	f*	WB
12094	**I**	IWCX		WB
12095	**I**	IWCX	r	WB
12096	**I**	IWCX		WB
12097	**I**	IWCX	r	WB
12098	**I**	IWCX	r	WB
12099	**I**	IWCX		WB
12100	**I**	IWCX		WB
12101	**I**	IWCX	rw	WB
12102	**I**	IWCX	f	WB
12103	**I**	IWCX	zw	WB
12104	**I**	IWCX	f	WB
12105	**I**	IWCX	f	WB
12106	**I**	IWCX	f	WB
12107	**I**	IWCX		WB

12108	I	IWCX	rw	WB
12109	I	IWCX	zw	WB
12110	I	IWCX	f	WB
12111	I	IWCX	r	WB
12112	I	IWCX	zw	WB
12113	I	IWCX	f	WB
12114	I	IWCX	f	WB
12115	I	IWCX	f	WB
12116	I	IWCX	f	WB
12117	I	PECX		EC
12118	I	IWCX	r	WB
12119	I	IWCX		WB
12120	I	IWCX		WB
12121	I	IWCX	f	WB
12122	I	IWCX	zw	WB
12123	I	IWCX		WB
12124	I	IWCX	f	WB
12125	I	IWCX	f	WB
12126	I	IWCX	r	WB
12127	I	IWCX	f	WB
12128	I	IWCX	rw	WB
12129	I	IWCX	f	WB
12130	I	IWCX	f	WB
12131	I	IWCX		WB
12132	I	IWCX	r	WB
12133	I	IWCX		WB
12134	I	IWCX	f	WB
12135	I	IWCX	f	WB
12136	I	IWCX		WB
12137	I	IWCX	f	WB
12138	I	IWCX	f	WB
12139	I	IWCX	r	WB
12140	I	IWCX	rw	WB
12141	I	IWCX	f	WB
12142	I	IWCX		WB
12143	I	IWCX	f	WB
12144	I	IWCX	rw	WB
12145	I	IWCX	f	WB
12146	I	IWCX		WB
12147	I	IWCX	r	WB
12148	I	IWCX	r	WB
12149	I	PECX	f	EC
12150	I	IWCX	f	WB
12151	I	IWCX	f	WB
12152	I	IWCX	f	WB
12153	I	IWCX	r	WB
12154	I	IWCX	r	WB
12155	I	IWCX	rw	WB
12156	I	IWCX	f	WB
12157	I	IWCX	f	WB
12158	I	IWCX		WB
12159	I	IWCX	f	WB
12160	I	IWCX	rw	WB
12161	I	IWCX		WB
12163	I	IWCX	f	WB
12164	I	IWCX	f	WB
12165	I	IWCX	f	WB
12166	I	IWCX	f	WB
12167	I	IWCX		WB
12168	I	IWCX	f	WB
12169		IWCX	rw	
12170		IWCX	rw	
12171		IWCX	rw	
12172		IWCX	rw	

12169–72 are being converted from 11908–10/22.

AI2J (TSOE) OPEN SECOND (END)

Dia. AI201. Mark 4. Air conditioned. 74S 2L. SIG bogies (BT41). ae. pa. ETH 6.

Lot No. 31047 Metro-Cammell 1989 onwards. 39.5 t.

12200	I	IECX	BN
12201	I	IECX	BN
12202	I	IECX	BN
12203	I	IECX	BN
12204	I	IECX	BN
12205			
12206			
12207			
12208			
12209			
12210			
12211			
12212			
12213			
12214			
12215			
12216			
12217			
12218			
12219			
12220			
12221			
12222			
12223			
12224			
12225			
12226			
12227			
12228			
12229			
12230			
12231			

AL2J (TSOD) OPEN SECOND (DISABLED ACCESS)

Dia. AL201. Mark 4. Air conditioned. 72S + wheelchair space 1L (suitable for a disabled person). SIG bogies (BT41). ae. pa. ETH 6.

Lot No. 31048 Metro-Cammell 1989 onwards. 39.4 t.

12300 I IECX BN
12301 I IECX BN
12302 I IECX BN
12303 I IECX BN
12304
12305
12306
12307
12308
12309
12310
12311
12312
12313
12314
12315
12316
12317
12318
12319
12320
12321
12322
12323
12324
12325
12326
12327
12328
12329
12330

AC2J (TSO) OPEN SECOND

Dia. AC214. Mark 4. Air conditioned. 74S 2L. SIG bogies (BT41). ae. pa. ETH 6.

Lot No. 31049 Metro-Cammell 1989 onwards. 39.9 t.

12400 I IECX BN
12401 I IECX BN
12402 I IECX BN
12403 I IECX BN
12404 I IECX BN
12405 I IECX BN
12406 I IECX BN
12407 I IECX BN
12408
12409
12410
12411
12412
12413
12414
12415
12416
12417
12418
12419
12420
12421
12422
12423
12424
12425
12426
12427
12428
12429
12430
12431
12432
12433
12434
12435
12436
12437
12438
12439
12440
12441
12442
12443
12444
12445
12446
12447
12448
12449
12450
12451
12452
12453
12454
12455
12456
12457
12458
12459
12460
12461
12462
12463
12464
12465
12466
12467
12468
12469
12470
12471
12472
12473
12474
12475
12476
12477
12478
12479
12480

12481	12492	12503
12482	12493	12504
12483	12494	12505
12484	12495	12506
12485	12496	12507
12486	12497	12508
12487	12498	12509
12488	12499	12510
12489	12500	12511
12490	12501	12512
12491	12502	

AA11 (FK) CORRIDOR FIRST

Dia. AA101. Mark 1. 42F 2L. ETH 3. d.

13225–13237. Lot No. 30381 Ashford/Eastleigh 1959. B4 bogies. 33 t.
13303–13344. Lot No. 30667 Swindon 1962. Commonwealth bogies. 36 t.

13225	**I**	ICHV	xk	BN
13227	**I**	ICHV	xk	BN
13228	**I**	ICHH	xk	BN
13229	**I**	ICHL	xk	PH
13230	**I**	IWRX	xk	OM
13233		IXXH	vk	CL
13236	**I**	IWRX	vk	OM
13237	**I**	IWRX	vk	OM
13303		IXXH	v	CL
13306		IXXH	v	CL
13308	**N**	NWRX	v	OM
13313		ICHH	v	CL
13314	**I**	IWRX	vk	OM
13316		IXXH	v	CL
13318	**I**	ICHV	a pa	BN
13320		IXXH	v	CL
13321		IXXH	v	CL
13324	**N**	NWRX	vf	OM
13325		IXXH	v	CL
13326	**N**	IXXH	v	CL
13328	**N**	NWRX	vf	OM
13329	**N**	NWRX	vf	OM
13331	**N**	NWRX	vf	OM
13335		NWRX	v	OM
13341	**I**	IWRX	vf	OM
13342		IWRX	v	OY
13343	**I**	IWRX	vk	OM
13344		ICHL	vf	PH

AA1A (FK) CORRIDOR FIRST

Dia. AA106. Mark 2A. Pressure ventilated. 42F 2L. B4 bogies. d. ETH 4.

13435–13463. Lot No. 30774 Derby 1968. 33 t.
13467–13475. Lot No. 30785 Derby 1968. 33 t.

13435	**N**	NWRX	v	OM
13436	**N**	NNEX	v	CA
13437	**N**	NWRX	v	OM
13440	**N**	NWRX	v	OM
13441	**N**	NWRX	v	OM
13442	**N**	NWRX	v	OM
13443	**N**	NWRX	v	OM
13444	**N**	NWRX	v	OM
13445	**P**	PHTX	a	HT
13446	**N**	NWRX	v	OM
13447	**N**	NNEX	v	CA
13450	**N**	NNEX	v	CA
13451	**P**	PHTX	a	HT
13456		PECX	a	EC
13459	**N**	NSSX	a	EH
13461		PECX	a	EC
13462	**N**	NSSX	a	EH
13463	**P**	PHTX	a	HT
13467	**N**	NSSX	a	EH
13470	**N**	NNEX	v	CA
13472		PDYX	a	DY
13473	**N**	NSSX	a	EH
13474	**N**	NNEX	v	CA
13475	**N**	NWRX	v	OM

AA1B (FK) CORRIDOR FIRST

Dia. AA107. Mark 2B. Pressure ventilated. 42F 2L. B4 bogies. ad. ETH 4.

Lot No. 30789 Derby 1969. 33 t.

13479	**N**	NSSX	EH
13482	**N**	NSSX	EH
13488	**N**	NWRX	OM
13493	**N**	NWRX	OM
13499		NSSX	EH
13502	**N**	NWRX	OM
13507	**N**	NSSX	EH
13513	**N**	NSSX	EH

AA1C (FK) CORRIDOR FIRST

Dia. AA108. Mark 2C. Pressure ventilated. 42F 2L. B4 bogies. ad. pa. ETH 4.

Lot No. 30797 Derby 1969–70. 33.5 t.

13514	**N**	NSSX		EH
13516	**P**	PHTX	pa	HT
13517	**N**	NWRX	pa	OM
13519	**P**	PHTX	pa	HT
13520	**P**	PHTX	pa	HT
13525	**N**	NSSX		EH
13526	**N**	NSSX	pa	EH
13527	**P**	PHTX	pa	HT
13532	**P**	PHTX	pa	HT

AA1D (FK) CORRIDOR FIRST

Dia. AA109. Mark 2D. Air conditioned (Stones). 42F 2L. B4 bogies. ae. pa. ETH 5. 13585–13610 require at least 800 V train heat supply.

Lot No. 30825 Derby 1971–2. 34.5 t.

13562	**I**	ICCX	MA
13563		IECX	BN
13564	**I**	IWRX	OM
13565	**I**	ICCX	MA
13566	**I**	ICCX	MA
13567	**I**	IWRX	OM
13568	**I**	ICCX	PC
13569	**I**	ICCX	PC
13571	**I**	ICCX	PC
13572	**I**	IWRL	OM
13574	**I**	IWRL	OM
13575		IECX	BN
13576		ICHM	MA
13577	**I**	IWRX	OM
13578	**I**	ICCX	PC
13579	**I**	IWRX	OM
13580	**I**	ICCX	MA
13581	**I**	ICCX	MA
13582	**I**	ICCX	PC
13583	**I**	IWRX	OM
13584	**I**	IWRX	OM
13585	**I**	IECX	BN
13586	**I**	IWRX	OM
13587	**I**	ICCX	MA
13588	**I**	ICCX	PC
13589	**I**	IWRX	OM
13590	**I**	ICHM	MA
13591	**I**	ICCX	MA
13592	**I**	ICCX	MA
13593	**I**	ICCX	MA
13594	**I**	IWRX	OM
13595	**I**	IWRX	OM
13596	**I**	ICCX	PC
13597	**I**	ICHM	MA
13598	**I**	ICCX	MA
13599		ICCX	MA
13600		IECX	BN
13601	**I**	ICCX	PC
13602	**I**	ICCX	OM
13603	**I**	IWRL	OM
13604	**I**	IECX	BN
13605		IECX	BN
13606	**I**	IWRX	OM
13607		IECX	BN
13608		IECX	BN
13609	**I**	IXXH	BN
13610	**I**	IWRL	OM

SPECIAL NOTE: All BFKs were formerly numbered in the 14xxx series. Subtract 3000 from present number to obtain former number.

AB11 (BFK) CORRIDOR BRAKE FIRST

Dia. AB101. Mark 1. 24F 1L. d. ETH 2.

17005. Lot No. 30218 Ashford/Eastleigh 1959–60. BR1 bogies. 34 t.
17007. Lot No. 30382 Ashford/Eastleigh 1960. BR1 bogies. 34 t.
17015–17021. Lot No. 30668 Swindon 1961. Commonwealth bogies. 36 t.
17023–17025. Lot No. 30718 Swindon 1963. Commonwealth bogies and metal window frames. 36 t.

Non-Standard Livery: Early BR Carmine & Cream.

17005		PISX	v	IS
17007	**0**	ICHX	v	OM
17015	**0**	ICHV	x	BN
17021	**I**	ICHL	vk	FW
17023	**I**	ICHV	x	BN
17025	**I**	ICHV	vf	BN

AB1Z (BFK) CORRIDOR BRAKE FIRST

Dia. AB102. Mark 2. Pressure ventilated. 24F 1L. B4 bogies. vd ETH 4.

Lot No. 30756 Derby 1966 31.5 t.

17031		IXXH		OY
17037	**I**	IWRX	k	OM
17039		PISX		IS
17040	**N**	NWRX		OM
17042	**I**	ICHV	k	OM
17043		IXXH		OY
17047		IXXH		OY
17051		IWRX		OM
17052		ICHC		OY
17054	**I**	ICHC		OY
17055		IXXH		OY

AB1A (BFK) CORRIDOR BRAKE FIRST

Dia. AB103. Mark 2A. Pressure ventilated. 24F 1L. B4 bogies. d. ETH 4.

17056–17077. Lot No. 30775 Derby 1967–8. 32 t.
17079–17103. Lot No. 30786 Derby 1968. 32 t.

17056	**N**	NSSX	a pt	LA
17057	**N**	NSSX	a pt	LA
17058	**N**	NWRX	a	OM
17059		PISX	v	IS
17060	**N**	NWRX	v	OM
17062	**N**	NWRX	a	OM
17063	**N**	NSSX	a pt	LA
17064	**N**	NWRX	v	OM
17066	**I**	ICHC	v	OY
17068	**N**	NWRX	v	OM
17069	**N**	NSSX	a pt	LA
17070	**N**	NSSX	a pt	LA
17072	**N**	NSSX	a pt	LA
17073	**N**	NSSX	a pt	LA
17074	**N**	NSSX	a pt	LA
17075	**I**	ICHD	a	OY
17076	**N**	NSSX	a pt	LA
17077	**N**	NWRX	a	OM
17079	**N**	NSSX	a	EH
17080	**N**	NWRX	a pt	LA
17081	**N**	NSSX	a pt	EH
17082	**N**	NSSX	a pt	LA
17085	**N**	NWRX	v	OM
17086	**N**	NSSX	a pt	LA
17088	**N**	NSSX	a pt	LA
17089	**N**	NWRX	v	OM
17090	**N**	NWRX	v	OM
17091	**N**	NWRX	v	OM
17092	**N**	NWRX	v	OM
17093	**I**	IECX	a	BN
17096	**N**	NSSX	a pt	LA
17097	**N**	NSSX	a pt	LA
17098	**I**	ICHD	a	OY
17099		PISX	v	IS
17100	**I**	ICHD	a	OY
17101		ICHC	v	OY
17102	**I**	ICHX	a	OY
17103		ICHC	v	OY

AB1C (BFK) CORRIDOR BRAKE FIRST

Dia. AB105. Mark 2C. Pressure ventilated. 24F 1L. B4 Bogies. ad. ETH 4.

Lot No. 30796 Derby 1969–70. 32.5 t.

17113	**I**	ICHD	pt	OY
17114	**N**	NSSX	pt	LA
17115	**N**	NSSX	pt	LA
17117	**I**	ICHX	pt	OM
17118	**N**	NWRX		OM
17119		IXXH	pt	PC
17120	**N**	NSSX	pt	LA
17121	**N**	NSSX	pt	LA
17122	**I**	ICCX	pt	PC
17123		ICCX	pt	PC
17124		ICCX	pt	MA
17125		ICCX	pt	PC
17126	**I**	ICCX	pt	MA
17127		PISX	pt	IS
17128		PISX		IS
17129	**I**	ICCX	pt	PC
17130	**I**	ICCX	pt	MA
17131		ICCX	k pt	PC
17132	**N**	NSSX		LA
17133	**N**	NSSX	k pt	EH
17134	**I**	ICCX	pt	MA
17135	**I**	IANX	pt	NC
17136		IECX		BN
17137		IWCD	k pt	MA
17138	**I**	ICCX	pt	MA

AB1D (BFK) CORRIDOR BRAKE FIRST

Dia. AB106. Mark 2D. Air conditioned (Stones equipment). 24F 1L. B4 Bogies. 17162–17172 require at least 800 V train heat supply. ae. pt. ETH 5.

Lot No. 30823 Derby 1971–2. 33.5 t.

17139	**I**	ICCX	pg	PC
17140	**I**	ICCX	pg	PC
17141	**I**	ICCX	pg	MA
17142	**I**	ICCX	pg	MA
17143	**I**	IANX	pg	NC
17144	**I**	ICCX		PC
17145	**I**	IANX	pg	NC
17146	**I**	ICCX	pg	MA
17147	**I**	ICCX	pg	MA
17148	**I**	IANX	pg	NC
17149	**I**	IANX	pg	NC
17150	**I**	ICCX	pg	PC
17151	**I**	ICCX	pg	MA
17152		ICCX	pg	PC
17153	**I**	IANX	pg	NC
17154	**I**	IANX	pg	NC
17155	**I**	ICCX		MA
17156	**I**	ICCX		MA
17157	**I**	ICCX	pg	MA
17158	**I**	IANX	pg	NC
17159		ICCX	pg	MA
17160	**I**	ICCX	pg	PC
17161	**I**	IANX	pg	NC
17162	**I**	ICCX	pg	PC
17163	**I**	ICCX		PC
17164	**I**	ICCX	pg	MA
17165	**I**	ICCX		MA
17166	**I**	ICCX	pg	PC
17167	**I**	ICCX	pg	PC
17168	**I**	ICCX		MA
17169	**I**	ICCX	pg	MA
17170	**I**	ICCX	pg	PC
17171	**I**	ICCX	pg	PC
17172	**I**	IWRX	pg	LA

AE1G (BFO) OPEN BRAKE FIRST

Dia. AE101. Mark 3B. Air conditioned. 36F 1L. BT10 bogies. ae. pg. ETH 5X. Fitted with hydraulic handbrake.

Lot No. 30990 Derby 1986.

17173	**I**	IWCX	WB
17174	**I**	IWCX	WB
17175	**I**	IWCX	WB

AA21 (SK) CORRIDOR SECOND

Dia. AA201 (AA202*). Mark 1. As with CKs there are two variants depending on whether the compartments have armrests. Each vehicle has eight compartments. 48S 2L (64S 2L*). ETH 4.

f-Facelifted with fluorescent lighting and wide mirrors. Blinds removed.

NOTE: The SKs below were formerly numbered in the 25xxx and 26xxx series. To find former number, add 7000 to present number.

18284–18288. Lot No. 30349 Wolverton 1957. Single bolster bogies. vd. 34 t.
18597–18627. Lot No. 30426 Wolverton 1957–8. B4 bogies. vd. 32 t.

18284		PISX		IS
18288		PISX		IS
18597		PLLX	*f	LL
18601	**N**	NWRX	*f	OM
18602	**N**	NWRX	*f	OM
18603	**N**	NWRX	*f	OM
18604	**N**	NWRX	*f	OM
18605	**N**	NWRX	*f	OM
18608	**N**	NWRX	*f	OM
18611	**N**	NWRX	*f	OM
18616	**N**	NWRX	*f	OM
18622	**N**	NWRX	*f	OM
18627	**N**	NWRX	*f	OM

The following lots of SK all have Commonwealth bogies, metal window frames and melamine interior panelling.

18704–18901. Lot No. 30685 Derby 1961–2. vd. 36 t.
18911–18970. Lot No. 30686 Derby 1962. vd. 36 t.
18977–19022. Lot No. 30719 Derby 1962. vd. 37 t.

18704	**N**	NWRX	*f	OM
18706	**N**	NWRX	*f	OM
18709	**N**	NWRX	*f	OM
18711	**N**	NWRX	*f	OM
18712		PLLX		LL
18714	**N**	NWRX	*f	OM
18715	**N**	NWRX	*f	OM
18716		PLLX	*f	LL
18717	**N**	NWRX	*f	OM
18727	**N**	NWRX	*f	OM
18728	**N**	NWRX		OM
18729	**N**	NWRX	*f	OM
18730	**N**	NWRX	*f	OM
18734		PLLX		LL
18738		PLLX		LL
18743	**N**	NWRX		OM
18749		PNLX		NL
18750		PLLX		LL
18752	**N**	NWRX		OM
18756	**I**	ICHL		MA
18762		PISX		IS
18769		NWRX		OM
18771	**N**	NWRX		OM
18774		PNLX		NL
18775	**N**	NWRX	*f	OM
18783		PLLX		LL
18784	**N**	NWRX	*f	OM
18788	**N**	NWRX	*f	OM
18789	**N**	NWRX	*f	OM
18792	**N**	NWRX	*f	OM
18794	**N**	NWRX	*f	OM
18805	**N**	NWRX	*f	OM
18806	**I**	ICHL		MA
18809		PLLX		LL
18813		PNLX		NL
18814	**N**	NWRX	*f	OM
18818	**N**	NWRX	*f	OM
18819		ICHH		OM
18830		PCLX		CL
18835		PLLX	*f	LL
18838		PNLX		NL
18849		PLLX		LL
18863	**N**	NWRX		OM
18869	**N**	NWRX		OM
18875	**N**	NWRX	*f	OM
18901		ICHH		OM
18911		PCFX	*f	CF
18921	**N**	NWRX	*f	OM
18925	**N**	NWRX	*f	OM
18931	**N**	NWRX	*f	OM
18933	**N**	NWRX	*f	OM
18935	**N**	NWRX	*f	OM

18936	**N** NWRX	*f	OM
18937	**N** NWRX	*f	OM
18938	**N** NWRX	*f	OM
18940	**N** NWRX	*f	OM
18941	**N** NWRX	*f	OM
18943	**N** NWRX	*f	OM
18945	**N** NWRX	*f	OM
18947	PLLX	*f	LL
18952	**N** NWRX	*f	OM
18955	**N** NWRX	*f	OM
18957	PLLX	*f	LL
18958	**N** NWRX	*f	OM
18959	**N** NWRX	*f	OM
18960	**N** NWRX	*f	OM
18967	**N** NWRX	*f	OM
18968	**N** NWRX	*f	OM
18970	**N** NWRX	*f	OM
18977	ICHH		OM
18987	PLLX		LL
18996	PNLX		NL
18997	**N** NWRX		OM
19011	ICHH		OM
19012	PLLX		LL
19014	ICHL		OM
19022	**N** NWRX	*f	OM

AA2A (SK) CORRIDOR SECOND

Dia. AA206. Mark 2A. Pressure ventilated. Renumbered from FK from 1985 onwards. 42S 2L. B4 bogies. ad. ETH 4.

19452–19460. Lot No. 30774 Derby 1968. 33 t.
19464–19465. Lot No. 30785 Derby 1968. 33 t.

19452	(13452)	PDYX	DY
19454	(13454)	NSSX	EH
19455	(13455)	PDYX	DY
19458	(13458)	PDYX	DY
19460	(13460)	IWRX	OM
19464	(13464)	NSSX	EH
19465	(13465)	IECX	BN

AA2B (SK) CORRIDOR SECOND

Dia. AA207. Mark 2B. Pressure ventilated. Renumbered from FK from 1985 onwards. 42S 2L. B4 bogies. ad. ETH 4.

Lot No. 30789 Derby 1969. 33 t.

19469	(13469)	IWCD	WB
19478	(13478)	IECX	BN
19486	(13486)	PDYX	DY
19512	(13512)	IWCD	WB

AA2C (SK) CORRIDOR SECOND

Dia. AA208. Mark 2C. Pressure ventilated. Renumbered from FK from 1985 onwards. 42S 2L. B4 bogies. ad. ETH 4.

Lot No. 30797 Derby 1969–70. 33.5 t.

19522	(13522)	PDYX	pa	DY
19523	(13523)	PDYX	pa	DY
19529	(13529)	PDYX	pa	DY
19534	(13534) **I**	IWCD	pa	EH
19536	(13536)	PDYX	pa	DY
19540	(13540)	IWCD	pa	WB
19545	(13545)	PDYX	pa	DY
19548	(13548)	PDYX	pa	DY
19549	(13549)	PDYX	pa	DY
19555	(13555)	PDYX	pa	DY
19556	(13556)	PDYX	pa	DY

AB31 (BCK) CORRIDOR BRAKE COMPOSITE

Dia.AB301 (AB302*). Mark 1. As with the CKs there are two variants depending upon whether the second class compartments have armrests.

Each vehicle has two first class and three second class compartments. 12F 18S 2L (12F 24S 2L*).

f-Fitted with fluorescent lighting and wide mirrors. Blinds removed.

21241–21247. Lot No. 30669 Swindon 1961–2. Commonwealth bogies. 36 t.
21265–21274. Lot No. 30732 Derby 1964. Commonwealth bogies. 37 t.

21241	**E**	ICHS	vd	BN
21242	**I**	ICHS	vd	BN
21246	**I**	ICHV	vd	BN
21247		ICHL	vd	OM
21265		NSSX	*ae	EH
21266	**I**	ICHV	*ae	BN
21268	**I**	ICHX	*ae	BN
21269		ICCX	*ad	EC
21274	**I**	ICHX	*ae	BN

AA21 (SK) CORRIDOR SECOND

Dia. AA201. Class continued from 19022. BR1 bogies. vd.

24895. Lot No. 30153 Derby 1955–6. 33 t. vd.
25008. Lot No. 30208 Derby 1956. 33 t. vs. For steam loco driver training.
26125. Lot No. 30153 Derby 1962–3. 36 t. Commonwealth bogies. vs. For steam loco driver training.

24895	PPCX	PC
25008	ICHS	FW
26125	ICHS	FW

AB21 (BSK) CORRIDOR BRAKE SECOND

Dia. AB201 (AB202*). Mark 1. As with other compartment vehicles there are two variants depending upon whether the compartments have armrests. each vehicle has four compartments. 24S 1L (32S 1L*). ETH 2 (2S•).

34538. Lot No. 30095 Wolverton 1955. BR1 bogies. vd. 34 t.
34668–34681. Lot No. 30156 Wolverton 1955. BR1 bogies. vd. 34 t.

34538	PLLX	LL
34668	PPCX	PC
34681	PNLX	NL

The following lots of BSK all have metal window frames and melamine interior panelling: (all vd).

35284–35290. Lot No. 30573 Gloucester 1960. B4 bogies. 33 t.
35306–35343. Lot No. 30669 Wolverton 1962–3. Commonwealth bogies. 37 t.
35447–35449. Lot No. 30728 Wolverton 1963. Commonwealth bogies. 37 t.
35450–35486. Lot No. 30721 Wolverton 1963. Commonwealth bogies. 37 t.

f–Facelifted with fluorescent lighting.

35284		PLAX		LA
35290	**N**	NNWX	pa	EN
35306		PLLX		LL
35309	**N**	NWRX		OM
35317	**N**	NNWX	pa	EN
35329	**N**	NNWX		EN
35333		PLLX		LL
35339	**N**	NNWX		EN
35340		PLLX		LL
35343		PLLX		LL
35447	**N**	NWRX	pa	OM
35448	**N**	NWRX		OM
35449		ICHL		OM
35450		ICHL		OM
35452	**N**	NNWX	pa	EN
35453		PLLX		LL

35454	IXXH	f	OY
35455 **N**	NNWX		EN
35456	PCLX		CL
35457	PNLX		NL
35459	ICHL		MA
35464 **N**	NWRX	pa	OM
35465	IXXH	f	OY
35468	ICHL		CL
35469	ICHS		OY
35470	PLLX		LL
35471 **N**	NWRX	*f	OM
35473	PLLX		LL
35474 **N**	NWRX		OM
35475 **N**	NWRX	*f	OM
35476	IXXH		OM
35479 **N**	NNWX	pa	EN
35482	PLAX	*f	LA
35483	PISX		IS
35486	ICHL		CL

AB2A (BSK) CORRIDOR BRAKE SECOND

Dia. AB204. Mark 2A. Pressure ventilated. Renumbered from BFK in 1985/6. 24S 1L. B4 bogies. vd. ETH 4.

Lot No. 30786 Derby 1968. 32 t.

35500 (17094) **N** NWRX v OM

HIGH SPEED TRAIN TRAILER CARS

HSTs run in formations of 7 or 8 trailer cars with a driving motor brake (power car) at each end. All vehicles are classified mark 3. All trailer cars have BT10 bogies with disc brakes. Heating is by a three-phase supply and vehicles have air conditioning. Max. Speed is 125 mph.

GN4G (TRB) TRAILER BUFFET FIRST

Dia. GN401. Converted from TRSB by fitting first class seats. Renumbered from 404xx series by subtracting 200. pt. 23F.

Lot No. 30883 Derby 1976–7. 36.12 t.

40204 **I**	IWRX	r	PM
40205 **I**	IWRX		PM
40206	IWRX		PM
40207 **I**	IWRX	r	PM
40208 **I**	IWRX	r	PM
40209 **I**	IWRX	r	PM
40210 **I**	IWRX	r	PM
40211 **I**	IWRX	r	PM
40212	IWRX		PM
40213 **I**	IWRX	r	PM
40221 **I**	IWRX	p	PM
40228 **I**	IWRX	r	PM
40231			
40232			
40233			
40234			

GK2G (TRSB) TRAILER BUFFET STANDARD

Dia. GK202. pt. Renumbered from 400xx series by adding 400. 35S.

40401–40428. Lot No. 30883 Derby 1976–7. 36.12 t.
40429–40437. Lot No. 30899 Derby 1978–9. 36.12 t.

40401 **I**	ICCX	rp	PM
40402 **I**	ICCX	rp	PM
40403 **I**	ICCX	rp	PM
40414 **I**	ICCX	fp	LA
40415 **I**	ICCX	fp	PM
40416 **I**	IWRX	r	LA
40417 **I**	IWRX	r	LA
40418 **I**	IWRX	r	LA
40419 **I**	IWRX	r	LA
40420 **I**	ICCX	rp	PM
40422 **I**	IWRX	r	PM
40423 **I**	IWRX	r	LA
40424 **I**	IWRX	rp	LA
40425 **I**	ICCX	rp	PM
40426 **I**	ICCX	rp	LA
40427 **I**	IWRX	r	LA
40429 **I**	IECX	fp	NL
40430 **I**	ICCX	fp	LA
40431 **I**	ICCX	p	LA
40432	ICCX	p	LA
40433 **I**	ICCX		LA

40434	ICCX	p	LA
40435I	IWRX	r	LA
40436I	IWRX	r	LA
40437I	IWRX	r	PM

GL1G (TRFK) TRAILER KITCHEN FIRST

Dia. GL101. pt. Reclassified from TRUK. Used in "Pullman" sets and as replacements for out of service TRFBs. pt. 24F.

*–Converted to lounge first seating 16F 1L. Dia. GM101.
Lot No. 30884 Derby 1976–7. 37 t.

40501I	IECX	f	HT
40505I	IECX	f	NL
40511I	IECX	f	NL
40513I	ICHV	*	BN

GK1G (TRFM) TRAILER MODULAR BUFFET FIRST

Dia. GK102. Converted to modular catering (Cuisine 2000) from 407xx series from which they are being renumbered by subtracting 100 from previous number. pt. 17F.

40619. Lot No. 30921 Derby 1978–9. 38.16 t.

40619I	IMLX	rp	NL

GK1G (TRFB) TRAILER BUFFET FIRST

Dia. GK101. These vehicles have larger kitchens than the 402xx and 404xx series vehicles, and are used in trains where full meal service is required. They have been renumbered from the 403xx series (in which the seats were unclassified) by adding 400 to previous number. pt. 17F.

40700–40721. Lot No. 30921 Derby 1978–9. 38.16 t.
40722–40735. Lot No. 30940 Derby 1979–80. 38.16 t.
40736–40753. Lot No. 30948 Derby 1980–1. 38.16 t.
40754–40757. Lot No. 30966 Derby 1982. 38.16 t.

40700I	IECX	fp	NL
40701I	IECX	fp	BN
40702I	IECX	fp	BN
40703I	IECX	fp	BN
40704I	IECX	rp	BN
40705I	IECX	fp	BN
40706I	IECX	fp	BN
40707I	IECX	fp	BN
40708I	IECX	fp	BN
40709I	IMLX	fp	NL
40710I	IECX	rp	BN
40711I	IECX	rp	NL
40712I	IMLX	rp	NL
40713I	IECX	rp	NL
40714I	IECX	rp	NL
40715I	IECX	rp	BN
40716I	IECX	r	BN
40717I	IECX	f	BN
40718I	IECX	rp	BN
40720I	IECX	rp	BN
40721I	IECX	rp	BN
40722I	IWRX	f	LA
40723I	IWRX	f	LA
40724I	IWRX	f	LA
40725I	IWRX	f	LA
40726I	IWRX	f	LA
40727I	IWRX	f	LA
40728I	IECX	fp	NL
40729I	IECX	fp	BN
40730I	IMLX	rp	NL
40731I	IWRX	f	LA
40732I	IWRX	f	LA
40733I	IECX	fp	BN
40734I	IECX	fp	BN
40735I	IMLX	rp	NL
40736I	IECX	rp	BN
40737I	IMLX	fp	NL
40738I	IMLX	fp	NL
40739I	IMLX	fp	BN
40740I	IECX	fp	BN
40741I	IECX	rp	HT
40742I	IECX	fp	BN
40743I	IECX	rp	NL
40744	IMLX	p	NL
40745I	IMLX	fp	NL
40746I	IMLX	fp	NL
40747	IMLX	p	NL
40748I	IMLX	r	NL
40749I	IMLX	r	NL
40750I	IMLX	rp	NL
40751I	IECX	fp	NL
40752	IECX	p	NL
40753	IECX	p	NL
40754I	IECX	fp	BN
40755I	IWRX	fp	LA
40756I	IECX	rp	BN
40757I	IWRX	r	LA

GH1G (TF) TRAILER FIRST

Dia. GH102. pa. 48F 2L.

41003–41056. Lot No. 30881 Derby 1976–7. 33.66 t.
41057–41120. Lot No. 30896 Derby 1977–8. 33.66 t.
41121–41148. Lot No. 30938 Derby 1979–80. 33.66 t.
41149–41166. Lot No. 30947 Derby 1980. 33.66 t.
41167–41169. Lot No. 30963 Derby 1982. 33.66 t.
41170–41174. Lot No. 30967 Derby 1982. Ex prototype vehicles. 33.66 t.
41175–41176. Lot No. 30897 Derby 1977. Converted 1982 from TS. 33.66 t.

Note: 41170–6 were converted from 41001, 42001/0/2, 41002, 42114/142 respectively.

41003I	IWRX	rp	PM
41004I	IWRX	r	PM
41005	IWRX	p	PM
41006	IWRX		PM
41007I	IWRX	rp	PM
41008I	IWRX	r	PM
41009I	IWRX	rp	PM
41010I	IWRX	r	PM
41011I	IWRX	rp	PM
41012I	IWRX	r	PM
41013I	IWRX	rp	PM
41014I	IWRX	r	PM
41015I	IWRX	rp	PM
41016I	IWRX	r	PM
41017I	IWRX	rp	PM
41018I	IWRX	r	PM
41019I	IWRX	rp	PM
41020I	IWRX	r	PM
41021I	IWRX	rp	PM
41022I	IWRX	r	PM
41023I	IWRX	fp	LA
41024I	IWRX	f	LA
41025I	IWRX	fp	LA
41026I	IWRX	f	LA
41027I	IWRX	fp	LA
41028I	IWRX	f	LA
41029I	IWRX	fp	LA
41030I	IWRX	f	LA
41031I	IWRX	fp	LA
41032I	IWRX	f	LA
41033I	IWRX	fp	LA
41034I	IWRX	f	LA
41035I	IWRX	fp	LA
41036I	IWRX	f	LA
41037I	IWRX	fp	LA
41038I	IWRX	f	LA
41039I	IMLX	r	NL
41040I	IMLX	r	NL
41041I	IECX	fp	NL
41042I	ICCX	f	LA
41043I	IMLX	f	NL
41044I	IMLX	f	NL
41045I	IMLX	f	NL
41046I	IECX	f	BN
41047	IMLX		NL
41048	IMLX		NL
41049I	IMLX	f	NL
41050I	IMLX	f	NL
41051I	IMLX	f	NL
41052I	IMLX	f	NL
41053	IMLX		NL
41054	IMLX		NL
41055I	IMLX	f	NL
41056I	IMLX	f	NL
41057I	IECX	f	BN
41058I	IECX	f	BN
41059I	IECX	f	BN
41060I	IECX	f	BN
41061I	IECX	f	BN
41062I	IECX	f	BN
41063I	IECX	f	BN
41064I	IECX	f	BN
41065I	IMLX	f	NL
41066I	IECX	fp	HT
41067I	IECX	r	NL
41068I	IECX	r	NL
41069I	IECX	f	NL
41070I	IECX	f	NL
41071I	IECX	f	NL
41072I	IECX	f	NL
41073	IECX		NL
41074	IECX		NL
41075I	IECX	r	BN
41076I	IECX	r	BN
41077	IECX		NL
41078	IECX		NL
41079I	IECX	f	NL
41080I	IECX	f	NL
41081I	IECX	f	BN
41082I	IECX	f	BN
41083I	IECX	f	BN
41084I	IECX	f	BN
41085I	IECX	f	BN
41086I	IECX	f	BN
41087I	IECX	f	BN
41088I	IECX	f	BN
41089I	IECX	f	BN
41090I	IECX	f	BN
41091I	IECX	r	BN
41092I	IECX	r	BN
41093I	IECX	f	BN
41094I	IECX	f	BN
41095I	IECX	r	BN
41096I	IECX	r	BN
41097I	IMLX	r	NL
41098I	IMLX	r	NL
41099I	IECX	r	NL
41100I	IECX	r	NL
41101I	IMLX	r	NL
41102I	IMLX	r	NL
41103I	IECX	r	NL
41104I	IECX	r	NL
41105I	IECX	r	NL
41106I	IECX	r	NL
41107	IECX		BN
41108I	IECX	r	BN
41109I	IECX	r	BN
41110I	IECX	r	BN
41111I	IMLX	r	NL
41112I	IMLX	r	NL
41113I	IECX	r	BN

No.	Type		Depot
41114I	IECX	r	BN
41115I	IECX	f	BN
41116I	IECX	f	BN
41117I	IMLX	r	NL
41118I	IECX	r	BN
41119I	IECX	f	BN
41120I	IECX	f	BN
41121I	IWRX	fp	LA
41122I	IWRX	f	LA
41123I	IWRX	rp	PM
41124I	IWRX	r	PM
41125I	IWRX	r	LA
41126I	IWRX	rp	LA
41127I	IWRX	rp	PM
41128I	IWRX	r	PM
41129I	IWRX	rp	LA
41130I	IWRX	r	LA
41131I	IWRX	rp	PM
41132I	IWRX	r	PM
41133I	IWRX	fp	LA
41134I	IWRX	f	LA
41135I	IWRX	fp	LA
41136I	IWRX	f	LA
41137I	IWRX	rp	LA
41138I	IWRX	r	LA
41139I	IWRX	rp	LA
41140I	IWRX	r	LA
41141I	IWRX	rp	LA
41142I	IWRX	r	LA
41143I	IWRX	rp	LA
41144I	IWRX	r	LA
41145I	IWRX	rp	LA
41146I	IWRX	r	LA
41147I	ICCX	r	PM
41148I	ICCX	r	PM
41149I	IECX	f	BN
41150I	IECX	f	BN
41151I	IECX	f	BN
41152I	IECX	f	BN
41153I	IECX	r	HT
41154I	IECX	r	HT
41155I	IMLX	r	NL
41156I	IMLX	r	NL
41157I	IECX	r	NL
41158I	IECX	r	NL
41159I	ICCX	r	PM
41160I	ICCX	r	PM
41161I	ICCX	r	PM
41162I	ICCX	f	LA
41163I	IMLX	f	NL
41164I	IWRX	f	PM
41165I	ICCX	f	LA
41166	ICCX		LA
41167	ICCX		LA
41168I	ICCX	r	LA
41169	ICCX		LA
41170I	IECX	f	BN
41171I	IECX	f	BN
41172I	IECX	r	BN
41173I	IECX	f	BN
41174I	IECX	r	BN
41175I	IMLX	r	NL
41176I	IMLX	r	NL

GH2G (TS) TRAILER STANDARD

Dia. GH202 (GH203*r). pa. 72S 2L. (76S 2L*r)

42003–42090. Lot No. 30882 Derby 1976–7. 33.60 t.
42091–42250. Lot No. 30897 Derby 1977–9. 33.60 t.
42251–42305. Lot No. 30939 Derby 1979–80. 33.60 t.
42306–42322. Lot No. 30969 Derby 1982. 33.60 t.
42323–42341. Lot No. 30983 Derby 1984–5. 33.60 t.
42342–42345. Converted from TGS.

Notes: 42158 was also numbered 41177. 42342–5 were converted from TGS 44082/95/92/96 respectively.

No.	Type		Depot
42003I	IWRX	r	PM
42004I	IWRX	r	PM
42005I	IWRX	r	PM
42006	IWRX		PM
42007	IWRX		PM
42008	IWRX		PM
42009I	IWRX	r	PM
42010I	IWRX	r	PM
42011I	IWRX	r	PM
42012I	IWRX	r	PM
42013I	IWRX	r	PM
42014I	IWRX	r	PM
42015I	IWRX	r	PM
42016I	IWRX	r	PM
42017I	IWRX	r	PM
42018I	IWRX	r	PM
42019I	IWRX	r	PM
42020	IWRX		PM
42021I	IWRX	r	PM
42022I	IWRX	r	PM
42023I	IWRX	r	PM
42024I	IWRX	r	PM
42025I	IWRX	r	PM
42026I	IWRX	r	PM
42027I	IWRX	r	PM
42028I	IWRX	r	PM
42029I	IWRX	r	PM
42030I	IWRX	r	PM
42031I	IWRX	r	PM
42032I	IWRX	f	PM
42033I	IWRX	f	LA
42034I	IWRX	f	LA
42035I	IWRX	f	LA
42036I	IWRX	r	LA
42037I	IWRX	r	LA
42038I	IWRX	r	LA
42039I	IWRX	r	LA
42040I	IWRX	r	LA
42041I	IWRX	r	LA
42042I	IWRX	r	LA
42043I	IWRX	r	LA
42044I	IWRX	r	LA
42045I	IWRX	r	LA
42046I	IWRX	r	LA
42047I	IWRX	r	LA
42048I	IWRX	r	LA
42049I	IWRX	r	LA
42050I	IWRX	r	LA
42051I	IWRX	f	LA
42052I	IWRX	f	LA
42053I	IWRX	f	LA

42054I	IWRX	f	LA
42055I	IWRX	f	LA
42056I	IWRX	f	LA
42057I	IMLX	r	NL
42058I	IMLX	r	NL
42059I	IMLX	r	NL
42060I	ICCX	f	LA
42061I	ICCX	f	LA
42062I	IMLX	f	NL
42063I	IMLX	f	NL
42064I	IMLX	f	NL
42065I	IMLX	f	NL
42066I	IMLX	f	NL
42067I	IMLX	f	NL
42068I	IMLX	f	NL
42069	IMLX		NL
42070	IMLX		NL
42071	IMLX		NL
42072I	IMLX	f	NL
42073I	IMLX	f	NL
42074I	IMLX	f	NL
42075I	IMLX	f*	NL
42076I	IMLX	f	NL
42077I	IECX	f	NL
42078I	IMLX	r	NL
42079	IMLX		NL
42080	IMLX		NL
42081I	IMLX	f	NL
42082I	IMLX	f	NL
42083I	IMLX	f	NL
42084I	ICCX	r	PM
42085I	ICCX	r	PM
42086I	ICCX	r	PM
42087I	ICCX	r	PM
42088I	ICCX	r	LA
42089I	ICCX	r	LA
42090I	ICCX	r	LA
42091I	ICCX	r	LA
42092I	IMLX	f	NL
42093I	IWRX	f	LA
42094I	IMLX	f	NL
42095I	IMLX	f	NL
42096I	IWRX	f	LA
42097I	IWRX	r	LA
42098I	IWRX	r	LA
42099I	IWRX	r	LA
42100I	IECX	r	NL
42101I	IECX	r	BN
42102I	IECX	r	BN
42103I	IECX	r	BN
42104I	IECX	r	BN
42105I	IECX	r	BN
42106I	IECX	r	BN
42107I	IMLX	r	BN
42108I	ICCX	f	LA
42109I	ICCX	f	LA
42110I	ICCX	f	LA
42111I	IECX	r	BN
42112I	IECX	r	BN
42113I	IECX	r	BN
42115I	IECX	r	BN
42116I	IECX	r	BN
42117I	IECX	r	BN
42118	IMLX		NL
42119I	IECX	f	BN
42120I	IECX	f	BN
42121I	IECX	f	BN
42122I	IECX	r	BN
42123I	IECX	f*	BN
42124I	IECX	f*	BN
42125I	IECX	f*	BN
42126I	IMLX	f	NL
42127I	IECX	f	BN
42128I	IECX	f	BN
42129I	IECX	r	BN
42130I	IECX	r	BN
42131I	IECX	r	NL
42132I	IECX	r	NL
42133I	IECX	r	NL
42134I	IECX	r	BN
42135I	IECX	f	NL
42136I	IECX	f	NL
42137I	IECX	f	NL
42138I	IECX	r	BN
42139I	IECX	r	NL
42140I	IECX	r	NL
42141I	IECX	r	NL
42143	IECX		NL
42144	IECX		NL
42145I	IECX	r	NL
42146I	IMLX	r	NL
42147I	IECX	r	BN
42148I	IECX	r	BN
42149I	IECX	r	BN
42150I	IMLX	r	NL
42151	IECX		NL
42152	IECX		NL
42153	IECX		NL
42154I	IMLX	r	NL
42155I	IECX	r	NL
42156I	IECX	r	NL
42157I	IECX	r	NL
42158I	IMLX	r	NL
42159I	IECX	r	BN
42160I	IECX	r	BN
42161I	IECX	r	BN
42162I	IECX	r	BN
42163I	IECX	r	BN
42164I	IECX	r	BN
42165I	IECX	r	BN
42166I	IECX	r	BN
42167I	IECX	r	BN
42168I	IECX	r	BN
42169I	IECX	r	BN
42170I	IECX	r	BN
42171I	IECX	f	BN
42172I	IECX	f	BN
42173I	IECX	f	BN
42174I	IECX	r	BN
42175I	IECX	r	BN
42176I	IECX	r	BN
42177I	IECX	r	BN
42178I	IECX	r	BN
42179I	IECX	r	BN
42180I	IECX	r	BN
42181I	IECX	r	BN
42182I	IECX	r	BN
42183I	IECX	r	BN
42184I	IECX	r	BN
42185I	IECX	r	BN
42186I	IECX	r	BN
42187I	IECX	r	BN
42188I	IECX	r	BN
42189I	IECX	r	BN
42190I	IECX	r	BN
42191I	IMLX	r	NL
42192I	IMLX	r	NL
42193I	IMLX	r	NL
42194I	IMLX	r	NL
42195I	IMLX	r	NL
42196I	ICCX	r	LA
42197I	ICCX	r	LA
42198I	IECX	r	NL
42199I	IECX	r	NL
42200I	IMLX	r	NL
42201I	IMLX	r	NL
42202I	IMLX	r	NL
42203I	IMLX	r	NL
42204I	IMLX	r	NL
42205I	IECX	r	HT
42206I	IECX	r	NL
42207I	IECX	r	NL
42208I	IECX	r	NL
42209I	IECX	r	NL
42210I	IECX	r	HT
42211I	IECX	r	NL

42212I	IECX	r	NL
42213I	IECX	r	NL
42214I	IECX	r	NL
42215I	IMLX	r	NL
42216I	IWRX	r	LA
42217I	IECX	r	BN
42218I	IECX	r	BN
42219I	IECX	r	BN
42220I	IMLX	r	NL
42221I	IWRX	r	LA
42222I	IECX	r	BN
42223I	IECX	r	BN
42224I	IECX	r	BN
42225I	IMLX	r	NL
42226I	IECX	r	BN
42227I	IMLX	r	NL
42228I	IMLX	r	NL
42229I	IMLX	r	NL
42230I	IMLX	r	NL
42231I	IECX	r	BN
42232I	IECX	r	BN
42233I	IECX	r	BN
42234I	IECX	r	BN
42235I	IECX	r	NL
42236	IMLX		NL
42237I	IECX	r	BN
42238I	IECX	r	BN
42239I	IECX	r	BN
42240I	IECX	r	NL
42241I	IECX	r	BN
42242I	IECX	r	BN
42243I	IECX	r	BN
42244I	IECX	r	BN
42245I	IECX	r	NL
42246I	IECX	r	BN
42247I	IECX	r	BN
42248I	IECX	r	BN
42249I	IECX	r	BN
42250	IECX		NL
42251I	IWRX	r	LA
42252I	IWRX	r	LA
42253I	IWRX	r	LA
42254I	ICCX	r	PM
42255I	IWRX	r	PM
42256I	IWRX	r	PM
42257I	IWRX	r	PM
42258I	ICCX	r	PM
42259I	IWRX	r	LA
42260I	IWRX	r	LA
42261I	IWRX	r	LA
42262I	ICCX	r	PM
42263I	IWRX	r	PM
42264I	IWRX	r	PM
42265I	IWRX	r	PM
42266I	ICCX	r	PM
42267I	IWRX	r	LA
42268I	IWRX	r	LA
42269I	IWRX	r	LA
42270I	ICCX	r	PM
42271I	IWRX	r	PM
42272I	IWRX	r	PM
42273I	IWRX	r	PM
42274I	ICCX	r	PM
42275I	IWRX	r	LA
42276I	IWRX	r	LA
42277I	IWRX	r	LA
42278I	ICCX	r	PM
42279I	IWRX	r	LA
42280I	IWRX	r	LA
42281I	IWRX	r	LA
42282I	ICCX	r	PM
42283I	IWRX	r	LA
42284I	IWRX	r	LA
42285I	IWRX	r	LA
42286I	ICCX	r	PM
42287I	IWRX	r	LA
42288I	IWRX	r	LA
42289I	IWRX	r	LA
42290I	ICCX	r	PM
42291I	IWRX	r	LA
42292I	IWRX	r	LA
42293I	IWRX	r	LA
42294I	ICCX	r	PM
42295I	IWRX	r	LA
42296I	IWRX	r	LA
42297I	IWRX	r	LA
42298I	ICCX	r	PM
42299I	IWRX	r	LA
42300I	IWRX	r	LA
42301I	IWRX	r	LA
42302I	ICCX	r	PM
42303I	ICCX	r	PM
42304I	ICCX	r	PM
42305I	ICCX	r	PM
42306	ICCX		LA
42307	ICCX		LA
42308	ICCX		LA
42309	ICCX		LA
42310	ICCX		LA
42311	ICCX		LA
42312	ICCX		LA
42313	ICCX		LA
42314I	ICCX	r	LA
42315I	ICCX	r	LA
42316I	ICCX	r	LA
42317I	ICCX	r	LA
42318	ICCX		LA
42319	ICCX		LA
42320	ICCX		LA
42321	ICCX		LA
42322I	ICCX	f	LA
42323I	IMLX	r*	NL
42324	IECX	*	NL
42325I	IMLX	f*	NL
42326I	IECX	r*	BN
42327I	IECX	r*	BN
42328I	IECX	r*	BN
42329I	IECX	r*	NL
42330I	IECX	r*	BN
42331I	IECX	r*	NL
42332	IMLX	*	NL
42333	IECX	*	NL
42334I	IECX	r*	BN
42335I	IECX	f*	BN
42336I	IECX	f*	BN
42337I	IECX	f*	BN
42338I	IECX	f*	BN
42339I	IECX	f*	NL
42340I	IMLX	f*	NL
42341I	IMLX	f*	NL
42342I	IWRX	r*	LA
42343I	IWRX	r*	PM
42344I	IWRX	r*	LA
42345I	IWRX	r*	LA

GJ2G (TGS) TRAILER GUARD'S STANDARD

Dia. GJ202. pg. 61S 1L. One tip-up seat is also being fitted.

44000. Lot No. 30953 Derby 1980. 33.47 t.
44001–44090. Lot No. 30949 Derby 1980–2. 33.47 t.

44091–44094. Lot No. 30964 Derby 1982. 33.47 t.
44095–44101. Lot No. 30970 Derby 1982. 33.47 t.

† Fitted with side buffers and drophead buckeye couplings.

No.	Type		Depot
44000I	ICCX	r	PM
44001I	IWRX	r	PM
44002	IWRX		PM
44003I	IWRX	r	PM
44004I	IWRX	r	PM
44005I	IWRX		PM
44006	IWRX		PM
44007I	IWRX		PM
44008I	IWRX		PM
44009I	IWRX		PM
44010I	IWRX		PM
44011I	IWRX	f	LA
44012I	IWRX	f	LA
44013I	IWRX	f	LA
44014I	IWRX	f	LA
44015I	IWRX	f	LA
44016I	IWRX	f	LA
44017I	IWRX	f	LA
44018I	IWRX	f	LA
44019I	IMLX	r	NL
44020I	ICCX	f	LA
44021I	IECX	f†	BN
44022I	IMLX	f	NL
44023	IMLX		NL
44024I	IMLX	f	NL
44025I	IMLX	f	NL
44026	IMLX		NL
44027I	IMLX	f	NL
44028I	IWRX	r	LA
44029I	IWRX	r	PM
44030I	IWRX	r	LA
44031I	IWRX	r	PM
44032I	IWRX	r	LA
44033I	IWRX	r	PM
44034I	IWRX	r	LA
44035I	IWRX	r	LA
44036	IWRX		LA
44037I	IWRX	r	LA
44038I	IWRX	r	LA
44039I	IWRX	r	LA
44040I	IWRX	r	LA
44041I	IECX	f	BN
44042I	IECX	f	BN
44043I	IECX	f	BN
44044I	IECX	r	BN
44045I	IMLX	f	NL
44046I	IECX	r	NL
44047I	IECX	f	NL
44048I	IECX	r	NL
44049	IECX		NL
44050I	IECX	r	BN
44051	IECX		NL
44052I	IECX	r	NL
44053I	IECX	r	BN
44054I	IECX	r	BN
44055I	IECX	r	BN
44056I	IECX†	f	BN
44057I	IECX	r	BN
44058I	IECX†	r	BN
44059I	IECX†	f	BN
44060I	IECX	r	BN
44061I	IMLX	r	NL
44062	ICCX		PM
44063I	IECX	r	NL
44064I	IMLX	r	NL
44065I	ICCX	r	PM
44066I	IECX	r	NL
44067I	IECX	r	NL
44068I	ICCX	r	PM
44069I	ICCX	r	PM
44070I	IECX	r	HT
44071I	IECX	r	NL
44072I	ICCX	f	LA
44073I	IMLX	r	NL
44074I	IECX	f	BN
44075I	IECX	r	BN
44076I	IMLX	f	NL
44077I	IECX	r†	BN
44078I	IECX	r	BN
44079I	IECX	r	BN
44080I	IECX	r	BN
44081I	IWRX	f	PM
44083I	IECX	r	HT
44084I	ICCX	r	LA
44085I	IMLX	r	NL
44086I	IECX	r†	NL
44087I	ICCX	f	LA
44088	ICCX		LA
44089	ICCX		LA
44090I	ICCX	r	LA
44091	ICCX		LA
44093I	IECX	r	BN
44094I	IMLX	r	NL
44097I	IECX	r†	BN
44098I	IECX	r	BN
44099I	IWRX	r	LA
44100	IECX		EC
44101I	IECX	r†	BN

2.2. HST FORMATIONS

Note: Set NL 45 is the "Master Cutler", Set HT 35 is the "Tees–Tyne Pullman" and Set NL 37 is a spare Pullman set.

The following cars are spare:

Spare BN: 40513 (ICHX), 40704 (IECX), 42129 (IECX).
Spare EC: 44100 (IECX).
Spare HT: 44070 (IECX).
Spare LA: 40433 (ICCX), 40755 (IWRX), 40757 (IWRX), 42093 (IWRX), 42096 (IWRX), 42344 (IWRX), 42345 (IWRX), 44084 (ICCX), 44099 (IWRX).
Spare NL: 40511 (IECX), 42126 (IMLX), 42195 (IECX), 44071 (IECX).
Spare PM: 41164 (IWRX), 42343 (IWRX), 44081 (IWRX).

WESTERN REGION UNITS

PM 1	41003	41004	40209		42003	42004	42005	44001	IWRX
PM 2	41005	41006	40212		42006	42007	42008	44002	IWRX
PM 3	41007	41008	40213		42009	42010	42011	44003	IWRX
PM 4	41009	41010	40204		42012	42013	42014	44004	IWRX
PM 5	41011	41012	40205		42015	42016	42017	44005	IWRX
PM 6	41013	41014	40206		42018	42019	42020	44006	IWRX
PM 7	41015	41016	40207		42021	42022	42023	44007	IWRX
PM 8	41017	41018	40208		42024	42025	42026	44008	IWRX
PM 9	41019	41020	40211		42027	42028	42029	44009	IWRX
PM 10	41021	41022	40210		42030	42031	42032	44010	IWRX
LA 11	41023	41024	40435		42033	42034	42035	44011	IWRX
LA 12	41025	41026	40723	42097	42036	42037	42038	44012	IWRX
LA 13	41027	41028	40724	42098	42039	42040	42041	44013	IWRX
LA 14	41029	41030	40726	42099	42042	42043	42044	44014	IWRX
LA 15	41031	41032	40727	42216	42045	42046	42047	44015	IWRX
LA 16	41033	41034	40731	42221	42048	42049	42050	44016	IWRX
LA 17	41035	41036	40732	42342	42051	42052	42053	44017	IWRX
LA 18	41037	41038	40725		42054	42055	42056	44018	IWRX
LA 28	41121	41122	40722		42251	42252	42253	44028	IWRX
PM 29	41123	41124	40422		42255	42256	42257	44029	IWRX
LA 30	41126	41125	40426		42259	42260	42261	44030	IWRX
PM 31	41127	41128	40221		42263	42264	42265	44031	IWRX
LA 32	41129	41130	40436		42267	42268	42269	44032	IWRX
PM 33	41131	41132	40437		42271	42272	42273	44033	IWRX
LA 34	41133	41134	40427		42275	42276	42277	44034	IWRX
LA 35	41135	41136	40423		42279	42280	42281	44035	IWRX
LA 36	41137	41138	40228		42283	42284	42285	44036	IWRX
LA 37	41139	41140	40416		42287	42288	42289	44037	IWRX
LA 38	41141	41142	40417		42291	42292	42293	44038	IWRX
LA 39	41143	41144	40418		42295	42296	42297	44039	IWRX
LA 40	41145	41146	40419		42299	42300	42301	44040	IWRX

NORTH EAST–SOUTH WEST ROUTE UNITS

PM 41	41147	40401	42254	42258	42262	42266	44000	ICCX
PM 42	41148	40402	42270	42274	42278	42282	44062	ICCX
PM 43	41159	40403	42286	42290	42294	42298	44065	ICCX
PM 44	41160	40425	42302	42303	42304	42305	44068	ICCX
PM 45	41161	40420	42084	42085	42086	42087	44069	ICCX
LA 46	41162	40414	42088	42089	42090	42091	44072	ICCX
LA 50	41166	40424	42196	42197	42060	42061	44020	ICCX
LA 51	41165	40430	42108	42109	42110	42322	44087	ICCX
LA 52	41042	40431	42306	42307	42308	42309	44088	ICCX
LA 53	41167	40432	42310	42311	42312	42313	44089	ICCX
LA 54	41168	40415	42314	42315	42316	42317	44090	ICCX
LA 55	41169	40434	42318	42319	42320	42321	44091	ICCX

EAST COAST & MIDLAND LINES UNITS

BN 01	41057	41058	40708	42335	42111	42112	42113	44041	IECX
BN 02	41059	41114	40733	42336	42115	42116	42117	44042	IECX
BN 03	41061	41062	40729	42337	42119	42120	42121	44043	IECX
BN 04	41063	41064	40754	42328	42123	42124	42125	44044	IECX
BN 05	41171	41173	40734	42338	42178	42127	42128	44021	IECX
NL 06	41067	41068	40730	42331	42131	42132	42133	44046	IECX
NL 07	41069	41070	40751	42339	42135	42136	42137	44047	IECX
NL 08	41071	41072	40728	42329	42139	42140	42141	44048	IECX
NL 09	41073	41074	40752	42333	42143	42144	42145	44049	IECX
BN 10	41075	41076	40756	42327	42147	42148	42149	44050	IECX
NL 11	41077	41078	40753	42324	42151	42152	42153	44051	IECX
NL 12	41079	41080	40700	42157	42156	42155	42100	44052	IECX
BN 13	41081	41082	40701	42159	42160	42161	42101	44053	IECX
BN 14	41083	41084	40702	42163	42164	42165	42102	44054	IECX
BN 15	41085	41086	40703	42169	42168	42167	42103	44055	IECX
BN 16	41087	41088	40706	42104	42171	42172	42173	44056	IECX
BN 17	41089	41090	40705	42175	42176	42177	42105	44057	IECX
BN 18	41091	41092	40717	42179	42180	42181	42106	44058	IECX
BN 19	41093	41094	40707	42183	42184	42185	42107	44059	IECX
BN 20	41095	41096	40736	42326	42187	42188	42189	44060	IECX
NL 21	41097	41098	40750	42158	42191	42192	42193	44061	IMLX
NL 22	41099	41100	40711	42235	42240	42198	42199	44063	IECX
NL 23	41101	41102	40712	42201	42202	42203	42204	44064	IMLX
NL 24	41103	41104	40713	42206	42207	42208	42209	44066	IECX
NL 25	41105	41106	40714	42211	42212	42213	42214	44067	IECX
BN 26	41113	41108	41107	40715	42217	42218	42219	44097	IECX
BN 27	41109	41110	40716	42334	42222	42223	42224	44101	IECX
NL 28	41111	41112	40749	42229	42194	42227	42228	44073	IMLX
BN 29	41046		40718	42231	42232	42233	42234	44074	IECX
BN 30	41115	41116	40710	42330	42237	42238	42239	44075	IECX
BN 31	41170	41118	40720	42241	42242	42243	42244	44098	IECX
BN 32	41119	41120	40721	42246	42247	42248	42249	44078	IECX
BN 33	41149	41150	40739	42162	42166	42170	42174	44079	IECX
BN 34	41151	41152	40740	42226	42182	42186	42190	44080	IECX
HT 35	41153	41066	40501	41154	40741	42205	42210	44083	IECX
NL 36	41155	41156	40748	42215	42220	42225	42230	44085	IMLX
NL 37	41157	41041	40505	41158	40743	42245	42250	44086	IECX
NL 38	41055	41056	40709	42341	42081	42082	42083	44027	IMLX
NL 39	41045	41065	40738	42062	42066	42067	42068	44022	IMLX
NL 40	41047	41048	40744	42069	42070	42071	42118	44023	IMLX
NL 41	41049	41050	40745	42325	42072	42073	42074	44024	IMLX
NL 42	41051	41052	40746	42075	42076	42077	42078	44025	IMLX
NL 43	41053	41054	40747	42332	42236	42079	42080	44026	IMLX
BN 44	41174	41172	40742	42130	42134	42138	42122	44093	IECX
NL 45	41117	41176	41175	40619	42146	42150	42154	44094	IMLX
NL 46	41039	41040	40735	42323	42057	42058	42059	44019	IMLX
NL 47	41043	41044	40737	42340	42063	42064	42065	44045	IMLX
NL 49	41163		40429	42200	42092	42094	42095	44076	IMLX

3.DIESEL MULTIPLE UNITS

NUMBERING

Diesel mechanical and diesel hydraulic multiple unit vehicles are numbered in the series 51000–59999. All vehicles numbered in the 53000–53999 series were originally numbered in the series 50000–50999, and were renumbered by having 3000 added to their original numbers. All vehicles in the series 54000–54504 were originally numbered in the series 56000–56505, and were renumbered by having 2000 subtracted from their original numbers.

Diesel electric multiple unit vehicles are numbered in the series 60108–60999. A number of vehicles which were numbered in the series 60000–60107 were renumbered during 1989 to avoid conficting with Class 60 locomotives. The previous numbers of these vehicles are given in the text where appropriate.

Regional prefix letters used to be carried preceeding the vehicle number, but these are now no longer used and many have been removed. The prefixes were: E–Eastern Region, M–London Midland Region, S–Southern Region, SC–Scottish Region. Prefixes are not shown in this book as they no longer officially form part of the vehicle number.
The codes used by the BR Operating Department to describe the various different types of diesel multiple unit vehicles and quoted in the class headings are as follows:

Diesel Mechanical & Diesel Hydraulic Units.

DMBC Driving Motor Brake Composite.
DMBS Driving Motor Brake Standard.
DMCL Driving Motor Composite with Lavatory.
DMPMV Driving Motor Parcels & Mails Van.
DMS Driving Motor Standard.
DMSL Driving Motor Standard with Lavatory.
DTCL Driving Trailer Composite with Lavatory
DTPMV Driving Trailer Parcels & Mails Van.
DTSL Driving Trailer Standard with Lavatory.
MS Motor Standard.
MSL Motor Standard with Lavatory.
TBSL Trailer Brake Standard with Lavatory.
TC Trailer Composite.
TCL Trailer Composite with Lavatory.
TS Trailer Standard.
TSL Trailer Standard with Lavatory.
TSLRB Trailer Standard with Lavatory & Miniature Buffet (all buffets are now disused).

It should be noted that as all vehicles are of an open configuration the letter 'O' is omitted for all vehicles.

The letters (A) and (B) may be added to the above codes to differentiate between two cars of the same operating type which have differences between them. Note that a consistent system is used, rather than the official operator codes which are sometimes inconsistent.

A composite is a vehicle containing both First and Standard class accommodation, and vehicles are described as such even though most first class accommodation has now been declassified on most vehicles. This is done so as to differentiate between the different styles of seat provided in standard and erstwhile first class areas of a vehicle. At the time of writing only composite vehicles allocated to BY, CA, OO and RG depots retained first class accommodation in use as such.

A brake vehicle is a vehicle containing seperate specific accommodation for the guard (as opposed to the use of spare driving cabs on second generation units).

Diesel Electric Units.

DMBSO Driving Motor Brake Standard (Open).
DTCsoL Driving Trailer Composite with Lavatory (Semi-Open).
DTSOL Driving Trailer Standard with Lavatory (Open).
DTSO Driving Trailer Standard (Open).
TCsoL Trailer Composite with Lavatory (Semi-Open).
TSO Trailer Standard (Open).
TSOL Trailer Standard with Lavatory (Open).

The notes as above apply regarding composite and brake vehicles. A semi-open composite vehicle has first class accommodation in compartments with a side corridor and standard class accommodation provided in an open saloon.

DIAGRAMS AND DESIGN CODES

For each type of vehicle, the official design code consists of a seven character code of two letters, four numbers and another letter, e.g. DP2010A. The first five characters of this are the diagram code and are given in the class heading or sub heading. These are explained as follows:

1st Letter

This is always 'D' for a diesel multiple unit vehicle.

2nd Letter

as follows for various vehicle types:

P Driving motor passenger vehicles without a brake compartment.
Q Driving motor passenger vehicles with a brake compartment.
R Non-Driving motor passenger vehicles.
S Driving trailer passenger vehicles.
T Trailer passenger vehicles without a brake compartment.
U Trailer passenger vehicles with a brake compartment.
X Parcels and Mails vehicles and single unit railcars.

1st Figure

as follows for various vehicle configurations:

2 Standard class accommodation (incl. declassified seats).
3 Composite accommodation.
5 No passenger accommodation.

3rd & 4th Figures

These denote distinguish between the different designs of vehicle, each different design being allocated a unique two digit number.

Special Note

Where vehicles have been declassified the correct design code for a declassified vehicle is given, even though this may be at variance with official records which do not show the reality of the current position. A declassified vehicle will be denoted by an asterisk (*) and will still retain the first class style seats in the erstwhile first class section of the vehicle.

LOT NUMBERS

Each batch of vehicles is allocated a Lot (or batch) number when ordered and these are quoted in class headings and sub headings.

BUILDERS

These are shown in class headings where the following abbreviations are used:

AlexanderWalter Alexander Ltd., Falkirk.
Barclay Andrew Barclay Ltd., Kilmarnock.
BRCW Birmingham Railway Carriage & Wagon Company Ltd.
Derby BR Derby Carriage Works or British Rail Engineering Limited, Derby Carriage Works.
GloucesterGloucester Railway Carriage & Wagon Company Ltd.
Leyland BusLeyland Bus Ltd., Workington.
Metro-CammellMetropolitan-Cammell Ltd.
Pressed SteelPressed Steel Ltd.
York British Rail Engineering Ltd., York.

Where a dual BR works builder is shown (e.g. Ashford/Eastleigh) the first named built the underframe and the last named built the body and assembled the vehicle. For second generation vehicles, the first name is that of the main contractor with the second name being the underframe and final assembly sub-contractor

ACCOMMODATION

This information is given in class headings and sub headings in the form nF nS nL, where n is the number of seats or lavatories and F, S & L denote first class seats, standard class seats and lavatories respectively. (e.g. 12F 54S 1L denotes 12 first class seats, 54 standard class seats and one lavatory). In declassified vehicles, the capacity is still shown in terms of first and standard class seats to differentiate between the two physically different seat types available, although all seats are officially standard class in such instances.

WEIGHTS

Approximate weights in working order are given in tons for all vehicle types in the class headings and sub headings as appropriate.

DIMENSIONS

The dimensions of each type of vehicle are given in metric units, with length followed by width by height. All lengths quoted are over buffers (1st generation vehicles) or couplings (2nd generation vehicles). All widths and heights quoted are maxima.

ABBREVIATIONS

The following are used throughout this section:

DEMU	diesel electric multiple unit.
DHMU	diesel hydraulic multiple unit.
DMMU	diesel mechanical multiple unit.
DMU	diesel multiple unit (general term).
F	First class seats.
f	Facelifted (Fitted with fluorescent lights and recovered interior panelling and seats).
g	Gangwayed where other members of the class are not.
h	Headlight fitted (where non standard for type).
hp	horsepower.
kW	Kilowatts.
L	Lavatories.
m	metres.
mph	miles per hour.
p	Passenger vehicle adapted for parcels use by removal of some or all of the seats.
r	Fitted with radio electronic token block apparatus.
S	Standard class seats.
t	tons.
(S)	Stored servicable. (Not necessarily at allocated depot).
(U)	Stored unservicable. (Not necessarily at allocated depot).
*	Composite vehicle now declassified to standard class only but retaining original seats.
§	Fitted with headlight and radio electronic token block apparatus.

3.1.1. FIRST GENERATION DMUS

Unit Types

There are four basic types of first generation vehicle as referred to in the class headings as follows:

(1) Standard (known as 'low density' by BR). Exterior doors are located in entrance vestibules. Largely unidirectional 3+2 seating is provided in standard class, with 2+2 seating in first class. Where no specific type of vehicle is quoted in class headings the vehicle may be assumed to be of this type.

(2) Suburban (known as 'high density' by BR). Exterior doors to each seating bay. Back to back seating arranged 3+2 in standard class with 2+1 seating in first class.

(3) Cross Country. Exterior doors are located in entrance vestibules. These were originally designed for longer distance services and have 2+2 seating in standard class, with 2+1 in first class.

(4) Parcels. Double or roller shutter exterior doors. No seats with whole of floor space devoted to parcels and mails stowage.

Transmission

All motor vehicles have mechanical transmission consisting of a cardan shaft and freewheel to a four-speed epicylic gearbox with a further cardan shaft to the final drive, each engine driving the inner axle of one bogie.

Gangways

Unless stated otherwise, all vehicles are gangwayed at non driving ends with British Standard gangways.

Doors

All units are equipped with slam doors unless otherwise stated.

Couplings

Screw couplings are used on all vehicles. All remaining first generation vehicles may be coupled together to work in multiple up to a maximum of 6 motor cars or 12 cars in total in a formation. First generation vehicles may not be coupled in multiple with second generation vehicles.

Maximum Speed

Maximum permitted speed of all vehicles is 70 mph.

3.1.2. SECOND GENERATION DMUS

Unit Types

There are five basic types of second generation vehicle as referred to in the class headings as follows:

●**Railbus** (known as 'Pacers' on BR). Folding power operated exterior doors. Bus-type 3+2 (2+2 on classes 140 and 141) largely unidirectional seating. Limited luggage space. Four wheel chassis.

●**Sprinter.** Sliding power operated exterior double doors to large entrance vestibules. High backed 3+2 seating. Limited luggage space.

●**Super Sprinter.** Sliding/sliding plug power operated exterior doors. High backed 2+2 largely unidirectional seating with some tables.

●**Express.** Sliding plug power operated exterior doors. Air conditioned. High backed 2+2 largely unidirectional seating with some tables.

Public Address System: All vehicles are equipped with public address, with transmission equipment for driving vehicles.

Transmission: All vehicles unless otherwise stated are fitted with Voith T211r hydraulic transmission and a Gmeinder final drive.

Gangways: Unless stated otherwise, all vehicles have flexible diaphragm gangways.

Couplings: Unless otherwise stated all vehicles are fitted with BSI automatic couplings at their outer ends. Railbus types are fitted with bar couplings at their inner ends, but all other types have BSI couplings at their inner ends unless otherwise stated.

51060–52065. FIRST GENERATION DMUS.

CLASS 119 GLOUCESTER CROSS COUNTRY

Engines: Two Leyland 1595 of 112 kW (150 hp).
Bogies: DD10.
Dimensions: 20.45 x 2.82 x 3.86 m.

DMBC. Dia. DQ302 (*DQ2??). Lot No. 30421 1959. 18F 16S. 37.5 t.

51060	**N**	NSBX	RG
51062	**N**	NWRX	RG
51065	**N**	NSBX	RG
51066	**N**	NSBX	RG
51073	**N**	NSBX	RG
51074	**N**	NSBX	RG
51076	**N**	NSBX	RG
51079	**N**	NSBX	RG

DMSL. Dia. DP203. Lot No. 30422 1958–59. 68S 2L. 38.5 t.

51086	**N**	NSBX	RG
51088	**N**	NSBX	RG
51090	**N**	NWRX	RG
51094	**N**	NSBX	RG
51099	**N**	NSBX	RG
51103	**N**	NSBX	RG
51104	**N**	NSBX	RG
51107	**N**	NSBX	RG

CLASS 116 DERBY SUBURBAN

Engines: Two Leyland 680/1 of 112 kW (150 hp).
Bogies: DD10.
Dimensions: 20.45 x 2.82 x 3.86 m.

DMBS. Dia. DQ230. Lot No. 30446 1958. 65S. 36.5 t.

51128	PCFX	fh	CF
51129	PTSX	f	TS
51130	PTSX		TS
51131	PTSX	f	TS
51132	PCFX	fh	CF
51133	PCFX		CF
51134	PCFX	fh	CF
51135	PCFX	fh	CF
51136	PCFX	f	CF
51138	PTSX	f	TS
51139	PCFX	fh	CF
51140	PCFX	fh	CF

DMS. Dia. DP220. Lot No. 30447 1958. 89S. 36.5 t.

51141	PCFX	fh	CF
51142	PTSX	f	TS
51143	PTSX	f	TS
51144	PTSX	f	TS
51145	PCFX	fh	CF
51146	PTSX		TS
51147	PCFX	fh	CF
51148	PCFX	fh	CF
51149	PTSX	f	TS
51151	PTSX	f	TS
51152	PCFX	fh	CF
51153	PCFX	fh	CF

CLASS 101 METRO-CAMMELL

Engines: Two Leyland or AEC of 112 kW (150 hp).
Bogies: DD15.
Dimensions: 18.49 x 2.82 x 3.85 m.

DMBS. Dia. DQ202 (†DQ232). Lot No. 30467 1958–59. 52S (†49S with additional luggage rack). 32.5 t.

51174		PNLX	f	NL	51213		PNCX	f	NC
51175		PNLX		NL	51215		NSBX	f	OO
51177		PNLX	f	NL	51218		PNCX	f	NC
51178		PNLX	f	NL	51219		PLAX	f	LA
51179		PLAX	f	LA	51220	**N**	NWRX	f	OO
51180		PNLX	f	NL	51221	**N**	NWRX	f	OO
51181		PLAX	f	LA	51222		PCHX	f	CH
51182		PHAX	f	HA	51223		PLAX	f	LA
51184		PNLX	f	NL	51224	**S**	PEDX	f	ED
51185		PNCX	f	NC	51225	**N**	NWRX	f	OO
51187		PNCX	f	NC	51226	**N**	NSBX	f†	RG
51188		PNLX	f	NL	51228		PHAX	f	HA
51189		PNCX	f	NC	51230		PNCX	f	NC
51190	**N**	NSBX		RG	51231		PHAX	f	HA
51191		PNLX		NL	51234		PHAX	f	HA
51192		PNCX	f	NC	51241		PHAX		HA
51194		PNLX	f	NL	51244		PHAX	f	HA
51201		PNCX	f	NC	51245		PCHX	f	CH
51205		PLAX	f	LA	51246		PLAX	f	LA
51207	**N**	NGEX	f	CA	51247		PNCX	f	NC
51208	**N**	NGEX	f	CA	51249	**S**	PEDX	f	ED
51210		PLOX	f	LO	51252		PNCX	f	NC
51211		NSBX	f†	RG	51253		PNLX	f	NL
51212	**N**	NWRX	f	OO					

CLASS 118 BRCW SUBURBAN

Engines: Two Leyland 680/1 of 112 kW (150 hp).
Bogies: DD10.
Dimensions: 20.45 x 2.82 x 3.86 m.

DMBS. Dia. DQ220. Lot No. 30543 1960. 65S. 36.5 t.

51306		PTSX	f	TS	51315		PCFX	h	CF
51312		PCFX		CF	51316		PTSX	f	TS
51314		PTSX		TS					

DMS. Dia. DP221. Lot No. 30545 1960. 89S. 36.5 t.

51319	**N**	NGBX		OO	51329		PTSX		TS
51321		PTSX	f	TS	51330		PCFX	h	CF
51327		PCFX		CF	51331		PTSX	f	TS

CLASS 117 PRESSED STEEL SUBURBAN

Engines: Two Leyland 680/1 of 112 kW (150 hp).
Bogies: DD10.
Dimensions: 20.45 x 2.82 x 3.86 m.

DMBS. Dia. DQ220. Lot No. 30546 1959–60. 65S. 36.5 t.

51332	**N**	NWRX	f	RG	51335	**N**	NWRX	f	RG
51333	**N**	NWRX	f	RG	51336	**N**	NWRX	f	RG
51334		PTSX	f	TS	51337	**N**	NWRX	f	RG

51338		PTSX	f	TS
51339		PTSX	f	TS
51340	**N**	NWRX	f	RG
51341	**N**	NWRX	f	RG
51342	**N**	NWRX	f	RG
51343	**N**	NWRX	f	RG
51344	**N**	NWRX	f	RG
51345	**N**	NWRX	f	RG
51346	**N**	NWRX	f	RG
51347	**N**	NWRX	f	RG
51348		PTSX	f	TS
51349	**N**	NWRX	f	RG
51350	**N**	NWRX	f	RG
51351	**N**	NWRX	f	RG
51352		PTSX	f	TS
51353		PTSX	f	TS
51354	**N**	NWRX	f	RG
51355	**N**	NWRX	f	RG
51356	**N**	NWRX	f	RG
51358	**N**	NWRX	f	RG
51359	**N**	NWRX	f	RG
51360		PTSX	f	TS
51361	**N**	NWRX	f	RG
51362	**N**	NWRX	f	RG
51363	**N**	NWRX	f	RG
51364		PTSX	f	TS
51365		PTSX	f	TS
51366	**N**	NWRX	f	RG
51367	**N**	NWRX	f	RG
51368	**0**	PTSX	f	TS
51369		PTSX	f	TS
51370		PTSX	f	TS
51371		PTSX	f	TS
51372		PTSX	f	TS
51373		PTSX	f	TS

DMS. Dia. DP221. Lot No. 30548 1959–60. 89S. 36.5 t.

51374	**N**	NWRX	f	RG
51375	**N**	NWRX	f	RG
51376		PTSX	f	TS
51377	**N**	NWRX	f	RG
51378	**N**	NWRX	f	RG
51379	**N**	NWRX	f	RG
51380		PTSX	f	TS
51381	**N**	NWRX	f	RG
51382		PTSX	f	TS
51383	**N**	NWRX	f	RG
51384	**N**	NWRX	f	RG
51385	**N**	NWRX	f	RG
51386	**N**	NWRX	f	RG
51387	**N**	NWRX	f	RG
51388	**N**	NWRX	f	RG
51389	**N**	NWRX	f	RG
51390		PTSX	f	TS
51391	**N**	NWRX	f	RG
51392	**N**	NWRX	f	RG
51393	**N**	NWRX	f	RG
51394		PTSX	f	TS
51395		PTSX	f	TS
51396	**N**	NWRX	f	RG
51397	**N**	NWRX	f	RG
51398	**N**	NWRX	f	RG
51399	**N**	NWRX	f	RG
51400	**N**	NWRX	f	RG
51401	**N**	NWRX	f	RG
51402		PTSX	f	TS
51404	**N**	NWRX	f	RG
51405	**N**	NWRX	f	RG
51406		PTSX	f	TS
51407		PTSX	f	TS
51408	**N**	NWRX	f	RG
51409	**N**	NWRX	f	RG
51410	**0**	PTSX	f	TS
51411		PTSX	f	TS
51412		PTSX	f	TS
51413		PTSX	f	TS
51414		PTSX	f	TS
51415		PTSX	f	TS

CLASS 108 DERBY 'LIGHTWEIGHT'

Engines: Two Leyland 680/1 or 680/13 of 112 kW (150 hp).
Bogies: DD7.
Dimensions: 18.49 x 2.82 x 3.87 m.

DMBS. Dia. DQ213. Lot No. 30498 1960. 52S. 29.5 t.

51416	PCFX	f	CF
51417	PCFX	fh	CF
51418	PCHX	f	CH
51419	PCFX	h	CF
51420	PNLX	f	NL
51421	PCHX	f	CH

51422	PCHX	f	CH
51424	PCHX	f	CH

CLASS 101 METRO-CAMMELL

Engines: Two Leyland or AEC of 112 kW (150 hp).
Bogies: DD15.
Dimensions: 18.49 x 2.82 x 3.85 m.

DMBS. Dia. DQ202 (†DQ232). Lot No. 30500 1959. 52S (†49S with additional luggage rack). 32.5 t.

51425	**N**	NSBX	f†	RG
51426		PCHX	f	CH
51427		PNCX	f	NC
51428		PNCX	f	NC
51429		PNCX	f	NC
51431	**N**	NSBX	f†	RG
51432	**N**	NSBX	f	RG
51434		NSBX	f†	RG
51435		PNLX	f	NL
51436		PCFX	f	CF
51437	**N**	NSBX	f	RG
51438	**N**	NGEX	f	CA
51442		PNCX	f	NC
51443	**N**	NWRX	f	OO
51444	**N**	NGEX	f	CA
51445	**N**	NGEX	fh	CA
51455		PNLX	f	NL
51462		PCHX	fh	CH
51463		PNLX	fh	NL
51468		PEDX	f	ED

DMCL. Dia. DP304(*DP210) or†DP317. Lot No. 30501 1959. 12F 53S 1L. (†12F 46S with additional luggage racks 1L). 32.5t.

51495		PNLX	*	NL
51496		PCHX	f*	CH
51498	**N**	NSBX	f	RG
51499	**N**	NSBX	f†	RG
51500		PLAX	f*	LA
51501	**N**	NSBX	f†	RG
51503		NSSX	f†	RG
51504	**N**	NSBX	f†	RG
51505		PCHX	f*	CH
51506		PNCX	f§*	NC
51508		PNCX	f§*	NC
51509		PCHX	*	CH
51511		PCHX	f*	CH
51512		PLAX	f*	LA
51513		PCHX	*	CH
51524		PNLX	f*	NL
51530		PLAX	fh*	LA
51531		PCHX	f*	CH
51533		PNLX	fh*	NL

CLASS 108 DERBY 'LIGHTWEIGHT'

Engines: Two Leyland 680/1 or 680/13 of 112 kW (150 hp).
Bogies: DD7.
Dimensions: 18.49 x 2.79 x 3.87 m.

DMCL. Dia. DP311 (*DP226, †DP227). Lot No. 30461 1959–60. 12F 52S 1L. (†12F 53S 1L) 28.5 t.

51561		PLEX	fh*	LE
51562		PLEX	h*	LE
51563		PLEX	fh*	LE
51565		PLEX	h*	LE
51566		PCHX	h*	CH
51567		PLEX	fh*	LE
51568		PCHX	f*	CH
51569		PCHX	f†	CH
51570		PLAX	f*	LA
51571		NMYX	f	BY
51572	**N**	NMYX		BY

CLASS 115 DERBY SUBURBAN

Engines: Two Leyland Albion of 149 kW (200 hp).
Gangways: Non gangwayed except as shown.
Bogies: DD13.
Dimensions: 20.45 x 2.82 x 3.87 m.
Notes: Fluorescent lighting as built. All BY allocated cars are fitted with power sanding equipment.

DMBS. Dia. DQ218 (g DQ233). Lot No. 30530 1960. 78S (g 74S). 38.5 t. (f 39.0 t.)

51651 **N** NMYX g BY
51652 NMYX g BY
51653 **N** NMYX f BY
51654 **N** NMYX g BY
51655 **N** NMYX f BY
51656 **N** NMYX g BY
51657 **N** NMYX g BY
51658 NMYX f BY
51659 **N** NMYX f BY
51660 **N** NMYX f BY
51661 **N** NMYX g BY
51662 PTSX fg TS
51663 **N** NMYX f BY
51664 **N** NMYX f BY
51665 **N** NMYX f BY
51666 **N** NMYX f BY
51667 **N** NMYX f BY
51668 **N** NMYX BY
51669 **N** NMYX f BY
51670 NMYX f BY
51671 **N** NMYX BY
51673 NMYX f BY
51674 **N** NMYX BY
51675 **N** NMYX BY
51676 **N** NMYX g BY
51677 **N** NMYX f BY
51678 **N** NMYX f BY
51679 **N** NMYX g BY
51680 **N** NMYX f BY

CLASS 101 METRO-CAMMELL

Engines: Two Leyland of 112 kW (150 hp).
Bogies: DD15.
Dimensions: 18.49 x 2.82 x 3.85 m.

DMBS. Dia. DQ202. Lot No. 30587 1959. 52S. 32.5 t.

51799 PNLX fh NL
51800 PCHX f CH

DMCL. Dia. DP304 (*DP210). Lot No. 30588 1959. 12F 53S 1L. 32.5 t.

51803 **S** PEDX f* ED
51808 PNLX fh* NL

CLASS 110 BRCW 'CALDER VALLEY'

Engines: Two Rolls-Royce C6 of 134 kW (180 hp).
Bogies: DD14.
Dimensions: 18.48 x 2.82 x 3.84 m.

DMBC. Dia. DQ301 (*DQ2??). Lot No. 30592 1961. 12F 33S. 32.5 t.

51813 PNLX f* NL
51817 PNLX f* NL
51823 PNLX f* NL

DMCL. Dia. DP313 (*DP2??). Lot No. 30593 1961. 12F 54S 1L. 32.5 t.

51829 PNLX f* NL
51830 PNLX f* NL

51834	PNLX	f*	NL
51840	PNLX	f*	NL
51842	PNLX	f*	NL
51843	PNLX	f*	NL
51847	PNLX	f*	NL

CLASS 115 DERBY SUBURBAN

Engines: Two Leyland Albion of 149 kW (200 hp).
Gangways: Non gangwayed except as shown.
Bogies: DD13.
Dimensions: 20.45 x 2.82 x 3.87 m.
Notes: Fluorescent lighting as built. All BY allocated cars are fitted with power sanding equipment.

51849–51860. DMBS. Dia. DQ218 (g DQ233). Lot No. 30595 1960. 78S (g 74S). 38.5 t. (f 39.0 t.).
51862–51900. DMBS. Dia. DQ218 (g DQ233). Lot No. 30598 1960. 78S (g 74S). 38.5 t. (f 39.0 t.).

51849	**N**	NMYX	g	BY
51851		PTSX	g	TS
51852		PTSX	fg	TS
51853		PTSX	fg	TS
51854		PTSX	fg	TS
51855	**N**	NMYX	g	BY
51856		PTSX	fg	TS
51857	**N**	NMYX	f	BY
51858		PTSX	fg	TS
51859		PTSX	fg	TS
51860		PTSX	fg	TS
51862		PTSX	fg	TS
51863	**N**	NMYX	f	BY
51865		PTSX	g	TS
51866	**N**	NMYX	g	BY
51867		PTSX	g	TS
51868		PTSX	fg	TS
51869		PTSX	fg	TS
51870		PTSX	g	TS
51871	**N**	NMYX	f	BY
51872		NMYX	f	BY
51873	**N**	NMYX	f	BY
51874		NMYX	f	BY
51875	**N**	NMYX	f	BY
51876		PTSX	g	TS
51877		PTSX	fg	TS
51878	**N**	NMYX	g	BY
51879	**N**	NMYX	f	BY
51880		PTSX	fg	TS
51883		NMYX	f	BY
51884		PTSX	fg	TS
51885	**N**	NMYX		BY
51886	**N**	NMYX		BY
51887	**N**	NMYX	f	BY
51888	**N**	NMYX		BY
51889	**N**	NMYX		BY
51890		NMYX	f	BY
51891	**N**	NMYX	f	BY
51892		PTSX	fg	TS
51893	**N**	NMYX	g	BY
51894	**N**	NMYX	f	BY
51895	**N**	NMYX	fg	BY
51896	**N**	NMYX	f	BY
51897		PTSX	g	TS
51898		PTSX	g	TS
51899	**N**	NMYX	f	BY
51900	**N**	NMYX	f	BY

Name: 51899 Aylesbury College SILVER JUBILEE 1987

CLASS 108 DERBY 'LIGHTWEIGHT'

Engines: Two Leyland 680/1 or 680/13 of 112 kW (150 hp).
Bogies: DD7.
Dimensions: 18.49 x 2.82 x 3.87 m.

DMBS. Dia. DQ213. Lot No. 30601 1960–61. 52S. 29.5 t.

51901	PCHX	f	CH
51902	PCHX	f	CH

51903		PCHX	f	CH
51904		PCHX	f	CH
51905		PCHX	f	CH
51906		PCHX	f	CH
51907		PCHX	f	CH
51908		PCHX	f	CH
51909	**N**	NMYX	f	BY
51911		PCHX	f	CH
51912	**N**	NMYX		BY
51913		PCHX	f	CH
51914	**N**	NMYX	f	BY
51916	**N**	NMYX		BY
51917		PCHX	f	CH
51919		PLEX	h	LE
51920		PCHX		CH
51922		PLEX	h	LE
51924		PLEX	fh	LE
51925		PLEX	fh	LE
51926		PLEX	fh	LE
51927		PLOX		LO
51928		PLEX	fh	LE
51930		PLEX	fh	LE
51931		PLEX	fh	LE
51932		PLAX	f	LA
51933		PLAX	f	LA
51935		PLOX	f	LO
51936		PLAX	f	LA
51937		PCHX	f	CH
51938		PCHX	f	CH
51939		PLAX	f	LA
51940		PLAX	f	LA
51941		PCHX	f	CH
51942	**N**	NMYX	f	BY
51943		PCHX	f	CH
51945		PCHX	f	CH
51947		PCHX	h	CH
51948		PCHX	f	CH
51950		PCHX	f	CH

CLASS 107 DERBY 'HEAVYWEIGHT'

Engines: Two Leyland 1595 of 112 kW (150 hp).
Bogies: DD10.
Dimensions: 18.49 x 2.82 x 3.87 m.

DMBS. Dia. DQ211. Lot No. 30611 1960–61. 52S. 35.0 t.

51985		PEDX	f	ED
51986	**S**	PEDX	f	ED
51987	**S**	PEDX	f	ED
51988		PEDX		ED
51989	**S**	PEDX	f	ED
51990	**S**	PEDX	f	ED
51991		PHAX	f	HA
51992		PHAX	f	HA
51993	**S**	PEDX	f	ED
51994		PEDX	f	ED
51996	**S**	PEDX	f	ED
51997	**S**	PEDX	f	ED
51998	**S**	PEDX	f	ED
51999	**S**	PEDX	f	ED
52000		PHAX	f	HA
52001	**S**	PEDX	f	ED
52004		PEDX	f	ED
52005	**S**	PEDX	f	ED
52006	**S**	PEDX	f	ED
52007	**S**	PEDX	f	ED
52008	**S**	PEDX	f	ED
52010	**S**	PEDX	f	ED

DMCL. Dia. DP3?? (*DP215). Lot No. 30612 1960–61. 12F 53S 1L. 35.5 t.

52011		PEDX	f*	ED
52012	**S**	PEDX	*	ED
52013		PEDX	*	ED
52015	**S**	PEDX	f*	ED
52016		PEDX	f*	ED
52018	**S**	PEDX	f*	ED
52019	**S**	PEDX	f*	ED
52020	**S**	PEDX	f*	ED
52021	**S**	PEDX	f*	ED
52023	**S**	PEDX	f*	ED
52024	**S**	PEDX	f*	ED
52025	**S**	PEDX	*	ED
52026		PHAX	f*	HA
52028		PEDX	f*	ED
52029	**S**	PEDX	f*	ED
52030	**S**	PEDX	f*	ED
52031	**S**	PEDX	f*	ED
52033	**S**	PEDX	f*	ED
52034	**S**	PEDX	f*	ED
52035		PHAX	f*	HA
52036		PHAX	f*	HA

CLASS 108 DERBY 'LIGHTWEIGHT'

Engines: Two Leyland 680/1 or 680/13 of 112 kW (150 hp).
Bogies: DD7.
Dimensions: 18.49 x 2.82 x 3.87 m.

DMCL. Dia. DP312 (*DP227). Lot No. 30660 1960–61. 12F 53S 1L. 28.5 t.

52038	PLEX	fh*	LE
52039	PCHX	f*	CH
52041	PLEX	fh*	LE
52042	PNLX	f*	NL
52043	PHAX	f*	HA
52044	PLEX	fh*	LE
52045	PLEX	fh*	LE
52046	PLEX	h*	LE
52047	PLAX	f*	LA
52048	PLEX	fh*	LE
52049	PCHX	f*	CH
52050	PCHX	f*	CH
52051	PLAX	f*	LA
52053	PCFX	f*	CF
52054	PLAX	f*	LA
52055	PLEX	fh*	LE
52056	PCHX	f*	CH
52057	PLAX	fh*	LA
52058	PCHX	f*	CH
52059	PCFX	f*	CF
52060	PCHX	f*	CH
52061	PLEX	h*	LE
52062	PCHX	*	CH
52063	PLAX	fh*	LA
52064	PCHX	f*	CH
52065	PCHX	h*	CH

CLASS 110 BRCW 'CALDER VALLEY'

Engines: Two Rolls-Royce C6 of 134 kW (180 hp).
Bogies: DD14.
Dimensions: 18.48 x 2.82 x 3.84 m.

DMBC. Dia. DQ301 (*DQ2??). Lot No. 30691 1961–62. 12F 33S. 32.5 t.

52066	PNLX	f*	NL
52069	PNLX	f*	NL
52071	PNLX	f*	NL
52072	PNLX	f*	NL
52075	PNLX	f*	NL

DMCL. Dia. DP313 (*DQ2??). Lot No. 30692 1961–62. 12F 54S 1L. 32.5 t.

52077	PNLX	f*	NL
52080	PNLX	f*	NL
52082	PNLX	f*	NL
52085	PNLX	f*	NL

52101–52808. SECOND GENERATION DMUS.

CLASS 150/1 BREL SPRINTER

Engine: One Cummins NT855R5 of 210 kW (285 hp).
Bogies: One BP38 and one BT38.
Gangways: Within unit only.
Seats: 2+3 facing (BR standard outer suburban seating).
Dimensions: 20.06 x 2.82 x 3.77 m.
Maximum Speed: 75 mph.

DMSL. Dia. DP238. Lot No. 31011 York 1985–6. 68S 1L. 36.5 t.

52101	**PO**	PDYX		DY
52102	**PO**	PDYX		DY
52103	**PO**	PDYX		DY
52104	**PO**	PDYX		DY
52105	**PO**	PDYX		DY
52106	**PO**	PDYX		DY
52107	**PO**	PDYX		DY
52108	**PO**	PDYX		DY
52109	**PO**	PDYX		DY
52110	**PO**	PDYX		DY
52111	**PO**	PDYX		DY
52112	**PO**	PDYX		DY
52113	**PO**	PDYX		DY
52114	**PO**	PDYX		DY
52115	**PO**	PDYX		DY
52116	**PO**	PDYX	r	DY
52117	**PO**	PDYX	r	DY
52118	**PO**	PDYX	r	DY
52119	**PO**	PDYX	r	DY
52120	**PO**	PDYX	r	DY
52121	**PO**	PDYX	r	DY
52122	**PO**	PDYX	r	DY
52123	**PO**	PDYX	r	DY
52124	**PO**	PDYX	r	DY
52125	**PO**	PDYX	r	DY
52126	**PO**	PDYX	r	DY
52127	**PO**	PDYX	r	DY
52128	**PO**	PDYX	r	DY
52129	**PO**	PDYX	r	DY
52130	**PO**	PDYX	r	DY
52131	**PO**	PDYX	r	DY
52132	**PO**	PDYX	r	DY
52133	**PO**	PDYX	r	DY
52134	**PO**	PDYX	r	DY
52135	**PO**	PDYX	r	DY
52136	**PO**	PDYX	r	DY
52137	**PO**	PDYX	r	DY
52138	**PO**	PDYX	r	DY
52139	**PO**	PDYX	r	DY
52140	**PO**	PDYX	r	DY
52141	**P**	PTSX	r	TS
52142	**P**	PTSX	r	TS
52143	**P**	PDYX	r	DY
52144	**P**	PTSX	r	TS
52145	**P**	PTSX	r	TS
52146	**PO**	PTSX	r	TS
52147	**P**	PDYX	r	DY
52148	**PO**	PTSX	r	TS
52149	**P**	PTSX	r	TS
52150	**P**	PDYX	r	DY

CLASS 150/2 YORK SPRINTER

Engine: One Cummins NT855R5 of 210 kW (285 hp).
Bogies: One BP38 and one BT38.
Gangways: Through gangwayed.
Seats: 2+3 mainly unidirectional ("Ashbourne seating").
Dimensions: 20.06 x 2.82 x 3.77 m.
Maximum Speed: 75 mph.

DMSL. Dia. DP242. Lot No. 31017 York 1986–87. 73S 1L. 35.8 t.

52201	**PN**	PNHX	NH
52202	**P**	PNHX	NH
52203	**PN**	PNHX	NH
52204	**P**	PNHX	NH

52205	**P**	PTSX	TS
52206	**P**	PNHX	NH
52207	**PN**	PNHX	NH
52208	**P**	PHAX	HA
52210	**P**	PNHX	NH
52211	**PN**	PNHX	NH
52213	**PN**	PNHX	NH
52214	**P**	PNHX	NH
52215	**PN**	PNHX	NH
52216	**P**	PNHX	NH
52217	**PN**	PNHX	NH
52218	**P**	PNHX	NH
52219	**P**	PNHX	NH
52220	**P**	PNHX	NH
52221	**P**	PNHX	NH
52222	**P**	PNHX	NH
52223	**P**	PNHX	NH
52224	**P**	PTSX	TS
52225	**P**	PNHX	NH
52226	**P**	PTSX	TS
52227	**P**	PNHX	NH
52228	**P**	PHAX	HA
52229	**P**	PNHX	NH
52230	**P**	PNLX	NL
52231	**P**	PNHX	NH
52232	**P**	PNLX	NL
52233	**P**	PNHX	NH
52234	**P**	PNLX	NL
52235	**P**	PNHX	NH
52236	**P**	PNLX	NL
52237	**P**	PNHX	NH
52238	**P**	PNLX	NL
52239	**P**	PNHX	NH
52240	**P**	PNLX	NL
52241	**P**	PTSX	TS
52242	**P**	PNLX	NL
52243	**P**	PNHX	NH
52244	**P**	PHAX	HA
52245	**P**	PHAX	HA
52246	**P**	PCFX	CF
52247	**P**	PCFX	CF
52248	**P**	PHAX	HA
52249	**P**	PCFX	CF
52250	**P**	PHAX	HA
52251	**P**	PCFX	CF
52252	**P**	PHAX	HA
52253	**P**	PTSX	TS
52254	**P**	PCFX	CF
52255	**P**	PHAX	HA
52256	**P**	PHAX	HA
52257	**P**	PHAX	HA
52258	**P**	PHAX	HA
52259	**P**	PHAX	HA
52260	**P**	PHAX	HA
52261	**P**	PCFX	CF
52262	**P**	PHAX	HA
52263	**P**	PCFX	CF
52264	**P**	PHAX	HA
52265	**P**	PCFX	CF
52266	**P**	PCFX	CF
52267	**P**	PCFX	CF
52268	**P**	PCFX	CF
52269	**P**	PCFX	CF
52270	**P**	PCFX	CF
52271	**P**	PCFX	CF
52272	**P**	PCFX	CF
52273	**P**	PCFX	CF
52274	**P**	PCFX	CF
52275	**P**	PCFX	CF
52276	**P**	PCFX	CF
52277	**P**	PCFX	CF
52278	**P**	PCFX	CF
52279	**P**	PCFX	CF
52280	**P**	PCFX	CF
52281	**P**	PCFX	CF
52282	**P**	PCFX	CF
52283	**P**	PHAX	HA
52284	**P**	PHAX	HA
52285	**P**	PHAX	HA

CLASS 155 LEYLAND BUS SUPER SPRINTER

Engine: One Cummins NT855R5 of 213 kW (285 hp).
Bogies: One P3-10 and one BT38.
Gangways: Through gangwayed.
Seats: 2+2 facing/unidirectional with wheelchair space in DMSL.
Dimensions: 23.21 x 2.70 x 3.75 m.
Maximum Speed: 75 mph.

52301–335. DMSL. Dia. DP248. Lot No. 31026 1987–8. 80S 1L (disabled persons toilet). 38.8 t.

52341–347. DMSL. Dia. DP248. Lot No. 31057 1988. 80S 1L (disabled persons toilet). 39.0 t.

52301	**P**	PCFX	CF
52302	**P**	PCFX	CF
52303	**P**	PCFX	CF
52304	**P**	PCFX	CF
52305	**P**	PCFX	CF
52306	**P**	PCFX	CF
52307	**P**	PCFX	CF
52308	**P**	PCFX	CF
52309	**P**	PCFX	CF
52310	**P**	PCFX	CF
52311	**P**	PCFX	CF
52312	**P**	PCFX	CF
52313	**P**	PCFX	CF
52314	**P**	PCFX	CF
52315	**P**	PCFX	CF
52316	**P**	PCFX	CF
52317	**P**	PCFX	CF
52318	**P**	PCFX	CF
52319	**P**	PCFX	CF
52320	**P**	PCFX	CF
52321	**P**	PCFX	CF
52322	**P**	PCFX	CF
52323	**P**	PCFX	CF
52324	**P**	PCFX	CF
52325	**P**	PCFX	CF
52326	**P**	PCFX	CF
52327	**P**	PCFX	CF
52328	**P**	PCFX	CF
52329	**P**	PCFX	CF
52330	**P**	PCFX	CF
52331	**P**	PCFX	CF
52332	**P**	PCFX	CF
52333	**P**	PCFX	CF
52334	**P**	PCFX	CF
52335	**P**	PCFX	CF
52341	**Y**	PNLX	NL
52342	**Y**	PNLX	NL
52343	**Y**	PNLX	NL
52344	**Y**	PNLX	NL
52345	**Y**	PNLX	NL
52346	**Y**	PNLX	NL
52347	**Y**	PNLX	NL

CLASS 156 METRO-CAMMELL SUPER SPRINTER

Engine: One Cummins NT855R5 of 210 kW (285 hp).
Bogies: One P3-10 and one BT38.
Gangways: Through gangwayed.
Seats: 2+2 facing/unidirectional with wheelchair space in DMSL.
Dimensions: 23.03 x 2.73 x 3.81 m.
Maximum Speed: 75 mph.

DMSL. Dia. DP244. Lot No. 31028 1988–9. 74S 1L (disabled persons toilet). 36.1 t.

52401	**P**	PNCX	NC
52402	**P**	PNCX	NC
52403	**P**	PNCX	NC
52404	**P**	PNCX	NC
52405	**P**	PNCX	NC
52406	**P**	PNCX	NC
52407	**P**	PNCX	NC
52408	**P**	PNCX	NC
52409	**P**	PNCX	NC
52410	**P**	PNCX	NC
52411	**P**	PNCX	NC
52412	**P**	PNCX	NC
52413	**P**	PNCX	NC
52414	**P**	PNCX	NC
52415	**P**	PNCX	NC
52416	**P**	PNCX	NC
52417	**P**	PNCX	NC
52418	**P**	PNCX	NC
52419	**P**	PNCX	NC
52420	**P**	PNCX	NC
52421	**P**	PNCX	NC
52422	**P**	PNCX	NC
52423	**P**	PNCX	NC
52424	**P**	PNCX	NC
52425	**P**	PNCX	NC
52426	**P**	PNCX	NC
52427	**P**	PNCX	NC
52428	**P**	PNCX	NC
52429	**P**	PNCX	NC
52430	**P**	PCKX	CK
52431	**P**	PCKX	CK
52432	**P**	PCKX	CK

52433	**P**	PCKX		CK
52434	**P**	PCKX		CK
52435	**P**	PCKX		CK
52436	**P**	PCKX		CK
52437	**P**	PCKX		CK
52438	**P**	PNLX		NL
52439	**P**	PCKX		CK
52440	**P**	PNCX		NC
52441	**P**	PNLX		NL
52442	**P**	PCKX		CK
52443	**P**	PNLX		NL
52444	**P**	PNLX		NL
52445	**P**	PHAX	r	HA
52446	**P**	PMOX	r	MO
52447	**P**	PHAX	r	HA
52448	**P**	PNLX		NL
52449	**P**	PHAX	r	HA
52450	**P**	PHAX	r	HA
52451	**P**	PNLX		NL
52452	**P**	PNCX		NC
52453	**P**	PHAX	r	HA
52454	**P**	PNLX		NL
52455	**P**	PNLX		NL
52456	**P**	PHAX	r	HA
52457	**P**	PMOX	r	MO
52458	**P**	PISX	r	IS
52459	**P**	PNLX		NL
52460	**P**	PNLX		NL
52461	**P**	PNCX		NC
52462	**P**	PNLX		NL
52463	**P**	PNLX		NL
52464	**P**	PNCX		NC
52465	**P**	PHAX		HA
52466	**P**	PNCX		NC
52467	**P**	PNLX		NL
52468	**P**	PNCX		NC
52469	**P**	PNLX		NL
52470	**P**	PNCX		NC
52471	**P**	PNLX		NL
52472	**P**	PNCX		NC
52473	**P**	PNCX		NC
52474	**P**	PMOX	r	MO
52475	**P**	PNLX		NL
52476	**P**	PHAX		HA
52477	**P**	PMOX	r	MO
52478	**P**	PMOX	r	MO
52479	**P**	PNLX		NL
52480	**P**	PNLX		NL
52481	**P**	PNLX		NL
52482	**P**	PNLX		NL
52483	**P**	PNLX		NL
52484	**P**	PNLX		NL
52485	**P**	PISX	r	IS
52486	**P**	PNCX		NC
52487	**P**	PNLX		NL
52488	**P**	PNLX		NL
52489	**P**	PNLX		NL
52490	**P**	PNLX		NL
52491	**P**	PNLX		NL
52492	**P**	PISX		IS
52493	**P**	PISX		IS
52494	**P**	PISX		IS
52495	**P**	PISX		IS
52496	**P**	PISX	r	IS
52497	**P**	PNLX		NL
52498	**P**	PNLX		NL
52499	**P**	PHAX	r	HA
52500	**P**	PHAX	r	HA
52501	**S**	PCKX		CK
52502	**S**	PCKX		CK
52503	**S**	PCKX		CK
52504	**S**	PCKX		CK
52505	**S**	PCKX		CK
52506	**S**	PCKX		CK
52507	**S**	PCKX		CK
52508	**S**	PCKX		CK
52509	**S**	PCKX		CK
52510	**S**	PCKX		CK
52511	**S**	PCKX		CK
52512	**S**	PCKX		CK
52513	**S**	PCKX		CK
52514	**S**	PCKX		CK

CLASS 158 — BREL EXPRESS

Engine: One Cummins NTA855R of 260 kW (350 hp) [† One Perkins 2006-TWH of 260 kW (350 hp)].
Bogies: BREL series 4.
Gangways: Through gangwayed.
Seats: 2+2 facing/unidirectional.
Dimensions: 23.21 x 2.82 x 3.80 m.
Maximum Speed: 90 mph.

DMS(A).. Dia. DP252. Lot No. 31051 Derby 1989– . 70S 1L (disabled persons toilet). t.

52701	**PE**	PHAX	HA
52702		P	X
52703		P	X
52704		P	X
52705		P	X
52706		P	X
52707		P	X
52708		P	X
52709		P	X
52710		P	X
52711		P	X
52712		P	X
52713		P	X
52714		P	X
52715		P	X
52716		P	X
52717		P	X
52718		P	X
52719		P	X
52720		P	X
52721		P	X
52722		P	X
52723		P	X
52724		P	X
52725		P	X
52726		P	X
52727		P	X
52728		P	X
52729		P	X
52730		P	X
52731		P	X
52732		P	X
52733		P	X
52734		P	X
52735		P	X
52736		P	X
52737		P	X
52738		P	X
52739		P	X
52740		P	X
52741		P	X
52742		P	X
52743		P	X
52744		P	X
52745		P	X
52746		P	X
52747		P	X
52748		P	X
52749		P	X
52750		P	X
52751		P	X
52752		P	X
52753		P	X
52754		P	X
52755		P	X
52756		P	X
52757		P	X
52758		P	X
52759		P	X
52760		P	X
52761		P	X
52762		P	X
52763		P	X
52764		P	X
52765		P	X
52766		P	X
52767		P	X
52768		P	X
52769		P	X
52770		P	X
52771		P	X
52772		P	X
52773		P	X
52774		P	X
52775		P	X
52776		P	X
52777		P	X
52778		P	X
52779		P	X
52780		P	X
52781		P	X
52782		P	X
52783		P	X
52784		P	X
52785		P	X
52786		P	X
52787		P	X
52788		P	X
52789		P	X
52790		P	X
52791		P	X
52792		P	X
52793		P	X
52794		P	X
52795		P	X
52796		P	X

On order to 52904.

53002–55034. FIRST GENERATION DMUS.

CLASS 114 DERBY 'HEAVYWEIGHT'

Engines: Two Leyland Albion of 149 kW (200 hp).
Bogies: DD9.
Dimensions: 20.45 x 2.82 x 3.86 m.

DMBS. Dia. DQ217 (p DX5??). Lot No. 30209 1956–57. 62S (p 0). 38.0 t.

53002 PTSX f TS
53005 PTSX f TS
53006 PTSX f TS
53019 PTSX f TS
53020 **X** RPXN fp CA
53021 PTSX f TS
53027 **X** RPXN fp CA
53030 PTSX f TS (S)
53036 PTSX f TS
53044 PTSX f TS
53046 **X** RPXN fp CA

CLASS 116 DERBY SUBURBAN

Engines: Two Leyland 680/1 of 112 kW (150 hp).
Bogies: DD10.
Dimensions: 20.45 x 2.82 x 3.86 m.

DMBS. Dia. DQ230. Lot No. 30211 1957. 65S. 36.5 t.

53050 PTSX f TS
53053 PTSX f TS
53054 PTSX TS
53055 PTSX TS
53056 PTSX f TS
53058 PTSX f TS
53060 PTSX TS
53061 PCHX f TS
53071 PTSX TS
53073 PTSX f TS
53079 PTSX f TS
53082 PTSX f TS
53083 **N** NWRX f OO
53086 PCFX fh CF
53090 PTSX f TS

DMS. Dia. DP220. Lot No. 30213 1957. 89S. 36.5 t.

53092 PTSX f TS
53093 PTSX f TS
53101 PTSX TS
53102 PTSX f TS
53106 PTSX TS
53114 PTSX TS
53116 PTSX TS
53124 PTSX f TS
53128 PCFX fh CF
53132 PTSX f TS

CLASS 101 METRO-CAMMELL

Engines: Two Leyland or AEC of 112 kW (150 hp).
Bogies: DD15.
Dimensions: 18.49 x 2.82 x 3.85 m.

DMCL. Dia. DP301 (*DP212). Lot No. 30249 1956. 12F 45S 1L. 32.5 t. 2+2 seating.

53139 PNCX f§* NC
53146 **S** PEDX f* ED
53149 PNCX f§* NC

DMBS. Dia. DQ202. Lot No. 30252 1956. 52S. 32.5 t.

53155 **N**	NWRX	f	OO	53157	PLAX	f	LA

DMCL. Dia. DP303 (*DP214). Lot No. 30253 1956. 12F 53S 1L. 32.5 t.

53158 **S**	PEDX	f*	ED	53163	PHAX	f*	HA
53160 **S**	PEDX	f*	ED				

DMBS. Dia. DQ202. Lot No. 30254 1957. 52S. 32.5 t.

53164	PNLX	f	NL	53165	PLAX	f	LA

53168–53171. DMCL. Dia. DP303 (*DP214). Lot No. 30255 1957. 12F 53S 1L. 32.5 t.

53176–53197. DMCL. Dia. DP303 (*DP214). Lot No. 30256 1957. 12F 53S 1L. 32.5 t.

53168	PNCX	f§*	NC	53180	PNCX	f§*	NC
53169	PCFX	f*	CF	53181	PNCX	f§*	NC
53170	PNCX	f§*	NC	53185 **S**	PEDX	f*	ED
53171	PCHX	f*	CH	53189	PHAX	f*	HA
53176	PHAX	f*	HA	53193	PNCX	f§*	NC
53177	PNCX	f§*	NC	53194 **S**	PEDX	f*	ED

53198–53208. DMBS. Dia. DQ202. Lot No. 30259 1957. 52S. 32.5 t.
53211–53231. DMBS. Dia. DQ202. Lot No. 30261 1957. 52S. 32.5t.

53198	PCHX	f	CH	53211	PNLX	f	NL
53200	PLAX	f	LA	53212	PLAX	f	LA
53201	PNCX	f	NC	53215	PHAX	f	HA (U)
53202	PNCX	f	NC	53216	PLOX	f	LO
53203	PNLX	f	NL	53223	PLAX	f	LA
53204	PNLX	f	NL	53224	PCHX	f	CH
53207 **N**	NWRX	f	OO	53228	PNCX	f	NC
53208	PNCX	f	NC	53231	PNCX	f	NC

DMCL. Dia. DP302 (*DP213). Lot No. 30263 1957. 12F 45S 1L. 32.5 t. 2+2 seating.

53238	PNCX	f§*	NC	53243	PHAX	f*	HA
53239	PHAX	f*	HA	53244	PCHX	f*	CH
53241 **S**	PEDX	f*	ED	53245	PNLX	f*	NL
53242	PNLX	f*	NL				

DMBS. Dia. DQ203. Lot No. 30339 1957. 44S. 32.5 t. 2+2 seating.

53247	PLAX	f	LA	53248	PLAX	f	LA

DMBS. Dia. DQ202. Lot No. 30266 1957. 52S. 32.5 t.

53250	PNLX	f	NL	53256	PLAX	f	LA
53253	PLOX	f	LO				

DMCL. Dia. DP304 (*DP210) or †DP317. Lot No. 30267 1957. 12F 53S 1L. (†12F 46S with additional luggage racks 1L). 32.5 t.

53260	PHAX	f*	HA	53266	PNCX	f§*	NC
53265	NSBX	f†	RG	53267	PNCX	f§*	NC

53268	PNLX	f*	NL	53269	PCHX	f*	CH

53290–53296. DMBS. Dia. DQ202. Lot No. 30270 1957. 52S. 32.5 t.
53305–53315. DMBS. Dia. DQ202 (†DQ232). Lot No. 30275 1958. 52S. (†49S with additional luggage rack) 32.5 t.

53291	PNLX	f	NL	53310	NSBX	f†	RG
53293	PNCX	f	NC	53311	NSBX	f	RG
53294	PLOX	f	LO	53312	NSBX	f†	RG
53296 **N**	NGBX	f	OO	53314	NSBX	f†	RG
53305	PNCX	f§	NC	53315	PLAX	f	LA
53308	NSBX	f†	RG				

DMCL. Dia. DP304 (*DP210) or †DP317. Lot No. 30276 1958. 12F 53S 1L (†12F 46S with additional luggage racks 1L) 32.5 t.

53321	PNCX	f§	NC	53330	PLAX	f	LA
53322	NSBX	f	RG	53331	NSBX	f†	RG
53326	NSBX	f†	RG	53332	NSBX	f†	RG
53327	NSBX	f†	RG	53333 **N**	NSBX	f	RG

CLASS 104 BRCW

Engines: Two Leyland 680/1 of 112 kW (150 hp).
Bogies: DD14.
Dimensions: 18.49 x 2.82 x 3.84 m.

DMBS. Dia. DQ205. Lot No. 30293 1957–58. 52S 1L. 31.5 t.

53429 **N**	NGBX	OO	53451 **B**	PCHX	CH
53431 **B**	PCHX	CH	53454 **B**	PNHX	NH
53437 **N**	NGBX	OO	53455 **N**	NGBX	OO
53442 **B**	PCHX	CH	53470 **N**	NGBX	OO
53444 **B**	PNHX	NH	53477 **N**	NGBX	OO
53447 **B**	PCHX	CH	53479 **N**	NGBX	OO

DMCL. Dia. DP305 (*DP209). Lot No. 30294 1957–58. 12F 54S 1L. 31.5 t.

53493 **B**	PNHX	*	NH	53517 **B**	PCHX	*	CH
53494 **B**	PCHX	*	CH	53528 **B**	PNHX	*	NH
53512 **B**	PCHX	*	CH	53529 **B**	PCHX	*	CH
53516 **B**	PCHX	*	CH	53531 **B**	PCHX	*	CH

DMBS. Dia. DQ205. Lot No. 30296 1958. 52S 1L. 31.5 t.

53534	PCHX	CH	53540 **N**	NGBX	OO
53539 **N**	NGBX	OO			

CLASS 108 DERBY 'LIGHTWEIGHT'

Engines: Two Leyland 680/1 or 680/13 of 112 kW (150 hp).
Bogies: DD7.
Dimensions: 18.49 x 2.82 x 3.88 m.

53599–53624. DMBS. Dia. DQ212. Lot No. 30406 1958. 52S. 29.5 t.
53625–53629. DMBS. Dia. DQ212. Lot No. 30407 1958–59. 52S. 29.5 t.

53599 **N**	NMYX	f	BY
53602	PBRX	f	BR
53608	PBRX	f	BR
53612	PBRX	f	BR
53614	PBRX	f	BR
53616	PBRX	f	BR
53617	PCFX	f	CF
53618	PBRX	f	BR
53619	PBRX	f	BR
53620	PBRX	f	BR
53621	PBRX	f	BR
53622	PBRX	f	BR
53624	PCFX	f	CF
53625	PLEX	f	LE
53626	PNLX	f	NL
53627	PCFX	fh	CF
53628 **N**	NMYX	f	BY
53629	PCFX	fh	CF

DMCL. Dia. DP310 (*DP219). Lot No. 30408 1958. 12F 50S 1L. 28.5 t. Luggage rack opposite toilet.
† Converted to DMSL (Dia. DP2??). 71S 1L. No luggage rack.

53630	PNLX	f*	NL
53631	PLEX	fh*	LE
53632	PCFX	f*	CF
53633	PLEX	fh*	LE
53634	PCHX	*	CH
53635	PCFX	f*	CF
53636	PNLX	f*	NL
53637	PLAX	f*	LA
53638	PLAX	f†	LA
53639	PLAX	f*	LA
53641	PNLX	f*	NL
53642	PCFX	f*	CF
53643	PLEX	fh*	LE
53644	PNLX	f*	NL
53645	PCHX	f*	CH
53646	PLAX	f*	LA

CLASS 101 METRO-CAMMELL

Engines: Two Leyland or AEC of 112 kW (150 hp).
Bogies: DD15.
Dimensions: 18.49 x 2.82 x 3.85 m.

DMCL. Dia. DP304 (*DP210). Lot No. 30271 1957. 12F 53S 1L. 32.5 t.

53746	PNLX	f*	NL
53751 **N**	NSBX	f	RG

CLASS 116 DERBY SUBURBAN

Engines: Two Leyland 680/1 of 112 kW (150 hp).
Bogies: DD10.
Dimensions: 20.45 x 2.82 x 3.86 m.

DMBS. Dia. DQ230. Lot No. 30363 1957–58. 65S. 36.5 t.

53818	PTSX	f	TS
53820 **N**	NGBX		OO
53822	PTSX	f	TS
53826	PTSX	f	TS
53827	PTSX	f	TS
53837	PTSX	f	TS
53838	PTSX		TS
53844	PTSX	f	TS
53849	PTSX		TS
53850	PTSX	f	TS
53853	PTSX	f	TS
53854	PTSX	f	TS
53863	PTSX		TS
53865	PTSX	f	TS

DMS. Dia. DP220. Lot No. 30364 1957–58. 89S. 36.5 t.

53873	PTSX	f	TS
53878	PTSX	f	TS
53880	PTSX	f	TS
53881	PTSX	f	TS
53886	PTSX	f	TS
53890	PTSX	f	TS
53891	PTSX	f	TS
53893	PTSX		TS

53894	PTSX	f	TS
53897	PTSX	f	TS
53902	PTSX	f	TS
53907	PTSX	f	TS
53916	PTSX		TS
53919	PTSX	f	TS
53921	PTSX	f	TS

CLASS 108 DERBY 'LIGHTWEIGHT'

Engines: Two Leyland 680/1 or 680/13 of 112 kW (150 hp).
Bogies: DD7.
Dimensions: 18.49 x 2.79 x 3.87 m.
Non-standard Livery: Original BR DMU green.

53924–53935. DMBS. Dia. DQ212. Lot No. 30460 1959–60. 52S. 29.5 t.
53938–53987. DMBS. Dia. DQ212. Lot No. 30465 1959. 52S. 29.5 t.

53924	PCHX	f	CH
53925	PCHX		CH
53926	PLEX	fh	LE
53927	PCHX	f	CH
53928	PLEX	fh	LE
53929	PCHX	fh	CH
53930	PCHX		CH
53931	PCHX		CH
53932	PCHX	fh	CH
53933	PCHX	h	CH
53934	PCHX	f	CH
53935	PLEX	fh	LE
53938	PCHX	f	CH
53939	PLEX	fh	LE
53940	PNLX	f	NL
53941	PBRX	f	BR
53942	PLEX	fh	LE
53943	PNLX	f	NL
53944	PNLX	f	NL
53945	PLEX	fh	LE
53947	PLEX	fh	LE
53948	PBRX	f	BR
53949	PNLX	f	NL
53950	PHTX	f	HT
53951	PHTX		HT
53952	PHTX		HT
53953	PHTX		HT
53954	PHTX		HT
53955	PHTX		HT
53956	PHTX		HT
53957	PHTX		HT
53958	PHTX		HT
53959	PHTX		HT
53960	PHTX		HT
53962	PHTX		HT
53963	PHTX		HT
53964 **0**	PHTX		HT
53965	PNLX	f	NL
53966	PNLX		NL
53968	PNLX		NL
53969	PNLX	f	NL
53970	PCHX	f	CH
53971	PNLX	f	NL
53973	PHTX	f	HT
53974	PNLX	f	NL
53975	PCHX	f	CH
53976	PCHX	f	CH
53977	PCHX	f	CH
53978	PCHX	f	CH
53980	PCHX		CH
53981	PCHX	f	CH
53982	PCHX	f	CH
53983	PCHX	f	CH
53986	PNLX	f	NL
53987	PNLX		NL

CLASS 114 DERBY 'HEAVYWEIGHT'

Bogies: DT9.
Dimensions: 20.45 x 2.82 x 3.86 m.
Non-standard Livery: Blue with red stripe.

DTCL. Dia. DS313 (*DS2??, p DX5??). Lot No. 30210 1956–57. 12F 62S 1L p 1L). 30.0 t.

54006	PTSX	f*	TS
54008	PTSX	f*	TS (S)

54010	PTSX	f*	TS
54011	PTSX	f*	TS (S)
54012	PTSX	f*	TS
54027	PTSX	f*	TS
54028 **0**	RPXN	fp	CA
54039	PTSX	f*	TS
54041 **0**	RPXN	fp	CA
54043	PTSX	f*	TS
54047	PTSX	f*	TS

CLASS 101 METRO-CAMMELL

Bogies: DT11.
Dimensions: 18.49 x 2.82 x 3.85 m.

54050–54061. DTCL. Dia. DS302 (*DS206). Lot No. 30260 1957. 12F 53S 1L. 25.5 t.
54062–54085. DTCL. Dia. DS302 (*DS206). Lot No. 30262 1957. 12F 53S 1L. 25.5 t.

54050	PNCX	f*	NC
54055	PNCX	f*	NC
54056	PNLX	f*	NL
54060	PNCX	f*	NC
54061	PNLX	f*	NL
54062	PNCX	f*	NC
54065	PNCX	f*	NC
54068 **N**	NGEX	f	CA
54070 **N**	NWRX	f	OO
54073	PNCX	f*	NC
54075	PNHX	f*	NH
54081	NSBX	f	OO
54085	PNLX	f*	NL

CLASS 108 DERBY 'LIGHTWEIGHT'

Bogies: DT8.
Dimensions: 18.49 x 2.79 x 3.87 m.

54191–54210. DTCL. Dia. DS314 (*DS207). Lot No. 30409 1958. 12F 53S 1L. 21.5 t.
54212–54214. DTCL. Dia. DS314 (*DS207). Lot No. 30410 1958. 12F 53S 1L. 21.5 t.

54191	PBRX	f*	BR
54194 **N**	NMYX	f	BY
54197	PBRX	f*	BR
54201	PBRX	f*	BR
54202	PBRX	f*	BR
54203	PBRX	f*	BR
54204	PBRX	f*	BR
54205	PBRX	f*	BR
54207	PBRX	f*	BR
54208	PBRX	f*	BR
54209	PBRX	f*	BR
54210	PCFX	fh*	CF
54212	PCFX	fh*	CF
54214	PCHX	f*	CH

CLASS 101 METRO-CAMMELL

Bogies: DT11.
Dimensions: 18.49 x 2.82 x 3.85 m.

DTCL. Dia. DS303 (*DS210). Lot No. 30340 1957. 12F 45S 1L. 25.5 t. 2+2 seating.

54218	PNCX	f*	NC
54219	PNLX	f*	NL
54220	PCHX	f*	CH

CLASS 108 DERBY 'LIGHTWEIGHT'

Bogies: DT8.
Dimensions: 18.49 x 2.79 x 3.87 m. (DS310), 18.49 x 2.82 x 3.87 m. (DS311).

Non-standard Livery: Original BR DMU green.

54221–54270. DTCL. Dia. DS310 (*DS207). Lot No. 30466 1959. 12F 53S 1L. 21.5 t.
54271–54279. DTCL. Dia. DS311 (*DS2??). Lot No. 30499 1960. 12F 53S 1L. 22.5 t.

54221		PNLX	f*	NL
54222	**N**	NMYX	f	BY
54223	**N**	NMYX	f	BY
54224	**N**	NMYX	f	BY
54226		PCHX	f*	CH
54227		PHTX	f*	HT
54228	**N**	NMYX	f	BY
54230		PBRX	f*	BR
54231		PCHX	f*	CH
54232		PCHX	f*	CH
54235		PHTX	*	HT
54236		PHTX	*	HT
54238		PHTX	*	HT
54239		PHTX	*	HT
54240		PHTX	*	HT
54241		PHTX	*	HT
54242		PHTX	*	HT
54243		PHTX	*	HT
54244		PHTX	*	HT
54245		PHTX	*	HT
54246		PCHX	f*	CH
54247	**0**	PHTX	*	HT
54248		PNLX	f*	NL
54249		PNLX	*	NL
54251		PNLX	*	NL
54252		PNLX	f*	NL
54253		PCHX	f*	CH
54256		PCHX	*	CH
54257	**N**	NMYX	f	BY
54258		PCHX	f*	CH
54259		NMYX	f	BY
54260		PCHX	f*	CH
54261		PCHX	f*	CH
54262		PHTX	f*	HT
54263		PHTX	f*	HT
54264		PCHX	f*	CH
54265		PHTX	*	HT
54266		PCHX	f*	CH
54267		PCFX	h*	CF
54268		PBRX	f*	BR
54269		PCHX	f*	CH
54270		PNLX	*	NL
54271	**N**	NMYX		BY
54272		PCHX	*	CH
54273		PCFX	f*	CF
54274	**N**	NMYX	f	BY
54275		PCHX	*	CH
54276		PCHX	*	CH
54277		PCHX	f*	CH
54278		PCHX	f*	CH
54279	**N**	NMYX	f	BY

CLASS 121 PRESSED STEEL SUBURBAN

Gangways: Not ganywayed except as shown. **Bogies**: DT9.
Dimensions: 20.45 x 2.82 x 3.86 m.

DTSL. Dia. DS201 (g DS2??). Lot No. 30519 1960–61. 91S (g 89S). 30.0 t.

54280	**N**	NWRX	f	OO
54283	**N**	NWRX	f	OO
54284	**N**	NMYX	f	BY
54287	**N**	NWRX	fg	OO
54289	**N**	NWRX	fg	OO

CLASS 101 METRO-CAMMELL

Bogies: DT11.
Dimensions: 18.49 x 2.82 x 3.85 m.

DTCL. Dia. DS302 (*DS206). Lot No. 30468 1958. 12F 53S 1L. 25.5 t.

54332		PNCX	f*	NC
54340		PNLX	*	NL
54342		PNLX	f*	NL
54343		PNCX	f*	NC
54345		PHTX	f*	HT (S)
54346		PNLX	f*	NL
54347		PNCX	f*	NC
54351		PNLX	f*	NL
54352		PNCX	f*	NC
54354		PNCX	f*	NC
54358	**N**	NMYX		BY
54362	**N**	NGEX	f	CA
54363	**N**	NMYX	f	BY
54365		PNLX	f*	NL
54368		PNCX	f*	NC
54369		PNCX	f*	NC
54370		PHTX	f*	HT (S)
54371	**N**	NWRX	f	OO
54372	**N**	NWRX	f	OO
54379		PNCX	f*	NC
54380		PNCX	f*	NC
54381	**N**	NMYX	f	BY
54382	**N**	NGEX	f	CA
54385	**N**	NWRX	f	OO
54387		PNCX	f*	NC
54388		PNCX	f*	NC
54393		PNCX	f*	NC
54396	**N**	NWRX	f	OO
54399		PNCX	f*	NC
54402	**N**	NGEX	f	CA
54405	**N**	NGEX	f	CA
54408		PNLX	f*	NL

CLASS 108 DERBY 'LIGHTWEIGHT'

Bogies: DT8.
Dimensions: 18.49 x 2.82 x 3.87.

DTCL. Dia. DS311 (*DS2??). Lot No. 30602 1960. 12F 53S 1L. 22.5 t.

54484		PCHX	f*	CH
54485		PCHX	f*	CH
54486		PCHX	f*	CH
54487		PCHX	f*	CH
54488		PCHX	f*	CH
54489		PCHX	f*	CH
54490		PCHX	f*	CH
54491	**N**	NMYX	f	BY
54492		PBRX	f*	BR
54493	**N**	NMYX		BY
54494		PCFX	fh*	CF
54495		NMYX		BY
54496		PCHX	f*	CH
54497		PCHX	f*	CH
54498		PCFX	f*	CF
54499		NMYX	f	BY
54500	**N**	NMYX	f	BY
54501		PCHX	f*	CH
54503		PNLX	f*	NL
54504		PNLX	*	NL

CLASS 114 DERBY 'HEAVYWEIGHT' PARCELS

Bogies: DT9.
Dimensions: 20.45 x 2.82 x 3.86 m.

DTPMV. Dia. DX514. Lot No. 30210 1956–57. 1L. 30.0 t.

54900	(55933)	**R**	RPXN	CA
54901	(54016)	**R**	RPXN	CA
54902	(55932)	**R**	RPXN	CA
54903	(54009)	**R**	RPXN	CA
54904	(54015)	**R**	RPXN	CA
54905	(540)			
54906	(540)			
54907	(540)			

CLASS 122 GLOUCESTER SUBURBAN

Engines: Two AEC 220 of 112 kW (150 hp).
Gangways: Non gangwayed single cars.
Bogies: DD10.
Dimensions: 20.45 x 2.82 x 3.87 m.

DMBS. Dia. DX202. Lot No. 30419 1958. 65S. 36.5 t.

55000		PLAX	LA
55003		PLAX	LA
55004		PLAX	LA
55005		PLAX	LA
55006		PLAX	LA
55009		PLAX	LA
55011		PLAX	LA
55012		PLAX	LA

CLASS 121 PRESSED STEEL SUBURBAN

Engines: Two Leyland 1595 of 112 kW (150 hp)
Gangways: Non gangwayed single cars.
Bogies: DD10.
Dimensions: 20.45 x 2.82 x 3.87 m.

DMBS. Dia. DX201. Lot No. 30518 1960. 65S. 38.0 t.

55020	**N**	NWRX	f	OO
55021	**N**	NWRX	f	OO
55022	**N**	NWRX	f	OO
55023	**N**	NWRX	f	OO
55024	**N**	NWRX	f	OO
55025	**N**	NWRX	f	OO
55026		PLAX	f	LA
55027	**N**	NWRX	f	OO
55028	**N**	NWRX	f	OO
55029	**N**	NWRX	f	OO
55030	**N**	NWRX	f	OO
55031	**N**	NWRX	f	OO
55032		PTSX	f	TS
55033	**PM**	PTSX		TS
55034		PTSX	f	TS

55200–52859. SECOND GENERATION DMUS.

CLASS 150/0 & 154†§ BREL PROTOTYPE SPRINTER

Class 154 vehicles were converted from class 150/0 at RTC Derby 1986.

Engine: One Cummins NT855R5 of 210 kW (285 hp) One Cummins NTA855R1 of 260 kW (350 hp)†. One Perkins 2006-TWH of 260 kW (350 hp)§.
Bogies: BX8.
Couplings: BSI at outer end of driving vehicles, bar non-driving ends.
Transmission: Hydraulic. Voith T211r with Gmeinder final drive.
Gangways: Within unit only.
Seats: Various types now fitted. (†§2+2)
Dimensions: 20.06 x 2.82 m (outer cars), 20.18 x 2.82 m (inner car).
Maximum Speed: 75 mph (*90 mph).

DMSL. Dia. DP230 (†DP246). Lot No. 30984 York 1984. 76S 1L (†58S 1L). 35.8 t (†37.9 t).

55200 **P** PTSX TS | 55201 **P** PDYX † DY

DMS. Dia. DP231 (†DP247). Lot No. 30985 York 1984. 79S (†60S). 35.6 t. (†37.9 t).

55300 **P** PTSX TS | 55301 **P** PDYX † DY

MS. Dia. DR202 (§DR206). Lot No. 30986 York 1984. 84S (§72S). 34.4 t. (§36.5 t).

55400 **P** PTSX TS | 55401 **P** PDYX § DY

CLASS 151 METRO-CAMMELL 3-CAR PROTOTYPE

DMSL–MS–DMS. Prototype units. Aluminium alloy bodies. Sliding doors. BSI couplings. PA.

Engine: One Cummins NT855R5 of 210 kW (285 hp).
Bogies: BX9.
Transmission: Hydraulic. Twin disc Tb33-1316 with Gmeinder final drive.
Gangways: Within unit only.
Dimensions: 20.10 x 2.81 m (outer cars), 19.86 x 2.81 m (inner car).
Seats: 2+3 facing.
Maximum Speed: 75 mph.
Non-standard livery: Grey with light blue stripe.

DMSL. Dia. DP233. Lot No. 30987. 1985. 68S 1L. 32.4 t.

55202 **0** PDYX DY (U) | 55203 **0** PDYX DY (U)

DMS. Dia. DP232. Lot No. 30988. 1985. 80S. 32.4 t.

55302 **0** PDYX DY (U) | 55303 **0** PDYX DY (U)

MS. Dia. DR204. Lot No. 30989. 1985. 84S. 32.1 t.

55402 **0** PDYX DY (U) | 55403 **0** PDYX DY (U)

CLASS 140 BR/LEYLAND BUS PROTOTYPE RAILBUS

Engine: One Leyland TL11 152 kW (205 hp).
Transmission: Mechanical. Self Changing Gears 4 speed gearbox.
Gangways: Within unit only. End doors.
Couplings: Automatic Tightlock (outer ends), bar (inner ends).
Dimensions: 16.20 x 2.47 x 3.91 m.
Maximum Speed: 75 mph.
Non-standard livery: Dark blue.

DMSL. Dia. DP222 Lot No. 30962 Derby 1981. 50S 1L. 23.2 t.

55500 **0** PNLX NL (U) I

DMS. Dia. DP223 Lot No. 30962 Derby 1981. 52S. 23.0 t.

55501 **0** PNLX NL (U) I

CLASS 141 LEYLAND BUS/BREL RAILBUS

Engine: One Leyland TL11 152 kW (205 hp) (v Cummins LTA 10R).
Transmission: Mechanical. Self Changing Gears. (v Hydraulic. Voith T211r).
Gangways: Within unit only.
Dimensions: 15.45 x 2.50 x 3.91 m.
Maximum Speed: 75 mph.

DMS. Dia. DP228 Lot No. 30977 Derby 1984. Modified by Barclay 1988–89. 50S. 26.0 t.

55502 **Y** PNLX NL
55503 **Y** PNLX NL
55504 **Y** PNLX NL
55505 **Y** PNLX NL
55506 **Y** PNLX NL
55507 **Y** PNLX NL
55508 **Y** PNLX NL
55509 **Y** PNLX NL
55510 **Y** PNLX NL
55511 **Y** PNLX NL
55512 **Y** PNLX NL
55513 **Y** PNLX v NL
55514 **Y** PNLX NL
55515 **Y** PNLX NL
55516 **Y** PNLX NL
55517 **Y** PNLX NL
55518 **Y** PNLX NL
55519 **Y** PNLX NL
55520 **Y** PNLX NL
55521 **Y** PNLX NL

DMSL. Dia. DP229 Lot No. 30978 1984. Modified by Barclay 1988–89.44S 1L. 26.5 t.

55522 **Y** PNLX NL
55523 **Y** PNLX NL
55524 **Y** PNLX NL
55525 **Y** PNLX NL
55526 **Y** PNLX NL
55527 **Y** PNLX NL
55528 **Y** PNLX NL
55529 **Y** PNLX NL
55530 **Y** PNLX NL
55531 **Y** PNLX NL
55532 **Y** PNLX NL
55533 **Y** PNLX v NL
55534 **Y** PNLX NL
55535 **Y** PNLX NL
55536 **Y** PNLX NL
55537 **Y** PNLX NL
55538 **Y** PNLX NL
55539 **Y** PNLX NL
55540 **Y** PNLX NL
55541 **Y** PNLX NL

CLASS 142/0 LEYLAND BUS/BREL RAILBUS

Engine: One Leyland TL11 152 kW (205 hp).
Transmission: Mechanical. Self Changing Gears (v Hydraulic. Voith T211r).
Gangways: Within unit only.
Dimensions: 15.55 x 2.80 x 3.86 m.
Maximum Speed: 75 mph.

DMS. Dia. DP234 Lot No. 31003 Derby 1985–6. 62S. 24.5 t.

55542	**M**	PNHX		NH
55543	**M**	PNHX		NH
55544	**M**	PNHX		NH
55545	**M**	PNHX		NH
55546	**M**	PNHX		NH
55547	**M**	PNHX	v	NH
55548	**M**	PNHX		NH
55549	**M**	PNHX		NH
55550	**M**	PNHX		NH
55551	**M**	PNHX		NH
55552	**M**	PNHX		NH
55553	**M**	PNHX		NH
55554	**M**	PNHX		NH
55555	**M**	PNHX		NH
55556	**0**	PNHX	v	NH
55557	**0**	PHTX	v	HT
55558	**0**	PNHX	v	NH
55559	**0**	PHTX	v	HT
55560	**0**	PHTX	v	HT
55561	**0**	PHTX	v	HT
55562	**0**	PHTX	v	HT
55563	**0**	PHTX	v	HT
55564	**0**	PNHX	v	NH
55565	**0**	PNHX	v	NH
55566	**0**	PHTX	v	HT
55567	**0**	PNHX	v	NH
55568	**0**	PNHX	v	NH
55569	**PR**	PNHX		NH
55570	**PR**	PNHX		NH
55571	**PR**	PNHX		NH
55572	**PR**	PNHX		NH
55573	**PR**	PNHX		NH
55574	**PR**	PNHX		NH
55575	**PR**	PNHX	v	NH
55576	**PR**	PNHX	v	NH
55577	**PR**	PNHX	v	NH
55578	**PR**	PNHX	v	NH
55579	**PR**	PNHX	v	NH
55580	**PR**	PNHX	v	NH
55581	**PR**	PNHX	v	NH
55582	**PR**	PNHX	v	NH
55583	**PR**	PNHX	v	NH
55584	**PR**	PNHX	v	NH
55585	**PR**	PNHX	v	NH
55586	**PR**	PNHX	v	NH
55587	**PR**	PNHX	v	NH
55588	**PR**	PNHX	v	NH
55589	**PR**	PNHX	v	NH
55590	**PR**	PNHX	v	NH
55591	**PR**	PHTX	v	HT

DMSL. Dia. DP235 Lot No. 31004 Derby 1985–6. 59S 1L. 25.0 t.

55592	**M**	PNHX		NH
55593	**M**	PNHX		NH
55594	**M**	PNHX		NH
55595	**M**	PNHX		NH
55596	**M**	PNHX		NH
55597	**M**	PNHX	v	NH
55598	**M**	PNHX		NH
55599	**M**	PNHX		NH
55600	**M**	PNHX		NH
55601	**M**	PNHX		NH
55602	**M**	PNHX		NH
55603	**M**	PNHX		NH
55604	**M**	PNHX		NH
55605	**M**	PNHX		NH
55606	**0**	PNHX	v	NH
55607	**0**	PHTX	v	HT
55608	**0**	PNHX	v	NH
55609	**0**	PHTX	v	HT
55610	**0**	PHTX	v	HT
55611	**0**	PHTX	v	HT
55612	**0**	PHTX	v	HT
55613	**0**	PHTX	v	HT
55614	**0**	PNHX	v	NH
55615	**0**	PNHX	v	NH
55616	**0**	PHTX	v	HT
55617	**0**	PNHX	v	NH
55618	**0**	PNHX	v	NH
55619	**PR**	PNHX		NH
55620	**PR**	PNHX		NH
55621	**PR**	PNHX		NH

55622	**PR**	PNHX		NH
55623	**PR**	PNHX		NH
55624	**PR**	PNHX		NH
55625	**PR**	PNHX	v	NH
55626	**PR**	PNHX	v	NH
55627	**PR**	PNHX	v	NH
55628	**PR**	PNHX	v	NH
55629	**PR**	PNHX	v	NH
55630	**PR**	PNHX	v	NH
55631	**PR**	PNHX	v	NH
55632	**PR**	PNHX	v	NH
55633	**PR**	PNHX	v	NH
55634	**PR**	PNHX	v	NH
55635	**PR**	PNHX	v	NH
55636	**PR**	PNHX	v	NH
55637	**PR**	PNHX	v	NH
55638	**PR**	PNHX	v	NH
55639	**PR**	PNHX	v	NH
55640	**PR**	PNHX	v	NH
55641	**PR**	PHTX	v	HT

CLASS 143 ALEXANDER/BARCLAY RAILBUS

Engine: One Leyland TL11 152 kW (205 hp).
Transmission: Mechanical. Self Changing Gears (v Hydraulic. Voith T211r).
Gangways: Within unit only.
Dimensions: 15.55 x 2.70 x 3.52 m.
Maximum Speed: 75 mph.

DMS. Dia. DP236 Lot No. 31005 Andrew Barclay 1985–6. 62S. 24.5 t.

55642	**PR**	PHTX	v	HT
55643	**PR**	PHTX	v	HT
55644	**PR**	PHTX	v	HT
55645	**PR**	PHTX	v	HT
55646	**PR**	PHTX	v	HT
55647	**PR**	PHTX	v	HT
55648	**PR**	PHTX	v	HT
55649	**PR**	PHTX	v	HT
55650	**PR**	PHTX	v	HT
55651	**PR**	PHTX	v	HT
55652	**PR**	PHTX	v	HT
55653	**PR**	PHTX	v	HT
55654	**PR**	PHTX	v	HT
55655	**PR**	PHTX	v	HT
55656	**PR**	PHTX	v	HT
55657	**PR**	PHTX	v	HT
55658	**PR**	PHTX	v	HT
55659	**PR**	PHTX	v	HT
55660	**PR**	PHTX	v	HT
55661	**T**	PHTX		HT
55662	**T**	PHTX		HT
55663	**T**	PHTX		HT
55664	**T**	PHTX		HT
55665	**T**	PHTX		HT
55666	**T**	PHTX		HT

DMSL. Dia. DP237 Lot No. 31006 Andrew Barclay 1985–6. 60S 1L. 25.0 t.

55667	**PR**	PHTX	v	HT
55668	**PR**	PHTX	v	HT
55669	**PR**	PHTX	v	HT
55670	**PR**	PHTX	v	HT
55671	**PR**	PHTX	v	HT
55672	**PR**	PHTX	v	HT
55673	**PR**	PHTX	v	HT
55674	**PR**	PHTX	v	HT
55675	**PR**	PHTX	v	HT
55676	**PR**	PHTX	v	HT
55677	**PR**	PHTX	v	HT
55678	**PR**	PHTX	v	HT
55679	**PR**	PHTX	v	HT
55680	**PR**	PHTX	v	HT
55681	**PR**	PHTX	v	HT
55682	**PR**	PHTX	v	HT
55683	**PR**	PHTX	v	HT
55684	**PR**	PHTX	v	HT
55685	**PR**	PHTX	v	HT
55686	**T**	PHTX		HT
55687	**T**	PHTX		HT
55688	**T**	PHTX		HT
55689	**T**	PHTX		HT
55690	**T**	PHTX		HT
55691	**T**	PHTX		HT

CLASS 142/1 LEYLAND BUS/BREL RAILBUS

Engine: One Leyland TL11 152 kW (205 hp).
Transmission: Mechanical. Self Changing Gears (v Hydraulic. Voith T211r).
Gangways: Within unit only.
Dimensions: 15.55 x 2.80 x 3.86 m.
Maximum Speed: 75 mph.

DMS. Dia. DP234 Lot No. 31013 Derby 1986–7. 62S and wheelchair space. 24.5 t.

55701	**PR**	PNHX	v	NH
55702	**PR**	PNHX		NH
55703	**PR**	PNHX		NH
55704	**PR**	PNHX		NH
55705	**PR**	PNHX		NH
55706	**PR**	PNHX		NH
55707	**PR**	PNHX		NH
55708	**PR**	PNHX		NH
55709	**PR**	PNHX		NH
55710	**PR**	PNHX	v	NH
55711	**PR**	PNHX	v	NH
55712	**PR**	PNHX	v	NH
55713	**PR**	PNHX	v	NH
55714	**PR**	PNHX	v	NH
55715	**PR**	PNHX	v	NH
55716	**PR**	PNHX		NH
55717	**PR**	PNHX	v	NH
55718	**PR**	PNHX	v	NH
55719	**PR**	PNHX	v	NH
55720	**PR**	PNLX		NL
55721	**PR**	PNHX	v	NH
55722	**PR**	PNLX	v	NL
55723	**PR**	PNLX	v	NL
55724	**PR**	PNLX		NL
55725	**PR**	PNLX		NL
55726	**PR**	PNLX		NL
55727	**PR**	PNLX		NL
55728	**PR**	PNLX		NL
55729	**PR**	PNLX	v	NL
55730	**PR**	PNLX	v	NL
55731	**PR**	PNLX	v	NL
55732	**PR**	PNLX	v	NL
55733	**PR**	PNLX	v	NL
55734	**PR**	PNLX	v	NL
55735	**PR**	PNLX	v	NL
55736	**PR**	PNLX	v	NL
55737	**PR**	PNLX	v	NL
55738	**PR**	PNLX	v	NL
55739	**PR**	PNLX	v	NL
55740	**PR**	PNLX	v	NL
55741	**PR**	PNLX	v	NL
55742	**PR**	PNLX	v	NL
55743	**PR**	PNLX	v	NL
55744	**PR**	PNLX	v	NL
55745	**PR**	PNLX	v	NL
55746	**PR**	PNLX	v	NL

DMSL. Dia. DP235 Lot No. 31014 Derby 1986–7. 59S 1L. 24.96 t.

55747	**PR**	PNHX	v	NH
55748	**PR**	PNHX		NH
55749	**PR**	PNHX		NH
55750	**PR**	PNHX		NH
55751	**PR**	PNHX		NH
55752	**PR**	PNHX		NH
55753	**PR**	PNHX		NH
55754	**PR**	PNHX		NH
55755	**PR**	PNHX		NH
55756	**PR**	PNHX	v	NH
55757	**PR**	PNHX	v	NH
55758	**PR**	PNHX	v	NH
55759	**PR**	PNHX	v	NH
55760	**PR**	PNHX	v	NH
55761	**PR**	PNHX	v	NH
55762	**PR**	PNHX		NH
55763	**PR**	PNHX	v	NH
55764	**PR**	PNHX	v	NH
55765	**PR**	PNHX	v	NH
55766	**PR**	PNLX		NL
55767	**PR**	PNHX	v	NH
55768	**PR**	PNLX	v	NL
55769	**PR**	PNLX	v	NL
55770	**PR**	PNLX		NL
55771	**PR**	PNLX		NL
55772	**PR**	PNLX		NL
55773	**PR**	PNLX		NL
55774	**PR**	PNLX		NL
55775	**PR**	PNLX	v	NL
55776	**PR**	PNLX	v	NL
55777	**PR**	PNLX	v	NL
55778	**PR**	PNLX	v	NL

55779	**PR**	PNLX	v	NL	55786	**PR**	PNLX	v	NL
55780	**PR**	PNLX	v	NL	55787	**PR**	PNLX	v	NL
55781	**PR**	PNLX	v	NL	55788	**PR**	PNLX	v	NL
55782	**PR**	PNLX	v	NL	55789	**PR**	PNLX	v	NL
55783	**PR**	PNLX	v	NL	55790	**PR**	PNLX	v	NL
55784	**PR**	PNLX	v	NL	55791	**PR**	PNLX	v	NL
55785	**PR**	PNLX	v	NL	55792	**PR**	PNLX	v	NL

CLASS 144 ALEXANDER/BREL RAILBUS

Engine: One Leyland TL11 152 kW (205 hp).
Transmission: Mechanical. Self Changing Gears. (v Hydraulic. Voith T211r).
Gangways: Within unit only.
Dimensions: 15.25 x 2.70 x 3.73 m.
Maximum Speed: 75 mph.

DMS. Dia. DP240 Lot No. 31015 Derby 1986–7. 62S and wheelchair space. 24.2 t.

55801	**Y**	PNLX		NL	55813	**Y**	PNLX	v	NL
55802	**Y**	PNLX		NL	55814	**Y**	PNLX		NL
55803	**Y**	PNLX		NL	55815	**Y**	PNLX		NL
55804	**Y**	PNLX		NL	55816	**Y**	PNLX		NL
55805	**Y**	PNLX		NL	55817	**Y**	PNLX		NL
55806	**Y**	PNLX		NL	55818	**Y**	PNLX		NL
55807	**Y**	PNLX		NL	55819	**Y**	PNLX		NL
55808	**Y**	PNLX		NL	55820	**Y**	PNLX		NL
55809	**Y**	PNLX	v	NL	55821	**Y**	PNLX		NL
55810	**Y**	PNLX	v	NL	55822	**Y**	PNLX		NL
55811	**Y**	PNLX	v	NL	55823	**Y**	PNLX		NL
55812	**Y**	PNLX	v	NL					

DMSL. Dia. DP241 Lot No. Derby 31016 1986–7. 60S 1L. 25.0 t.

55824	**Y**	PNLX		NL	55836	**Y**	PNLX	v	NL
55825	**Y**	PNLX		NL	55837	**Y**	PNLX		NL
55826	**Y**	PNLX		NL	55838	**Y**	PNLX		NL
55827	**Y**	PNLX		NL	55839	**Y**	PNLX		NL
55828	**Y**	PNLX		NL	55840	**Y**	PNLX		NL
55829	**Y**	PNLX		NL	55841	**Y**	PNLX		NL
55830	**Y**	PNLX		NL	55842	**Y**	PNLX		NL
55831	**Y**	PNLX		NL	55843	**Y**	PNLX		NL
55832	**Y**	PNLX	v	NL	55844	**Y**	PNLX		NL
55833	**Y**	PNLX	v	NL	55845	**Y**	PNLX		NL
55834	**Y**	PNLX	v	NL	55846	**Y**	PNLX		NL
55835	**Y**	PNLX	v	NL					

MS. Dia. DR205 Lot No. Derby 31037 1987. 73S. 22.6 t.

55850	**Y**	PNLX	NL	55855	**Y**	PNLX	NL
55851	**Y**	PNLX	NL	55856	**Y**	PNLX	NL
55852	**Y**	PNLX	NL	55857	**Y**	PNLX	NL
55853	**Y**	PNLX	NL	55858	**Y**	PNLX	NL
55854	**Y**	PNLX	NL	55859	**Y**	PNLX	NL

55928–55995. FIRST GENERATION DMUS.

CLASS 114/1 DERBY 'HEAVYWEIGHT' PARCELS

Engines: Two Leyland TL11 of 152 kW (205 hp).
Bogies: DD9.
Dimensions: 20.45 x 2.82 x 3.86 m.
Carrying Capacity: 10 tons.

DMPMV. Dia. DX513. Lot No. 30209 1956–57. 38.0 t.

55928 (53018)	**R**	RPXN	CA	55957 (530)
55929 (53015)	**R**	RPXN	CA	55958 (530)
55930 (53010)	**R**	RPXN	CA	55959 (530)
55931 (53040)	**R**	RPXN	CA	55960 (530)
55932 (53032)	**R**	RPXN	CA	55961 (530)

CLASS 128 GLOUCESTER PARCELS

Engines: Two Leyland Albion 149 kW (200 hp) [†Leyland TL11 of 152 kW (205 hp)].

Gangways: Non gangwayed single cars.
Bogies:
Dimensions: 20.45 x 2.74 x 3.87 m.
Carrying Capacity: **DMPMV**. Dia. DX501. Lot No. 30551 1959–60. 41.5 t.

55991	**R**	RPXN		CA
55992	**R**	RPXN	†	CA
55993	**R**	RPXN		CA
55994	**R**	RPXN	†	CA
55995	**R**	RPXN	†	CA

57101–58866. SECOND GENERATION DMUS.

CLASS 150/1 BREL SPRINTER

For technical details see 52101–050.

DMS. Dia. DP239. Lot No. 31012 York 1985–6. 70S. 38.45 t.

57101	**PO**	PDYX		DY
57102	**PO**	PDYX		DY
57103	**PO**	PDYX		DY
57104	**PO**	PDYX		DY
57105	**PO**	PDYX		DY
57106	**PO**	PDYX		DY
57107	**PO**	PDYX		DY
57108	**PO**	PDYX		DY
57109	**PO**	PDYX		DY
57110	**PO**	PDYX		DY
57111	**PO**	PDYX		DY
57112	**PO**	PDYX		DY
57113	**PO**	PDYX		DY
57114	**PO**	PDYX		DY
57115	**PO**	PDYX		DY
57116	**PO**	PDYX	r	DY
57117	**PO**	PDYX	r	DY
57118	**PO**	PDYX	r	DY
57119	**PO**	PDYX	r	DY
57120	**PO**	PDYX	r	DY
57121	**PO**	PDYX	r	DY
57122	**PO**	PDYX	r	DY
57123	**PO**	PDYX	r	DY
57124	**PO**	PDYX	r	DY
57125	**PO**	PDYX	r	DY
57126	**PO**	PDYX	r	DY
57127	**PO**	PDYX	r	DY
57128	**PO**	PDYX	r	DY
57129	**PO**	PDYX	r	DY
57130	**PO**	PDYX	r	DY
57131	**PO**	PDYX	r	DY
57132	**PO**	PDYX	r	DY
57133	**PO**	PDYX	r	DY
57134	**PO**	PDYX	r	DY
57135	**PO**	PDYX	r	DY
57136	**PO**	PDYX	r	DY
57137	**PO**	PDYX	r	DY
57138	**PO**	PDYX	r	DY
57139	**PO**	PDYX	r	DY
57140	**P**	PDYX	r	DY
57141	**P**	PTSX	r	TS
57142	**P**	PTSX	r	TS
57143	**P**	PDYX	r	DY
57144	**P**	PTSX	r	TS
57145	**P**	PTSX	r	TS
57146	**PO**	PTSX	r	TS
57147	**P**	PDYX	r	DY
57148	**PO**	PTSX	r	TS
57149	**P**	PTSX	r	TS
57150	**P**	PDYX	r	DY

CLASS 150/2 BREL SPRINTER

For technical details see 52201–285.

DMS. Dia. DP243. Lot No. 31018 York 1986–7. 73S and luggage space. 34.90 t.

57201	**PN**	PNHX	NH
57202	**P**	PNHX	NH
57203	**PN**	PNHX	NH
57204	**P**	PNHX	NH
57205	**P**	PTSX	TS
57206	**P**	PNHX	NH
57207	**PN**	PNHX	NH
57208	**P**	PHAX	HA
57209	**P**	PNHX	NH
57210	**P**	PNHX	NH
57211	**PN**	PNHX	NH
57212	**P**	PTSX	TS
57213	**PN**	PNHX	NH
57214	**P**	PNHX	NH
57215	**PN**	PNHX	NH
57216	**P**	PNHX	NH
57217	**PN**	PNHX	NH
57218	**P**	PNHX	NH
57219	**P**	PNHX	NH
57220	**P**	PNHX	NH
57221	**P**	PNHX	NH
57222	**P**	PNHX	NH

57223	**P**	PNHX	NH
57224	**P**	PTSX	TS
57225	**P**	PNHX	NH
57226	**P**	PTSX	TS
57227	**P**	PNHX	NH
57228	**P**	PHAX	HA
57229	**P**	PNHX	NH
57230	**P**	PNLX	NL
57231	**P**	PNHX	NH
57232	**P**	PNLX	NL
57233	**P**	PNHX	NH
57234	**P**	PNLX	NL
57235	**P**	PNHX	NH
57236	**P**	PNLX	NL
57237	**P**	PNHX	NH
57238	**P**	PNLX	NL
57239	**P**	PNHX	NH
57240	**P**	PNLX	NL
57241	**P**	PTSX	TS
57242	**P**	PNLX	NL
57243	**P**	PNHX	NH
57244	**P**	PHAX	HA
57245	**P**	PHAX	HA
57246	**P**	PCFX	CF
57247	**P**	PCFX	CF
57248	**P**	PHAX	HA
57249	**P**	PCFX	CF
57250	**P**	PHAX	HA
57251	**P**	PCFX	CF
57252	**P**	PHAX	HA
57253	**P**	PTSX	TS
57254	**P**	PCFX	CF
57255	**P**	PHAX	HA
57256	**P**	PHAX	HA
57257	**P**	PHAX	HA
57258	**P**	PHAX	HA
57259	**P**	PHAX	HA
57260	**P**	PHAX	HA
57261	**P**	PCFX	CF
57262	**P**	PHAX	HA
57263	**P**	PCFX	CF
57264	**P**	PHAX	HA
57265	**P**	PCFX	CF
57266	**P**	PCFX	CF
57267	**P**	PCFX	CF
57268	**P**	PCFX	CF
57269	**P**	PCFX	CF
57270	**P**	PCFX	CF
57271	**P**	PCFX	CF
57272	**P**	PCFX	CF
57273	**P**	PCFX	CF
57274	**P**	PCFX	CF
57275	**P**	PCFX	CF
57276	**P**	PCFX	CF
57277	**P**	PCFX	CF
57278	**P**	PCFX	CF
57279	**P**	PCFX	CF
57280	**P**	PCFX	CF
57281	**P**	PCFX	CF
57282	**P**	PCFX	CF
57283	**P**	PHAX	HA
57284	**P**	PHAX	HA
57285	**P**	PHAX	HA

CLASS 155 LEYLAND BUS SUPER SPRINTER

For details see 52301–335/341–347.

57301–335. DMS. Dia. DP249. Lot No. 31027 1987–8. 80S and parcels area. 38 t.

57341–347. DMS. Dia. DP249. Lot No. 31058 1988. 80S and parcels area. 38.7 t.

57301	**P**	PCFX	CF
57302	**P**	PCFX	CF
57303	**P**	PCFX	CF
57304	**P**	PCFX	CF
57305	**P**	PCFX	CF
57306	**P**	PCFX	CF
57307	**P**	PCFX	CF
57308	**P**	PCFX	CF
57309	**P**	PCFX	CF
57310	**P**	PCFX	CF
57311	**P**	PCFX	CF
57312	**P**	PCFX	CF
57313	**P**	PCFX	CF
57314	**P**	PCFX	CF
57315	**P**	PCFX	CF
57316	**P**	PCFX	CF
57317	**P**	PCFX	CF
57318	**P**	PCFX	CF
57319	**P**	PCFX	CF
57320	**P**	PCFX	CF
57321	**P**	PCFX	CF
57322	**P**	PCFX	CF

57323	**P**	PCFX		CF
57324	**P**	PCFX		CF
57325	**P**	PCFX		CF
57326	**P**	PCFX		CF
57327	**P**	PCFX		CF
57328	**P**	PCFX		CF
57329	**P**	PCFX		CF
57330	**P**	PCFX		CF
57331	**P**	PCFX		CF
57332	**P**	PCFX		CF
57333	**P**	PCFX		CF
57334	**P**	PCFX		CF
57335	**P**	PCFX		CF
57341	**Y**	PNLX		NL
57342	**Y**	PNLX		NL
57343	**Y**	PNLX		NL
57344	**Y**	PNLX		NL
57345	**Y**	PNLX		NL
57346	**Y**	PNLX		NL
57347	**Y**	PNLX		NL

CLASS 156 METRO-CAMMELL SUPER SPRINTER

For technical details see 52401–514.

DMS. Dia. DP246. Lot No. 31029 1987–9. 35.5 t. 76S and parcels area.

57401	**P**	PNCX		NC
57402	**P**	PNCX		NC
57403	**P**	PNCX		NC
57404	**P**	PNCX		NC
57405	**P**	PNCX		NC
57406	**P**	PNCX		NC
57407	**P**	PNCX		NC
57408	**P**	PNCX		NC
57409	**P**	PNCX		NC
57410	**P**	PNCX		NC
57411	**P**	PNCX		NC
57412	**P**	PNCX		NC
57413	**P**	PNCX		NC
57414	**P**	PNCX		NC
57415	**P**	PNCX		NC
57416	**P**	PNCX		NC
57417	**P**	PNCX		NC
57418	**P**	PNCX		NC
57419	**P**	PNCX		NC
57420	**P**	PNCX		NC
57421	**P**	PNCX		NC
57422	**P**	PNCX		NC
57423	**P**	PNCX		NC
57424	**P**	PNCX		NC
57425	**P**	PNCX		NC
57426	**P**	PNCX		NC
57427	**P**	PNCX		NC
57428	**P**	PNCX		NC
57429	**P**	PNCX		NC
57430	**P**	PCKX		CK
57431	**P**	PCKX		CK
57432	**P**	PCKX		CK
57433	**P**	PCKX		CK
57434	**P**	PCKX		CK
57435	**P**	PCKX		CK
57436	**P**	PCKX		CK
57437	**P**	PCKX		CK
57438	**P**	PNLX		NL
57439	**P**	PCKX		CK
57440	**P**	PNCX		NC
57441	**P**	PNLX		NL
57442	**P**	PCKX		CK
57443	**P**	PNLX		NL
57444	**P**	PNLX		NL
57445	**P**	PHAX	r	HA
57446	**P**	PMOX	r	MO
57447	**P**	PHAX	r	HA
57448	**P**	PNLX		NL
57449	**P**	PHAX	r	HA
57450	**P**	PHAX	r	HA
57451	**P**	PNLX		NL
57452	**P**	PNCX		NC
57453	**P**	PHAX	r	HA
57454	**P**	PNLX		NL
57455	**P**	PNLX		NL
57456	**P**	PHAX	r	HA
57457	**P**	PMOX	r	MO
57458	**P**	PISX	r	IS
57459	**P**	PNLX		NL
57460	**P**	PNLX		NL
57461	**P**	PNCX		NC
57462	**P**	PNLX		NL
57463	**P**	PNLX		NL
57464	**P**	PNCX		NC
57465	**P**	PHAX		HA
57466	**P**	PNCX		NC
57467	**P**	PNLX		NL
57468	**P**	PNCX		NC
57469	**P**	PNLX		NL
57470	**P**	PNCX		NC
57471	**P**	PNLX		NL
57472	**P**	PNCX		NC

57473	**P**	PNCX		NC
57474	**P**	PMOX	r	MO
57475	**P**	PNLX		NL
57476	**P**	PHAX		HA
57477	**P**	PMOX	r	MO
57478	**P**	PMOX	r	MO
57479	**P**	PNLX		NL
57480	**P**	PNLX		NL
57481	**P**	PNLX		NL
57482	**P**	PNLX		NL
57483	**P**	PNLX		NL
57484	**P**	PNLX		NL
57485	**P**	PISX	r	IS
57486	**P**	PNCX		NC
57487	**P**	PNLX		NL
57488	**P**	PNLX		NL
57489	**P**	PNLX		NL
57490	**P**	PNLX		NL
57491	**P**	PNLX		NL
57492	**P**	PISX		IS
57493	**P**	PISX		IS
57494	**P**	PISX		IS
57495	**P**	PISX		IS
57496	**P**	PISX	r	IS
57497	**P**	PNLX		NL
57498	**P**	PNLX		NL
57499	**P**	PHAX	r	HA
57500	**P**	PHAX	r	HA
57501	**S**	PCKX		CK
57502	**S**	PCKX		CK
57503	**S**	PCKX		CK
57504	**S**	PCKX		CK
57505	**S**	PCKX		CK
57506	**S**	PCKX		CK
57507	**S**	PCKX		CK
57508	**S**	PCKX		CK
57509	**S**	PCKX		CK
57510	**S**	PCKX		CK
57511	**S**	PCKX		CK
57512	**S**	PCKX		CK
57513	**S**	PCKX		CK
57514	**S**	PCKX		CK

CLASS 158 BREL EXPRESS UNITS

For technical details see 52701–904.

DMSL. Dia. DP251. Lot No. 31052 Derby 1989– . t. 70S 1L and parcels area.

57701	**PE**	PHAX	HA
57702		P X	
57703		P X	
57704		P X	
57705		P X	
57706		P X	
57707		P X	
57708		P X	
57709		P X	
57710		P X	
57711		P X	
57712		P X	
57713		P X	
57714		P X	
57715		P X	
57716		P X	
57717		P X	
57718		P X	
57719		P X	
57720		P X	
57721		P X	
57722		P X	
57723		P X	
57724		P X	
57725		P X	
57726		P X	
57727		P X	
57728		P X	
57729		P X	
57730		P X	
57731		P X	
57732		P X	
57733		P X	
57734		P X	
57735		P X	
57736		P X	
57737		P X	
57738		P X	
57739		P X	
57740		P X	
57741		P X	
57742		P X	
57743		P X	
57744		P X	
57745		P X	
57746		P X	
57747		P X	
57748		P X	

57749	P	X
57750	P	X
57751	P	X
57752	P	X
57753	P	X
57754	P	X
57755	P	X
57756	P	X
57757	P	X
57758	P	X
57759	P	X
57760	P	X
57761	P	X
57762	P	X
57763	P	X
57764	P	X
57765	P	X
57766	P	X
57767	P	X
57768	P	X
57769	P	X
57770	P	X
57771	P	X
57772	P	X
57773	P	X
57774	P	X
57775	P	X
57776	P	X
57777	P	X
57778	P	X
57779	P	X
57780	P	X
57781	P	X
57782	P	X
57783	P	X
57784	P	X
57785	P	X
57786	P	X
57787	P	X
57788	P	X
57789	P	X
57790	P	X
57791	P	X
57792	P	X
57793	P	X
57794	P	X
57795	P	X
57796	P	X

On order to 57904.

MSL. Dia. DR207. Lot No. 31050 Derby 1989– . t. 70S 2L.

58701	P	X
58702	P	X
58703	P	X
58704	P	X
58705	P	X
58706	P	X
58707	P	X
58708	P	X
58709	P	X
58710	P	X
58711	P	X
58712	P	X
58713	P	X
58714	P	X
58715	P	X
58716	P	X
58717	P	X
58718	P	X
58719	P	X
58720	P	X
58721	P	X
58722	P	X
58723	P	X
58724	P	X
58725	P	X
58726	P	X
58727	P	X
58728	P	X
58729	P	X
58730	P	X
58731	P	X
58732	P	X
58733	P	X
58734	P	X
58735	P	X
58736	P	X
58737	P	X
58738	P	X
58739	P	X

59032–59817. FIRST GENERATION DMUS.

CLASS 116 DERBY SUBURBAN

Bogies: DT9.
Dimensions: 20.45 x 2.82 x 3.87 m.

TS. Dia. DT209. Lot No. 30385 1957. 98S. 29.5 t.

59032	PTSX	f	TS	59040	PHTX	f	HT (S)

CLASS 101 METRO-CAMMELL

Bogies: DT11.
Dimensions: 18.49 x 2.82 x 3.85 m.

TSL. Dia. DT201. Lot No. 30250 1956. 61S 1L. 25.5 t. 2+2 seating.

59042	PHAX	f	HA				

TBSL. Dia. DU202. Lot No. 30251 1956. 45S 1L. 25.5 t. 2+2 seating.

59049 **S**	PEDX	f	ED	59055	PNCX	f	NC

TSL. Dia. DT202. Lot No. 30257 1957. 71S 1L. 25.5 t.

59061	PHAX	f	HA	59072	PNLX	f	NL
59065	PLOX	f	LO				

TBSL. Dia. DU203. Lot No. 30258 1957. 53S 1L. 25.5 t.

59074 **S**	PEDX	f	ED	59080	PEDX	f	ED
59077	PNCX	f	NC	59084	PNCX	f	NC
59079	PNCX	f	NC				

TSL. Dia. DT201 (†DT228). Lot No. 30264 1957. 61S 1L (†58S plus additional luggage racks 1L). 25.5 t. 2+2 seating.

59086	PHAX	f	HA	59091	NSBX	f†	RG
59090 **S**	PEDX	f	ED				

TBSL. Dia. DU202. Lot No. 30265 1957. 45S 1L. 25.5 t. 2+2 seating.

59092	PNCX	f	NC	59095	PNCX	f	NC
59093	PLAX		LA	59097	PEDX	f	ED

TSL. Dia. DT202 (†DT228). Lot No. 30269 1957–58. 71S 1L (†58S plus additional luggage racks 1L). 25.5 t.

59101	NSBX	f†	RG	59105 **N**	NSBX	f	RG
59104	PHAX	f	HA	59107	PCHX	f	CH (S)

TCL. Dia. DT301 (*DT220) or †DT228. Lot No. 30277 1958. 12F 53S 1L (†58S 1L) 25.5 t.

59115	NSBX	f†	RG	59124	PHAX	f*	HA
59116	PHAX	f*	HA	59125	NSBX	f†	RG
59117	NSBX	f	RG	59128 **N**	NSBX		RG
59118	PLAX	f*	LA	59130	PLAX	f*	LA

CLASS 104 BRCW

Bogies: DT10.
Dimensions: 20.45 x 2.82 x 3.87 m.

TCL. Dia. DT302 (*DT218). Lot No. 30295 1957–58. 12F 54S 1L. 24.5 t.

59163 **N** NWRX RG
59187 **B** PNLX * NL

TSL. Dia. DT203. Lot No. 30299 1958. 69S 1L. 24.5 t. Luggage rack opposite toilet.

59206 **N** NMYX BY

TBSL. Dia. DU204. Lot No. 30300 1958. 51S 1L. 25.5 t. Luggage rack opposite toilet.

59228 **B** PNHX NH (U)

CLASS 108 DERBY 'LIGHTWEIGHT'

Bogies: DT8.
Dimensions: 18.49 x 2.79 x 3.87 m.

TBSL. Dia. DU201. Lot No. 30412 1958. 50S 1L. 23.5 t.

59245 PCFX f CF
59246 PNLX f NL
59248 PCFX f CF
59249 PNLX f NL
59250 PNLX f NL

CLASS 101 METRO-CAMMELL

Bogies: DT11.
Dimensions: 18.49 x 2.82 x 3.85 m.

TS. Dia. DT202 (†DT228). Lot No. 30273 1957. 71S 1L. (†64S plus additional luggage racks 1L). 25.5 t.

59302 PHAX f HA
59303 PLOX f LO
59304 PHAX f HA
59306 NSSX f† RG

CLASS 116 DERBY SUBURBAN

Bogies: DT9.
Dimensions: 20.45 x 2.82 x 3.87 m.

TC. Dia. DT3?? (*DT219). Lot No. 30365 1957–58. 20F 68S. 29.0 t.

59335 PTSX f* TS
59344 PTSX f* TS
59353 PTSX f* TS
59367 PTSX f* TS

CLASS 108 DERBY 'LIGHTWEIGHT'

Bogies: DT8.
Dimensions: 18.49 x 2.79 x 3.87 m.

TSL. Dia. DT205. Lot No. 30411 1958. 68S 1L. 22.5 t. Luggage rack opposite toilet.

59380	PLEX	f	LE	59383	PLEX	f	LE
59381	PLEX	f	LE	59384	PLEX	f	LE
59382	PLEX	f	LE	59385	PCFX	f	CF

TSL. Dia. DT206. Lot No. 30493 1958. 68S 1L. 23.5 t. Luggage rack opposite toilet.

59386	PLAX	f	LA	59389	PCHX	f	CH
59387	PNLX	f	NL	59390	PCHX	f	CH
59388	PLOX		LO				

CLASS 119 GLOUCESTER CROSS-COUNTRY

Bogies: DT9.
Dimensions: 20.45 x 2.82 x 3.86 m.
Notes: Buffets removed and converted into additional luggage area.

TSLRB. Dia. DT216. Lot No. 30423 1958–59. 60S 2L. 31.5 t.

59416 **N**	NSBX	RG	59425 **N**	NSBX	RG
59419 **N**	NSBX	RG	59430 **N**	NSBX	RG
59421 **N**	NWRX	RG	59435 **N**	NSBX	RG
59424 **N**	NSBX	RG	59437 **N**	NSBX	RG

CLASS 116 DERBY SUBURBAN

Bogies: DT9.
Dimensions: 20.45 x 2.82 x 3.87 m.

TC. Dia. DT3?? (*DT219). Lot No. 30448 1957–58. 20F 68S. 29.0 t.

59442	PTSX	f*	TS	59446	PTSX	f*	TS
59444	PTSX	f*	TS	59448	PTSX	f*	TS
59445	PTSX	f*	TS				

CLASS 118 BRCW SUBURBAN

Bogies: DT9.
Dimensions: 20.45 x 2.82 x 3.87 m.

TCL. Dia. DT305 (*DT230). Lot No. 30544 1960. 22F 48S 2L. 30.5 t.

59473	PTSX	f*	TS	59483	PTSX	f*	TS
59481	PTSX	*	TS				

CLASS 117 PRESSED STEEL SUBURBAN

Bogies: DT9.
Dimensions: 20.45 x 2.82 x 3.87 m.

TCL. Dia. DT305 (*DT230). Lot No. 30547 1959–60. 22F 48S 2L. 30.5 t.

59484 **N**	NWRX	f	RG	59486	PTSX	f*	TS
59485 **N**	NWRX	f	RG	59487 **N**	NWRX	f	RG

59488 **N**	NWRX	f	RG
59489 **N**	NWRX	f	RG
59490	PTSX	f*	TS
59491 **N**	NWRX	f	RG
59492	PTSX	f*	TS
59493 **N**	NWRX	f	RG
59494 **N**	NWRX	f	RG
59495 **N**	NWRX	f	RG
59496 **N**	NWRX	f	RG
59497 **N**	NWRX	f	RG
59498 **N**	NWRX	f	RG
59499 **N**	NWRX	f	RG
59500	PTSX	f*	TS
59501 **N**	NWRX	f	RG
59502 **N**	NWRX	f	RG
59503 **N**	NWRX	f	RG
59504	PTSX	f*	TS
59505	PTSX	f*	TS
59506 **N**	NWRX	f	RG
59507 **N**	NWRX	f	RG
59508 **N**	NWRX	f	RG
59509	PTSX	f*	TS
59510 **N**	NWRX	f	RG
59511 **N**	NWRX	f	RG
59512	PTSX	f*	TS
59513 **N**	NWRX	f	RG
59514 **N**	NWRX	f	RG
59515 **N**	NWRX	f	RG
59516	PTSX	f*	TS
59517	PTSX	f*	TS
59518 **N**	NWRX	f	RG
59519 **N**	NWRX	f	RG
59520 **0**	PTSX	f*	TS
59521	PTSX	f*	TS
59522	PTSX	f*	TS

CLASS 101 METRO-CAMMELL

Bogies: DT11.
Dimensions: 18.49 x 2.82 x 3.85 m.

TCL. Dia. DT301 (*DT220) or †DT228. Lot No. 30502 1959. 12F 53S 1L (†58S 1L). 25.5 t.

59525	PCHX	f*	CH (S)
59526 **N**	NSBX	f†	RG
59530 **N**	NSBX	f	RG
59532	PHAX	f*	HA
59536	PNCX	f*	NC
59539	PLAX	f*	LA
59540	NSBX	f	RG
59542 **S**	PEDX	f*	ED
59543 **N**	NSBX	f†	RG
59561	PLAX	f*	LA

TSL. Dia. DT202 (†DT228). Lot No. 30510 1959. 71S 1L (†58S 1L). 25.5 t.

59570 **N**	NSBX	f†	RG
59571	PHAX	f	HA

CLASS 127 DERBY SUBURBAN

Bogies: DT9.
Gangways: Lot 30523 are not gangwayed except as shown below.
Dimensions: 20.45 x 2.82 x 3.86 m.
Note: Fluorescent lighting as built.

TSL. Dia. DT226. Lot No. 30522 1959. 86S 2L. 30.5 t.

59589	PTSX	f	TS
59590	PTSX		TS
59591	PTSX	f	TS
59592	PTSX	f	TS
59593	PTSX	f	TS
59594	PTSX	f	TS
59595	PTSX		TS
59596	PTSX		TS
59597	PTSX	f	TS
59598	PTSX		TS
59600	PTSX		TS
59602	PTSX	f	TS
59603	PTSX		TS
59604	PTSX	f	TS
59606	PTSX		TS
59607	PTSX		TS
59608	PTSX	f	TS
59609	PTSX		TS

59610	PTSX		TS	59614	PTSX		TS
59611	PTSX		TS	59615	PTSX		TS
59612	PTSX		TS	59616	PTSX		TS
59613	PTSX		TS	59617	PTSX		TS

TS. Dia. DT208 (g DT229). Lot No. 30523 1959. 106S (g 96S). 29.5 t.

59621	PCPX	f	CP (U)	59631 **B**	PCPX		CP (U)
59622	PCPX		CP (U)	59632	PTSX	fg	TS
59625	PTSX	fg	TS	59641	PTSX	fg	TS
59627	PCPX		CP (U)	59643	PTSX	fg	TS
59629	PTSX	fg	TS	59648	PTSX	fg	TS

CLASS 115 — DERBY SUBURBAN

Bogies: DT9.
Gangways: Not gangwayed except as shown below.
Dimensions: 20.45 x 2.82 x 3.86 m.
Note: Fluorescent lighting as built.

TS. Dia. DT208 (g DT229). Lot No. 30531 1960. 106S (g 96S). 29.5 t.

59651	NMYX	f	BY	59658	PTSX	g	TS
59652 **N**	NMYX	f	BY	59659 **N**	NMYX	f	BY
59653	PCPX		CP (U)	59660	NMYX	f	BY
59654 **N**	NMYX	f	BY	59661	PTSX	g	TS
59655 **N**	NMYX		BY	59662	NMYX	f	BY
59656 **N**	NMYX		BY	59663 **N**	NMYX	f	BY
59657 **N**	NMYX	f	BY				

TCL. Dia. DT303 (*DT2??). Lot No. 30532 1960. 30F 40S 2L (g 30F 38S 2L). 30.5 t.

59664	NMYX		BY	59672	PTSX	fg*	TS
59665	NMYX	f	BY	59673	PTSX	g*	TS
59667 **N**	NMYX	f	BY	59674	PTSX	g*	TS
59668	PTSX	g*	TS	59675	NMYX	f	BY
59669 **N**	NMYX	f	BY	59676 **N**	NMYX	f	BY
59670	PTSX	g*	TS	59677	PTSX	g*	TS
59671 **N**	NMYX	f	BY	59678	NMYX	f	BY

CLASS 101 — METRO-CAMMELL

Bogies: DT11.
Dimensions: 18.49 x 2.82 x 3.85 m.

TCL. Dia. DT301 (*DT220). Lot No. 30589 1960. 12F 53S 1L. 25.5 t.

59688	PHAX	f*	HA

CLASS 110 — BRCW 'CALDER VALLEY'

Bogies: DT10.
Dimensions: 18.49 x 2.82 x 3.88 m.

TSL. Dia. DT207. Lot No. 30594 1961. 72S 1L. 24.5 t.

Class 101 unit No. L832 (51226/59570/51488) is at Godstone on the 08.06 Redhill–Tonbridge service. The date is 8th March 1989. *Rodney Lissenden*

Class 108 power twin P956 (51939/52057) waits at Plymouth with the 13.50 to Gunnislake on 19th August 1988. *John Augustson*

Three Class 107 units in Strathclyde PTE livery await their next duties at Eastfield TMD on 27th August 1989.
John Augustson

Class 110 BRCW 'Calder Valley' 2 car unit Nos. 52082/52071 wait to depart from Doncaster with the 14.20 shuttle to Goole on 23rd January 1989. This service was required as Hull services were being diverted via Selby due to temporary closure of Goole Swing Bridge. *Barry Nicolle*

Network SouthEast-liveried Class 119 set No. L576 (51076/59421/51090) stands at Oxford having arrived with a service from Reading on 8th August 1989. *John Augustson*

A Class 115 DMU, led by driving vehicle No. 51890, is seen at High Wycombe with the 11.44 Aylesbury–Marylebone service on 13th June 1989. This train manages a fairly typical average speed of just 31 mph and regular travellers will no doubt be looking forward to the completion of 'total route modernisation' and introduction of Class 165 Network Turbo trains. *N.J. Joynson*

Class 121 single car unit No. T132 (55032) with Welsh Valleys lines adornments stands under the semaphore at Worcester Shrub Hill on 6th May 1989.

Steve Widdowson

Class 122 single car DMU No. 109 (55009) in blue/grey livery stands at Exeter St. Davids on 24th June 1989. *John Augustson*

Class 128 DPU in Royal Mail livery passes Broxbourne with a Liverpool Street–Cambridge working on 23rd February 1989. *Chris Wilson*

Class 114 DPU vehicles, 54900 and 55930 stand in the Scarborough bay at York with the 17.02 parcels service to Leeds on 13th August 1989.
Barry Nicolle

Class 141 No. 141 114 in crimson and white West Yorkshire PTE livery stands at York on 30th September 1988.
John Augustson

Renumbered Class 143 No. 143 320 in Tyne and Wear PTE livery runs into the new Platform 15 at Newcastle with the 15.57 Hexham–Sunderland on 7th October 1989. *John Augustson*

Class 144 3-car DMU No. 144 018, in West Yorkshire PTE livery, calls at Shipley station after traversing Shipley Bingley Junction with a Skipton–Leeds working on 4th February 1989. As there is no platform on the up line here, trains calling at Shipley must cross to the down line. *Les Nixon*

Greater Manchester PTE-liveried Class 142 No. 142 011 is seen at Dore in the south western suburbs of Sheffield on a Manchester Piccadilly–Sheffield stopping service. All classes of Pacer (140–144) have been criticised for their spartan interiors, poor interior lighting, bus-type seating and lack of tinted glass for both driver and passengers, the latter feature offered on all buses and most new cars on the market! *Duncan Street*

Class 156 No. 156 503, in Strathclyde PTE livery, leaves Newcastle with the 17.15 to Girvan on 19th June 1989. *John Augustson*

The Class 150 Sprinters have not been the gold at the end of BR's rainbow with reliability being poor during 1989. Class 150/1 DMU No. 150 102 just east of the Railway Technical Centre, Derby, with the 09.10 Birmingham–Lincoln service on 14th October 1989. *Peter Fox*

Class 150/2 unit No. 150 211 in Network NorthWest livery is seen approaching Buxton on 29th May 1989. *Paul Shannon*

Class 155/0 units Nos. 155 325 and 155 332 power past Lymington Junction on a driver training trip from Eastleigh to Weymouth on the 19th April 1989. *Nick Bartlett*

Provincial's newest design, the Class 158, is now several months behind schedule and introduction on Express routes has been postponed. 158 701 is pictured here east of the Railway Technical Centre, Derby, on 14th October 1989. It is working a return Bedford–Derby demonstration special. *Peter Fox*

Eastleigh-based Class 205 DEMU No. 205 033 passes Worting Junction, just west of Basingstoke, with the 13.55 Reading–Portsmouth Harbour service on 20th April 1989. The Waterloo–Exeter main line diverges here, passing beneath the flyover. *Rodney Lissenden*

Class 207 ('3D') DEMU No. 207 005, departs from St. Denys with the 16.28 Portsmouth Harbour–Southampton stopping service. This route is due to be electrically-operated from May 1990. The date is 24th June 1989. *David Brown*

No.		Pool		Depot
59694		PNLX	f	NL
59701		PNLX	f	NL

CLASS 115 DERBY SUBURBAN

Bogies: DT9.
Gangways: Not gangwayed except as shown below.
Dimensions: 20.45 x 2.82 x 3.86 m.
Note: Fluorescent lighting as built.

TS. Dia. DT 229. Lot No. 30596 1960. 96S. 29.5 t.

No.		Pool		Depot
59713		PTSX	g	TS
59715		PCPX	fg	CP (U)
59716		PCPX	fg	CP (U)
59717		PCPX	g	CP (U)

TCL. Dia. DT303 (*g DT2??). Lot No. 30597 1960. 30F 40S 2L (g 30F 38S 2L). 30.5 t.

No.		Pool		Depot
59719		PTSX	fg*	TS
59720		PTSX	g*	TS
59721		PTSX	g*	TS
59722		PTSX	g*	TS
59723		PTSX	g*	TS
59724		PTSX	g*	TS

TS. Dia. DT208 (g DT229). Lot No. 30599 1960. 106S (g 96S). 29.5 t.

No.		Pool		Depot
59725		PCPX	f	CP (S)
59726		PTSX	fg	TS
59727	**N**	NMYX	f	BY
59728	**N**	NMYX	f	BY
59729	**N**	NMYX	f	BY
59731		NMYX	f	BY
59732	**N**	NMYX	f	BY
59733		NMYX	f	BY
59734		NMYX	g	BY
59735	**N**	NMYX	g	BY
59736	**N**	NMYX	f	BY
59737	**N**	NMYX		BY
59738		NMYX		BY
59740		NMYX	f	BY
59741		PTSX	fg	TS
59743		PTSX	fg	TS
59744		PCPX		CP (U)

TCL. Dia. DT303 (*g DT2??). Lot No. 30600 1960. 30F 40S 2L. (g 30F 38S 2L). 30.5 t.

No.		Pool		Depot
59745		PTSX	g*	TS
59746		NMYX	f	BY
59747		NMYX	f	BY
59749		NMYX	f	BY
59750		NMYX	f	BY
59751		PTSX	g*	TS
59752	**N**	NMYX	f	BY
59753		PTSX	g*	TS
59754		NMYX		BY
59755	**N**	NMYX	f	BY
59756		PTSX	g*	TS
59757		PTSX	g*	TS
59758		NMYX	f	BY
59759	**N**	NMYX	f	BY
59760		PTSX	fg*	TS
59761	**N**	NMYX	f	BY
59762	**N**	NMYX	f	BY
59763	**N**	NMYX	f	BY
59764		NMYX	f	BY

CLASS 107 DERBY 'HEAVYWEIGHT'

Bogies: DT9.
Dimensions: 18.49 x 2.82 x 3.87 m.

TSL. Dia. DT204. Lot No. 30613 1960–61. 71S 1L. 28.5 t.

No.		Pool		Depot
59782	**S**	PEDX	f	ED
59783	**S**	PEDX	f	ED
59784	**S**	PEDX	f	ED
59785		PHAX	f	HA

59786	PEDX		ED
59789	PHAX	f	HA
59790 **S**	PEDX	f	ED
59791	PEDX	f	ED
59792 **S**	PEDX		ED
59794	PNLX		NL
59795 **S**	PEDX	f	ED
59796 **S**	PEDX	f	ED
59797	PEDX	f	ED
59800	PHAX	f	HA
59801 **S**	PEDX	f	ED
59802 **S**	PEDX	f	ED
59803 **S**	PEDX	f	ED
59804 **S**	PEDX	f	ED
59805	PEDX	f	ED
59806 **S**	PEDX	f	ED
59807 **S**	PEDX	f	ED

CLASS 110 BRCW 'CALDER VALLEY'

Bogies: DT10.
Dimensions: 18.49 x 2.82 x 3.88 m.

TSL. Dia. DT207. Lot No. 30693 1961–62. 72S 1L. 25.5 t.

59809	PHAX	f	HA
59817	PNLX	f	NL

3.3. FORMATIONS OF OLD DMUS

SCOTTISH REGION

101 304	ED	51224	59090	53241
101 306	HA	51228	59086	53243
101 307	HA	51231	59688	53189
101 310	HA	51234	59302	53163
101 315	HA	51241	59042	53260
101 318	HA	51244	59061	53176
101 321	ED	51249	59542	53146
101 350	HA	51182	59124	53239
101 360	ED	53158	59074	53160
101 363	ED	53185	59049	53194
107 725	ED	51985	59797	52028
107 726	ED	51986	59803	52018
107 727	ED	51987		52020
107 728	ED	51988	59786	52013
107 729	ED	51989	59795	52023
107 730	ED	51990	59802	51803
107 731	HA	51991	59789	52035
107 732	HA	51992	59800	52026
107 733	ED	51993	59790	52012
107 734	ED	51994	59791	52011
107 735	ED	51996		52019
107 736	ED	51997	59784	52025
107 737	ED	51998	59782	52021
107 738	ED	51999	59796	52030
107 739	HA	52000	59785	52036
107 740	ED	52001	59783	52024
107 743	ED	52004	59805	52016
107 744	ED	52005	59804	52031
107 745	ED	52006	59801	52015
107 746	ED	52007	59806	52033
107 747	ED	52008	59792	52029
107 749	ED	52010	59807	52034

WESTERN REGION

P100	55000		
P103	55003		
P104	55004		
P105	55005		
P106	55006		
P109	55009		
P111	55011		
P112	55012		
L120	55020		
L121	55021		
L122	55022		
L123	55023		
L124	55024		
L125	55025		
P126	55026		
L127	55027		
L128	55028		
L129	55029		
L130	55030		
L131	55031		
L201	51212	54070	
L202	51225	54372	
L204	51220	54371	
L205	51443	54385	
L206	51215	54081	
L207	51221	54396	
L210	53083	54289	
L211	53155	54287	
L280	54280		
L283	54283		
C362	53086		53128
C393	51135		51148
C394	51139		51152
C395	51140		51153
L400	51332	59484	51374
L401	51333	59485	51375
L402	51335	59487	51377
L403	51336	59488	51378
L404	51337	59489	51379
L405	51340	59491	51381
L406	51341	59493	51383
L407	51342	59494	51384
L408	51343	59495	51385
L409	51344	59496	51386
L410	51345	59497	51387
L411	51346	59498	51388
L412	51347	59499	51389
L413	51349	59501	51391
L414	51350	59502	51392
L415	51351	59503	51393
L417	51354	59506	51396
L418	51355	59507	51397

L419	51356	59508	51398
L420	51358	59510	51400
L421	51359	59511	51401
L423	51361	59513	51399
L424	51362	59514	51404
L425	51363	59515	51405
L428	51366	59518	51408
L429	51367	59519	51409
C472	51315		51327
C480	51312	59419	51088
L575	51060	59419	51088
L576	51062	59421	51090
L579	51065	59424	51103
L580	51066	59425	51094
L588	51079	59437	51107
L594	51073	59435	51104
L595	51074	59430	51086
L596	51076	59416	51099
L700	53207	53296	
L701	53437		53479
L702	53455		53539
L703	53429		53470
L704	53477		53540
L705	51319		53820
P825	53165	59118	51246
P826	53223	59561	51219
P827	53212	59130	51223
P828	51936	59386	52051
P829	51500	59093	53157
L830	51190	59105	53333
L831	51211	59115	53265
L832	51226	59570	51499
L833	51425	59543	51504
L834	51431	59526	51501
L835	51432	59530	51498
L836	51434	59540	51503
L837	51437	59128	53751
L838	53308	59125	53331
L839	53310	59306	53326
L840	53311	59117	53322
L841	53312	59091	53332
L842	53314	59101	53327
P860	51181		51530
P861	51205		51512
P862	51179		51570
P870	53200		53638
P871	53256		53639

P872	53247		53646
P873	53248		53637
P874	53315		53330
C901	51128	53632	51141
C902	53642	51132	51145
C903	51134	52053	51147
C904	51133	52059	51136
S939	51926		53631
S940	51919		52048
S941	51922		51562
S942	51928	59382	52044
S943	51930		51561
S944	51931		52055
S945	51925	59380	53643
S946	53939		52046
S947	53942		51567
S948	53928		51565
S949	53945		51563
S950	53926		52061
S951	53935		52041
S952	53625		52045
P954	51932		52047
P955	51933		52054
P956	51939		52063
P957	51940		52057
S958	51924	59384	52038
S959	53947		53633
B960	53602	54197	
B961	53608	54208	
B962	53612	54492	
B963	53948	54268	
B964	53614	54204	
B965	53616	54209	
B966	53618	54202	
B967	53619	54230	
B968	53620	54203	
B969	53621	54201	
B970	53622	54207	
B971	53941	54205	
C972	53627	54210	
C973	53629	54267	
C974	51416	54498	
C975	53624	54273	
C976	51417	54212	
C977	51419	54494	
C983	53617	53635	
C995	51436		53169

BY BLETCHLEY

Units at this depot do not maintain fixed formations.

CA CAMBRIDGE

Set numbers not carried.

220	51207	54405
221	51208	54402
222	51438	54382
223	51444	54068
224	51445	54028
902	53027	54041
911	53020	53046
928	55928	54903
929	55929	54904
930	55930	54900
931	55931	54901
932	55932	54902

CH CHESTER

CH250	51418	54485	
CH252	53975	54258	
CH253	51421	54275	
CH254	51422	54277	
CH255	51424	54232	
CH256	51901	54231	
CH257	53970	54264	
CH258	51903	54486	
CH260	51905	54488	
CH261	51906	54489	
CH263	51908	54501	
CH264	51911	54496	
CH265	51913	54497	
CH267	51920	54276	
CH268	53924	54278	
CH269	53927	54214	
CH270	53929	54269	
CH271	53931	54260	
CH272	53938	54484	
CH273	53976	54225	
CH274	53978	54261	
CH275	53980	54256	
CH276	53981	54253	
CH277	53982	54246	
CH278	53983	54266	
CH351	51937	59390	51943

CH610	53447		53531
CH611	53451		53517
CH615	53534		51505
CH620	51948		51568
CH621	51902		52049
CH622	51904		52050
CH623	51907		52058
CH624	51917		53516
CH625	51938		52065
CH626	51941		52064
CH627	51945		52060
CH628	51947		53634
CH629	51950		52062
CH631	53925		52039
CH632	53930		52056
CH633	53932		51569
CH634	53933		51566
CH635	53934		53512
CH637	53977		53645
CH640	51800		51513
CH641	51245		53269
CH642	51426		53171
CH643	51462		51511
CH645	53224		51509
CH646	53198		53244

HT HEATON

Only HT's 108 units maintain fixed formations. Set numbers not carried.

HT001	53950	54227
HT002	53951	54235
HT003	53952	54244
HT004	53953	54242
HT005	53954	54265
HT006	53955	54262
HT007	53956	54238
HT008	53957	54236
HT009	53958	54239
HT010	53959	54243
HT011	53960	54241
HT012	53962	54245
HT013	53963	54240
HT014	53964	54247
HT015	53973	54263

LO LONGSIGHT

LO347	53216	59303	53294
LO348	51210	59065	53253
LO350	51927	59388	51935

NL NEVILLE HILL

Units at this depot do not maintain fixed formations.

NC NORWICH CROWN POINT

51	51218	54399	
53	51247	54387	
54	51230	54369	
55	51427	54368	
56	51429	54393	
58	51213	54380	
59	51442	54073	
63	51428	54218	
65	51189	54343	
66	51192	54354	
67	51187	54332	
68	53293	54062	
70	53202	54388	
72	51201	54347	
76	51252	54352	
78	53201	54379	
81	53231	54050	
85	51185	54065	
86	53228	54055	
88	53208	54060	
100	51508	59095	53181
101	51506	59055	53266
102	53149	59084	53193
103	53305	59536	53180
106	53139	59092	53238
107	53177	59079	53267
108	53168	59077	53170

TS TYSELEY

TS012	55032		
TS013	55033		
TS014	55034		
TS021	53021	54012	
TS022	53002	54047	
TS023	53036	54006	
TS024	53044	54043	
TS025	53005	54039	
TS026	53006	54010	
TS027	53019	54027	
TS301	51353	59505	51395
TS302	51360	59512	51402
TS303	51364	59516	51406
TS304	51365	59517	51407
TS305	51368	59520	51410
TS306	51369	59521	51411
TS307	51370	59522	51412
TS308	51371	59509	51413
TS309	51372	59490	51414
TS310	51373	59486	51415
TS311	51334	59500	51376
TS312	51338	59504	51380
TS313	51339	59492	51382
TS314	51348	59677	51390
TS315	51352	59745	51394
TS316	51306	59473	51321
TS317	51314	59481	51329
TS318	51316	59483	51331
TS319	51129	59756	51149
TS320	51138	59720	53890
TS321	53854	59723	53907
TS322	53071	59760	53880
TS323	53818		53093
TS324	53053	59672	53101
TS325	53060	59757	51142
TS326	53822	59444	53881
TS332	53865	59598	53073
TS333	53082	59617	51131
TS334	53054	59596	53850
TS335	53844	59674	53827
TS336	53849	59721	51130
TS337	53853	59751	53838
TS338	53050	59722	53058
TS339	53090	59724	53863
TS340	53826	59719	53837
TS341	53061	59753	53079

TS342	53055	59673	53056	
TS401	51662	59610	59743	53886
TS402	51851	59658	59614	51146
TS403	51852	59641	59616	53106
TS404	51853	59604	59445	53919
TS405	51854	59615	59335	51143
TS406	51856	59606	59344	53893
TS407	51858	59594	59442	53916
TS408	51859	59448	59592	53114
TS409	51860	59600	59367	53894
TS410	51862	59602	59713	53124
TS411	51898	59625	59597	53873
TS412	51867	59446	59591	53116
TS413	51868	59632	59593	53902
TS414	51869	59661	59613	51144
TS415	51870	59589	59353	53921
TS416	51876	59609	59648	53891
TS417	51877	59603	59726	53897
TS419	51884	59629	59668	51151
TS420	51892	59607	59032	53092
TS421	51897	59741	59612	53878

3.4. FORMATIONS OF SECOND GENERATION DMUS.

CLASS 140 PROTOTYPE RAILBUS

140 001	NL	55500	55501

CLASS 141 LEYLAND/BREL RAILBUS

141 101	NL	55521	55541
141 102	NL	55502	55522
141 103	NL	55503	55523
141 104	NL	55504	55524
141 105	NL	55505	55525
141 106	NL	55506	55526
141 107	NL	55507	55527
141 108	NL	55508	55528
141 109	NL	55509	55529
141 110	NL	55510	55530
141 111	NL	55511	55531
141 112	NL	55512	55532
141 114	NL	55514	55534
141 115	NL	55515	55535
141 116	NL	55516	55536
141 117	NL	55517	55537
141 118	NL	55518	55538
141 119	NL	55519	55539
141 120	NL	55520	55540
141 813	NL	55513	55533

CLASS 142 LEYLAND/BREL RAILBUS

142 001	NH	55542	55592
142 002	NH	55543	55593
142 003	NH	55544	55594
142 004	NH	55545	55595
142 005	NH	55546	55596
142 006	NH	55547	55597
142 007	NH	55548	55598
142 008	NH	55549	55599
142 009	NH	55550	55600
142 010	NH	55551	55601
142 011	NH	55552	55602
142 012	NH	55553	55603
142 013	NH	55554	55604
142 014	NH	55555	55605
142 015	NH	55556	55606
142 017	NH	55558	55608
142 023	NH	55564	55614
142 024	NH	55565	55615
142 026	NH	55567	55617
142 027	NH	55568	55618
142 028	NH	55569	55619
142 029	NH	55570	55620
142 030	NH	55571	55621
142 031	NH	55572	55622
142 032	NH	55573	55623
142 033	NH	55574	55624
142 034	NH	55575	55625
142 035	NH	55576	55626
142 036	NH	55577	55627
142 037	NH	55578	55628
142 038	NH	55579	55629
142 039	NH	55580	55630
142 040	NH	55581	55631
142 041	NH	55582	55632
142 042	NH	55583	55633
142 043	NH	55584	55634
142 044	NH	55585	55635
142 045	NH	55586	55636
142 046	NH	55587	55637
142 047	NH	55588	55638
142 048	NH	55589	55639
142 049	NH	55590	55640
142 050	HT	55591	55641
142 051	NH	55701	55747
142 052	NH	55702	55748
142 053	NH	55703	55749
142 054	NH	55704	55750
142 055	NH	55705	55751
142 056	NH	55706	55752
142 057	NH	55707	55753
142 058	NH	55708	55754
142 059	NH	55709	55755
142 060	NH	55710	55756
142 061	NH	55711	55757
142 062	NH	55712	55758
142 063	NH	55713	55759

142 064	NH	55714	55760	
142 065	NH	55715	55761	
142 066	NH	55716	55762	
142 067	NH	55717	55763	
142 068	NH	55718	55764	
142 069	NH	55719	55765	
142 070	NL	55720	55766	
142 071	NH	55721	55767	
142 072	NL	55722	55768	
142 073	NL	55723	55769	
142 074	NL	55724	55770	
142 075	NL	55725	55771	
142 076	NL	55726	55772	
142 077	NL	55727	55773	
142 078	NL	55728	55774	
142 079	NL	55729	55775	
142 080	NL	55730	55776	
142 081	NL	55731	55777	
142 082	NL	55732	55778	
142 083	NL	55733	55779	
142 084	NL	55734	55780	
142 085	NL	55735	55781	
142 086	NL	55736	55782	
142 087	NL	55737	55783	
142 088	NL	55738	55784	
142 089	NL	55739	55785	
142 090	NL	55740	55786	
142 091	NL	55741	55787	
142 092	NL	55742	55788	
142 093	NL	55743	55789	
142 094	NL	55744	55790	
142 095	NL	55745	55791	
142 096	NL	55746	55792	
142 516	HT	55557	55607	
142 518	HT	55559	55609	
142 519	HT	55560	55610	
142 520	HT	55561	55611	
142 521	HT	55562	55612	
142 522	HT	55563	55613	
142 525	HT	55566	55616	

CLASS 143 ALEXANDER/BARCLAY RAILBUS

143 320	HT	55661	55686
143 321	HT	55662	55687
143 322	HT	55663	55688
143 323	HT	55664	55689
143 324	HT	55665	55690
143 325	HT	55666	55691
143 601	HT	55642	55667
143 602	HT	55643	55668
143 603	HT	55644	55669
143 604	HT	55645	55670
143 605	HT	55646	55671
143 606	HT	55647	55672
143 607	HT	55648	55673
143 608	HT	55649	55674
143 609	HT	55650	55675
143 610	HT	55651	55676
143 611	HT	55652	55677
143 612	HT	55653	55678
143 613	HT	55654	55679
143 614	HT	55655	55680
143 615	HT	55656	55681
143 616	HT	55657	55682
143 617	HT	55658	55683
143 618	HT	55659	55684
143 619	HT	55660	55685

CLASS 144 ALEXANDER/BREL RAILBUS

144 001	NL	55801	55824	
144 002	NL	55802	55825	
144 003	NL	55803	55826	
144 004	NL	55804	55827	
144 005	NL	55805	55828	
144 006	NL	55806	55829	
144 007	NL	55807	55830	
144 008	NL	55808	55831	
144 009	NL	55809	55832	
144 010	NL	55810	55833	
144 011	NL	55811	55834	
144 012	NL	55812	55835	
144 013	NL	55813	55836	
144 014	NL	55814	55850	55837
144 015	NL	55815	55851	55838
144 016	NL	55816	55852	55839
144 017	NL	55817	55853	55840
144 018	NL	55818	55854	55841
144 019	NL	55819	55855	55842
144 020	NL	55820	55856	55843
144 021	NL	55821	55857	55844
144 022	NL	55822	55858	55845
144 023	NL	55823	55859	55846

CLASS 150/0 BREL PROTOTYPE 3-CAR DMU

150 001	TS	55200	55300	55400

CLASS 150/1 SPRINTER BREL 2/3-CAR DMU

150 101	DY	52101		57101
150 102	DY	52102		57102
150 103	DY	52103		57103
150 104	DY	52104		57104
150 105	DY	52105		57105
150 106	DY	52106		57106
150 107	DY	52107		57107
150 108	DY	52108		57108
150 109	DY	52109		57109
150 110	DY	52110		57110
150 111	DY	52111		57111
150 112	DY	52112		57112
150 113	DY	52113		57113
150 114	DY	52114		57114
150 115	DY	52115		57115
150 116	DY	52116		57116
150 117	DY	52117		57117
150 118	DY	52118		57118
150 119	DY	52119		57119
150 120	DY	52120		57120
150 121	DY	52121		57121
150 122	DY	52122		57122
150 123	DY	52123		57123
150 124	DY	52124		57124
150 125	DY	52125		57125
150 126	DY	52126		57126
150 127	DY	52127		57127
150 128	DY	52128		57128
150 129	DY	52129		57129
150 130	DY	52130		57130
150 131	DY	52131		57131
150 132	DY	52132		57132
150 133	DY	52133		57133
150 134	DY	52134		57134
150 135	DY	52135		57135
150 136	DY	52136		57136
150 137	DY	52137		57137
150 138	DY	52138		57138
150 139	DY	52139		57139
150 140	DY	52140		57140
150 141	TS	52141	52205	57141
150 142	TS	52142	57212	57142
150 143	DY	52143		57143
150 144	TS	52144	57205	57144
150 145	TS	52145	57253	57145
150 146	TS	52146	52253	57146
150 147	DY	52147		57147
150 148	TS	52148	52241	57148
150 149	TS	52149	57241	57149
150 150	DY	52150		57150

CLASS 150/2 SPRINTER BREL 2-CAR DMU

150 201	NH	52201	57201	
150 202	NH	52202	57202	52220
150 203	NH	52203	57203	
150 204	NH	52204	57204	57220
150 206	NH	52206	57206	52222
150 207	NH	52207	57207	
150 208	HA	52208	57208	
150 210	NH	52210	57210	57222
150 211	NH	52211	57211	
150 213	NH	52213	57213	
150 214	NH	52214	57214	52243
150 215	NH	52215	57215	
150 216	NH	52216	57216	57243
150 217	NH	52217	57217	
150 218	NH	52218	57218	57209
150 219	NH	52219	57219	
150 221	NH	52221	57221	
150 223	NH	52223	57223	
150 224	TS	52224	57224	
150 225	NH	52225	57225	
150 226	TS	52226	57226	
150 227	NH	52227	57227	
150 228	HA	52228	57228	
150 229	NH	52229	57229	
150 230	NL	52230	57230	
150 231	NH	52231	57231	
150 232	NL	52232	57232	
150 233	NH	52233	57233	
150 234	NL	52234	57234	
150 235	NH	52235	57235	
150 236	NL	52236	57236	
150 237	NH	52237	57237	
150 238	NL	52238	57238	
150 239	NH	52239	57239	
150 240	NL	52240	57240	
150 242	NL	52242	57242	

Unit	Depot	Vehicles
150 244	HA	52244 57244
150 245	HA	52245 57245
150 246	CF	52246 57246
150 247	CF	52247 57247
150 248	HA	52248 57248
150 249	CF	52249 57249
150 250	HA	52250 57250
150 251	CF	52251 57251
150 252	HA	52252 57252
150 254	CF	52254 57254
150 255	HA	52255 57255
150 256	HA	52256 57256
150 257	HA	52257 57257
150 258	HA	52258 57258
150 259	HA	52259 57259
150 260	HA	52260 57260
150 261	CF	52261 57261
150 262	HA	52262 57262
150 263	CF	52263 57263
150 264	HA	52264 57264
150 265	CF	52265 57265
150 266	CF	52266 57266
150 267	CF	52267 57267
150 268	CF	52268 57268
150 269	CF	52269 57269
150 270	CF	52270 57270
150 271	CF	52271 57271
150 272	CF	52272 57272
150 273	CF	52273 57273
150 274	CF	52274 57274
150 275	CF	52275 57275
150 276	CF	52276 57276
150 277	CF	52277 57277
150 278	CF	52278 57278
150 279	CF	52279 57279
150 280	CF	52280 57280
150 281	CF	52281 57281
150 282	CF	52282 57282
150 283	HA	52283 57283
150 284	HA	52284 57284
150 285	HA	52285 57285

CLASS 151 METRO-CAMMELL 3-CAR PROTOTYPE

Unit	Depot	Vehicles
151 003	DY	55202 55302 55402
151 004	DY	55203 55303 55403

CLASS 154 BREL PROTOTYPE 3-CAR DMU

Unit	Depot	Vehicles
154 002	DY	55201 55301 55401

CLASS 155 SUPER SPRINTER LEYLAND BUS

Unit	Depot	Vehicles
155 301	CF	52301 57301
155 302	CF	52302 57302
155 303	CF	52303 57303
155 304	CF	52304 57304
155 305	CF	52305 57305
155 306	CF	52306 57306
155 307	CF	52307 57307
155 308	CF	52308 57308
155 309	CF	52309 57309
155 310	CF	52310 57310
155 311	CF	52311 57311
155 312	CF	52312 57312
155 313	CF	52313 57313
155 314	CF	52314 57314
155 315	CF	52315 57315
155 316	CF	52316 57316
155 317	CF	52317 57317
155 318	CF	52318 57318
155 319	CF	52319 57319
155 320	CF	52320 57320
155 321	CF	52321 57321
155 322	CF	52322 57322
155 323	CF	52323 57323
155 324	CF	52324 57324
155 325	CF	52325 57325
155 326	CF	52326 57326
155 327	CF	52327 57327
155 328	CF	52328 57328
155 329	CF	52329 57329
155 330	CF	52330 57330
155 331	CF	52331 57331
155 332	CF	52332 57332
155 333	CF	52333 57333
155 334	CF	52334 57334
155 335	CF	52335 57335
155 341	NL	52341 57341
155 342	NL	52342 57342
155 343	NL	52343 57343
155 344	NL	52344 57344
155 345	NL	52345 57345

155 346	NL	52346 57346			155 347	NL	52347 57347	

CLASS 156 SUPER SPRINTER METRO-CAMMELL

156 401	NC	52401	57401	156 449	HA	52449		57449
156 402	NC	52402	57402	156 450	HA	52450		57450
156 403	NC	52403	57403	156 451	NL	52451		57451
156 404	NC	52404	57404	156 452	NC	52452		57452
156 405	NC	52405	57405	156 453	HA	52453		57453
156 406	NC	52406	57406	156 454	NL	52454		57454
156 407	NC	52407	57407	156 455	NL	52455		57455
156 408	NC	52408	57408	156 456	HA	52456		57456
156 409	NC	52409	57409	156 457	MO	52457		57457
156 410	NC	52410	57410	156 458	IS	52458		57458
156 411	NC	52411	57411	156 459	NL	52459		57459
156 412	NC	52412	57412	156 460	NL	52460		57460
156 413	NC	52413	57413	156 461	NC	52461		57461
156 414	NC	52414	57414	156 462	NL	52462		57462
156 415	NC	52415	57415	156 463	NL	52463		57463
156 416	NC	52416	57416	156 464	NC	52464		57464
156 417	NC	52417	57417	156 465	HA	52465		57465
156 418	NC	52418	57418	156 466	NC	52466		57466
156 419	NC	52419	57419	156 467	NL	52467		57467
156 420	NC	52420	57420	156 468	NC	52468		57468
156 421	NC	52421	57421	156 469	NL	52469		57469
156 422	NC	52422	57422	156 470	NC	52470		57470
156 423	NC	52423	57423	156 471	NL	52471		57471
156 424	NC	52424	57424	156 472	NC	52472		57472
156 425	NC	52425	57425	156 473	NC	52473		57473
156 426	NC	52426	57426	156 474	MO	52474		57474
156 427	NC	52427	57427	156 475	NL	52475		57475
156 428	NC	52428	57428	156 476	HA	52476		57476
156 429	NC	52429	57429	156 477	MO	52477		57477
156 430	CK	52430	57430	156 478	MO	52478		57478
156 431	CK	52431	57431	156 479	NL	52479		57479
156 432	CK	52432	57432	156 480	NL	52480		57480
156 433	CK	52433	57433	156 481	NL	52481		57481
156 434	CK	52434	57434	156 482	NL	52482		57482
156 435	CK	52435	57435	156 483	NL	52483		57483
156 436	CK	52436	57436	156 484	NL	52484		57484
156 437	CK	52437	57437	156 485	IS	52485	57485	52496
156 438	NL	52438	57438	156 486	NC	52486		57486
156 439	CK	52439	57439	156 487	NL	52487		57487
156 440	NC	52440	57440	156 488	NL	52488		57488
156 441	NL	52441	57441	156 489	NL	52489		57489
156 442	CK	52442	57442	156 490	NL	52490		57490
156 443	NL	52443	57443	156 491	NL	52491		57491
156 444	NL	52444	57444	156 492	IS	57492	52492	57495
156 445	HA	52445	57445	156 493	IS	57493	52493	57496
156 446	MO	52446	57446	156 494	IS	57494	52494	52495
156 447	HA	52447	57447	156 497	NL	52497		57497
156 448	NL	52448	57448	156 498	NL	52498		52500

156 499	HA	52499	57499	156 507	CK	52507	57507
156 500	HA	52500	57500	156 508	CK	52508	57508
156 501	CK	52501	57501	156 509	CK	52509	57509
156 502	CK	52502	57502	156 510	CK	52510	57510
156 503	CK	52503	57503	156 511	CK	52511	57511
156 504	CK	52504	57504	156 512	CK	52512	57512
156 505	CK	52505	57505	156 513	CK	52513	57513
156 506	CK	52506	57506	156 514	CK	52514	57514

CLASS 158 EXPRESS UNITS BREL

158 701	HA	52701	57701	158 742
158 702				158 743
158 703				158 744
158 704				158 745
158 705				158 746
158 706				158 747
158 707				158 748
158 708				158 749
158 709				158 750
158 710				158 751
158 711				158 752
158 712				158 753
158 713				158 754
158 714				158 755
158 715				158 756
158 716				158 757
158 717				158 758
158 718				158 759
158 719				158 760
158 720				158 761
158 721				158 762
158 722				158 763
158 723				158 764
158 724				158 765
158 725				158 766
158 726				158 767
158 727				158 768
158 728				158 769
158 729				158 770
158 730				158 771
158 731				158 772
158 732				158 773
158 733				158 774
158 734				158 775
158 735				158 776
158 736				158 777
158 737				158 778
158 738				158 779
158 739				158 780
158 740				158 781
158 741				158 782

3.5. DIESEL ELECTRIC MULTIPLE UNITS

All SR diesel-electric multiple unit power cars have above-floor-mounted engines and all vehicles are equipped with buckeye couplings and were built at Eastleigh with frames laid at Ashford.

CLASS 203 4L

DMBSO–2TSOL–DMBSO. Built to former loading gauge of Hastings Line.

Engine: English Electric 4SRKT engines of 370 kW (500 hp).
Transmission: Two EE 507 traction motors on the inner bogie.
Gangways: Within unit only.
Dimensions: 20.24 x 2.50 x 3.87 m.
Maximum Speed: 75 mph.
Non-standard livery: Southern Region green.

DMBSO. Dia. DB202. Lot No. 30395 Ashford/Eastleigh 1957–58. 30S. 55 t.
TSOL. Dia. DH202. Lot No. 30397 Ashford/Eastleigh 1957–58. 60S 2L. 30 t.

60152/3 renumbered from 60014/5 respectively).

203 001 (1011)	**0**	NKCX		SU	60152	60522	60523	60153

Name: 60152 St. Leonards

CLASS 205 3H

DMBSO–TSOL–DTCsoL. Buckeye couplings.

Engine: English Electric 4SRKT engines of 450 kW (600 hp).
Transmission: Two EE 507 traction motors on the inner bogie.
Gangways: Non-gangwayed.
Dimensions: 20.28 x 2.82 x 3.77 m.
Maximum Speed: 75 mph.

60108–117/154/5. DMBSO. Dia DB203. Lot No. 30332 1957. 52S. 56 t.
60120. DMBSO. Dia DB203. Lot No. 30398 1957. 52S. 56 t.
60122–125. DMBSO. Dia DB203. Lot No. 30540 1958–59. 52S. 56 t.
60145–151. DMBSO. Dia DB204. Lot No. 30671 1960–62. 42S. 56 t.
60650–671. TSO. Dia DH203. Lot No. 30542 1958–59. 104S. 30 t.
60672–678. TSO. Dia DH203. Lot No. 30672 1960–62. 104S. 30 t.
60800–817. DTCsoL. Dia DE302 (†DE301). Lot No. 30333 1956–57. 19F 50S 2L
†–One compartment converted to luggage compartment. 13F 50S 2L.
60820. DTCsoL. Dia DE302. Lot No. 30399 1957. 19F 50S 2L. 32 t.
60822–825. DTCsoL. Dia DE302 (†DE301). Lot No. 30541 1958–59. 19F 50S 2L
†–One compartment converted to luggage compartment. 13F 50S 2L.
60826–832. DTCsoL. Dia DE303. Lot No. 30673 1960–62. 13F 62S 2L. 32 t.

60154/5 renumbered from 60100/1 respectively.

205 001 (1101)	**N**	NKCX	†	SU	60154	60650	60800
205 002 (1102)		NKCX	†	SU	60155	60651	60801
205 008 (1108)		NSXX		SU	60120	60657	60820
205 009 (1109)	**N**	NSXX		SU	60108	60658	60808
205 012 (1112)	**N**	NSXX		SU	60111	60661	60811

205 015 (1115)	**N**	NSXX		SU	60114	60664	60814
205 016 (1116)	**N**	NSXX		SU	60115	60665	60815
205 018 (1118)		NSXX		SU	60117	60667	60817
205 023 (1123)		NSXX		SU	60122	60668	60822
205 024 (1124)	**N**	NSXX	†	SU	60123	60669	60823
205 025 (1125)		NSSX	†	EH	60124	60670	60824
205 026 (1126)	**N**	NSSX	†	EH	60125	60671	60825
205 027 (1127)		NSSX		EH	60145	60672	60826
205 028 (1128)		NSSX		EH	60146	60673	60827
205 029 (1129)	**N**	NSSX		EH	60147	60674	60828
205 030 (1130)	**N**	NSSX		EH	60148	60675	60829
205 031 (1131)		NSSX		EH	60149	60676	60830
205 032 (1132)		NSSX		EH	60150	60677	60831
205 033 (1133)		NSSX		EH	60151	60678	60832

CLASS 205/1 — 3H

DMBSO–TSO–DTSOL. Refurbished 1980. Fluorescent lighting. PA.

Engine: English Electric 4SRKT engines of 450 kW (600 hp).
Transmission: Two EE 507 traction motors on the inner bogie.
Gangways: Within unit only.
Dimensions: 20.28 x 2.82 x 3.77 m.
Maximum Speed: 75 mph.

DMBSO. Dia DB203. Lot No. 30332 1957. 39S. 57 t.
TSO. Dia DH203. Lot No. 30542 1959. 98S. 30 t.
DTSOL. Dia DE204. Lot No. 30333 1957. 76S 2L. 32 t.

205 101 (1111)	**N**	NKCX	SU	60110	60660	60810

CLASS 207 — 3D

DMBSO–TCsoL–DTSO. Buckeye couplings.

Engine: English Electric 4SRKT engines of 450 kW (600 hp).
Transmission: Two EE 507 traction motors on the inner bogie.
Gangways: Non-gangwayed.
Dimensions: 20.34 x 2.74 x 3.83 m. (DMBSO), 20.32 x 2.74 x 3.73 m. (DTS), 20.34 x 2.74 x 3.73 m. (TCsoL).
Maximum Speed: 75 mph.

DMBSO. Dia DB205. Lot No. 30625 1962. 42S. 56 t.
TCsoL. Dia DH301. Lot No. 30626 1962. 24F 42S 1L. 31 t.
DTSO. Dia DE201. Lot No. 30627 1962. 76S. 32 t.

207 001 (1301)		NSXX	SU	60126	60600	60900
207 002 (1302)	**N**	NSXX	SU	60127	60601	60901
207 004 (1304)		NSXX	SU	60129	60603	60903
207 005 (1305)		NSSX	EH	60130	60604	60904
207 010 (1310)		NSSX	EH	60135	60609	60909
207 011 (1311)	**N**	NKCX	SU	60139	60610	60910
207 013 (1313)		NSSX	EH	60138	60612	60912
207 017 (1317)		NSSX	EH	60142	60616	60916
Spare		NKCX	SU (U)			60918

3.6. DEPARTMENTAL DMU CARS

DB 975007	(M 79018)	RG	Ultrasonic test train coach.
DB 975008	(M 79612)	RG	Ultrasonic test train coach.
RDB 975010	(M 79900)	RTC	Lab coach 19 (Iris).
TDB 975023	(W 55001)	CH	Route learning car. (L101)
TDB 975025	(S 60755)	SL	SR general manager's saloon.
ADB 975042	(M 55019)	BY	Sandite car.
RDB 975089	(M 50396)	RTC	Rail top trim. Lab coach 5.
RDB 975090	(M 56162)	RTC	Rail top trim. Lab coach 5.
DB 975349	(E 51116)	CA	Inspection saloon.
DB 975539	(E 56101)	CA	Inspection saloon.
TDB 975540	(W 55016)	TS	Sandite car.
DB 975637	(E 56300)	CA	ER general manager's saloon.
TDB 975659	(W 55035)	OO	Route learning car. (L135)
DB 975664	(E 51122)	CA	ER general manager's saloon.
TDB 977123	(E 51286)	CA	Route learning car.
TDB 977124	(E 51296)	CA	Route learning car.
TDB 977125	(E 56444)	CA	Route learning car.
TDB 977126	(E 56445)	CA	Route learning car.
ADB 977376	(S 60002)	SU	Sandite unit 066.
ADB 977377	(S 60003)	SU	Sandite unit 066.
ADB 977379	(S 60504)	SU	Sandite unit 066.
DB 977391	(E 51433)	RG	Ultrasonic test train car.
DB 977392	(E 53167)	RG	MTA Pool. Ultrasonic test car.
DB 977393	(E 53246)	Cathays	Ultrasonic test car. (pending conversion)
TDB 977466	(W 54286)	LA	Sandite vehicle.
TDB 977486	(W 54285)	CF	Sandite vehicle.
TDB 977535	(E 53259)	HT	Sandite vehicle.
TDB 977536	(E 53295)	HT	Sandite vehicle.
TDB 977554	(M 54182)	BX	Sandite vehicle.
TDB 977555	(M 54183)	CH	Sandite vehicle.
TDB 977607	(SC51464)	ED	Route learning and sandite vehicle.
TDB 977608	(SC51525)	ED	Route learning and sandite vehicle.
ADB 977611	(E 51824)	NL	Sandite vehicle.
ADB 977612	(E 51838)	NL	Sandite vehicle.
ADB 977613	(E 51826)	NL	Sandite vehicle.
ADB 977614	(E 52076)	NL	Sandite vehicle.
ADB 977615	(E 54281)	OO	Sandite vehicle.
RDB 977645	(60450)	RTC	Proposed replacement for Lab. 6.
ADB 977649	(60200)		De-icing and Sandite unit
ADB 977650	(60201)		(awaiting conversion).
TDB 977651	(53290)	ScR	Sandite & Route learning Unit.
TDB 977652	(53197)	ScR	Sandite & Route learning Unit.

New build.

RDB 999507		RTC	Wickham Self-propelled laboratory.
DB 999600	DB 999601	RTC	Civil Engineers track recording train.

4. ELECTRIC MULTIPLE UNITS

NUMBERING

BR design electric multiple unit vehicles are numbered in the series 61000–78999. SR vehicles carry numbers in the SR 12000–15999 number series, and there are two units kept for special duties with vehicles numbered in the LNER 6xxxx number series.

Each BR multiple unit vehicle used to carry a regional prefix letter preceeding the number, but these prefixes are now being removed (see section 4). In addition, SR number series vehicles carry the suffix S (also being removed).

GENERAL

This section contains details of all BR electric multiple units. Unless stated otherwise, they are of BR design, and have buckeye couplings and tread brakes. Seating is 3+2 in standard class open vehicles, 2+2 in first class open vehicles, 12 to a non-corridor standard class compartment, 8 to a corridor standard class compartment and 6 to a corridor first class compartment. In express stock, open standards have 2+2 seating and open firsts have 2+1 seating.

ABBREVIATIONS

The following abbreviations are used in this book:

Operator Codes:

Each vehicle is referred to by an operator code as follows:

M Motor
DM Driving Motor
BDM Battery Driving Motor
T Trailer
DT Driving Trailer
BDT Battery Driving Trailer
B Brake, i.e. vehicle with luggage space and guards compartment.

F First
S Standard
C Composite
U Unclassified
RB Buffet Car
H handbrake fitted.
RSM Buffet Standard (Modular)
PMV Parcels and Mails Van
LV Luggage Van
O Open vehicle
K Side corridor with lavatory
so Semi-open vehicle.
L Open or semi-open Vehicle with lavatory
MLS Motor Luggage Standard (an MBS with guards equipment removed)

The letters (A) and (B) may be added to the above codes to differentiate between two cars of the same operating type which have differences between them. Note that a consistent system is used, rather than the official operator codes which are sometimes inconsistent.

Notes:
(1) Compartment Stock (non-corrridor) has no suffix.

(2) Semi-open composites generally have the first class accommodation in compartments and the standard class in open saloons.
(3) Unless stated otherwise, it is assumed that motor vehicles are fitted with pantographs. If the pantograph is on a trailer, then the trailer has the prefix 'P', e.g. PTSO – Pantograph trailer open standard.

Builder Codes:

AEI	Associated Electrical Industries Ltd.
BRCW	The Birmingham Railway Carriage & Wagon Co. Ltd.
BTH	The British Thomson Houston Electrical Co. Ltd.
CP	Crompton-Parkinson Ltd.
EE	The English Electric Co. Ltd.
GEC	The General Electric Company Ltd.
Gloucester	The Gloucester Railway Carriage and Wagon Co. Ltd.
Metro-Cammell	The Metropolitan Cammell Railway Carriage and Wagon Co. Ltd.

Other Abbreviations:

(S)	Stored Serviceable
(U)	Stored Unserviceable
LNER	London & North Eastern Railway
SR	Southern Railway

DIAGRAMS AND DESIGN CODES

See section 3 for an explanation of these. Codes applicable to EMUs are:

1st Letter

This is always 'E' for an electric multiple unit vehicle.

2nd Letter

as follows for various vehicle types:

A Driving motor passenger vehicles.
B Driving motor passenger vehicles with a brake compartment.
C Non-Driving motor passenger vehicles.
D Non-Driving trailer passenger vehicles with a brake compartment.
E Driving Trailer passenger vehicles.
F Battery Driving Trailer passenger vehicles.
G Driving Trailer passenger vehicles with a brake compartment.
H Trailer passenger vehicles.
I Battery Driving Motor passenger vehicles.
J Trailer passenger vehicles with a brake compartment.
N Trailer passenger vehicles with a buffet compartment.
O Battery Driving Trailer passenger vehicles with a brake compartment.
P Trailer passenger vehicles with a handbrake.
X Driving Motor Luggage Vans.
Y Non-driving Motor Luggage Van.

1st Figure

1 First class accommodation.
2 Standard class accommodation (incl. declassified seats).
3 Composite accommodation.
5 No passenger accommodation.

4.1. 25 kV a.c. EMUS

CLASS 302

BDTSOL–MBSO (MBS†)–TCsoL–DTSO (DTS†). (BDTCOL–MBSO–TSOL–DTSO. r).
System: 25 kV/6.25 kV a.c. overhead.
Bogies: Gresley (refurbished units have B4/B5 on trailers).
Gangways: Non-gangwayed [Gangwayed within sets only (r)].
Traction Motors: Four EE536A 143.5 kW.
Dimensions: 19.50 x 2.82 m (outer cars), 19.36 x 2.82 m (inner cars).
Maximum Speed: 75 mph.
Non-Standard Livery: BR Green.

BDTSOL. Dia. EF201. 80S 2L. 36.5 t.
MBSO (MBS†). Dia. ED205 (ED217†). 82S (96S†). 55.0 t.
TCsoL. Dia. EH302. 19F 60S 2L. 31.0 t.
DTSO (DTS†). Dia. EE205 (EE223†). 92S (108S†). 32.5 t.
r–Refurbished with new seats and PA.
BDTCOL. Dia. EF303. 24F 52S 1L. 39.5 t. B5 bogies.
MBSO. Dia. ED216. 76S. 55.3 t.
TSOL. Dia. EH223. 86S 1L. 34.4 t. B4 bogies.
DTSO. Dia. EE219. 88S. 33.4 t. B5 bogies.

75085–75207. BDTSOL/BDTCOL. Lot No. 30436. York/Doncaster 1958–59.
75287–75359. BDTSOL/BDTCOL. Lot No. 30440 York/Doncaster 1959.
61060–61093. MBS/MBSO. Lot No. 30434. York 1958–59.
61098–61226. MBS/MBSO. Lot No. 30438. York 1960.
70060–70093. TCsoL/TSOL. Lot No. 30437. York/Doncaster 1958–59.
70098–70226. TCsoL/TSOL. Lot No. 30441. York 1959–61.
75033–75084. DTS/DTSO. Lot No. 30435. York 1958–59.
75212–75280. DTS/DTSO. Lot No. 30439. York 1959–60.

Note: The unrefurbished sets are in a state of flux, and many cars have recently been withdrawn. A list of spare cars is given after the formations of the known sets.

Set		Pool		Depot				
302 200	(302 207) **0**	NTSX		EM	75091	61066	70066	75039
302 201		NTSX	r	EM	75085	61060	70060	75033
302 202		NTSX	r	EM	75086	61061	70061	75034
302 203	(302 263)	NTSX	r	EM	75311	61122	70122	75236
302 204		NTSX	r	EM	75088	61063	70063	75036
302 205		NTSX	r	EM	75089	61064	70064	75037
302 206		NTSX	r	EM	75090	61065	70065	75038
302 207	(302 310)	NTSX	r	EM	75358	61226	70226	75283
302 208	(302 308)	NTSX	r	EM	75356	61224	70224	75281
302 209		NTSX	r	EM	75093	61068	70068	75041
302 210		NTSX	r	EM	75094	61069	70069	75042
302 211		NTSX	r	EM	75095	61070	70070	75043
302 212		NTSX	r	EM	75096	61071	70071	75044
302 213		NTSX	r	EM	75097	61072	70072	75060
302 214	(302 304)	NTSX	r	EM	75352	61220	70220	75277

302 215			NTSX	r	EM	75099	61074	70074	75062
302 216			NTSX	r	EM	75100	61075	70075	75063
302 217		**N**	NTSX	r	EM	75190	61076	70076	75064
302 218			NTSX	r	EM	75191	61077	70077	75065
302 219			NTSX	r	EM	75192	61078	70078	75066
302 220			NTSX	r	EM	75193	61079	70079	75067
302 221			NTSX	r	EM	75194	61080	70080	75068
302 222			NTSX	r	EM	75195	61081	70081	75069
302 223	(302 293)		NTSX	r	EM	75341	61209	70209	75266
302 224			NTSX	r	EM	75197	61083	70083	75071
302 225			NTSX	r	EM	75198	61084	70084	75072
302 226			NTSX	r	EM	75199	61085	70085	75073
302 227	(302 277)		NTSX	r	EM	75325	61193	70193	75250
302 228			NTSX	r	EM	75201	61087	70087	75075
302 229			NTSX	r	EM	75202	61088	70088	75076
302 230	(302 232)		NTSX	r	EM	75205	61091	70091	75079
302 234			NTSX		EM	75207	61093	70120	75081
302 239			NTSX		EM	75287	61098	70125	75272
302 250	(302 300)		NTSX		EM	75348	61216	70216	75273
302 251			NTSX		EM	75299	61110	70110	75224
302 252			NTSX		EM	75300	61111	70111	75225
302 253	(302 301)		NTSX		EM	75349	61217	70217	75274
302 255			NTSX		EM	75092	61219	70219	75276
302 259			NTSX		EM	75329	61118	70198	75232
302 261			NTSX		EM	75309	61120	70073	75234
302 262			NTSX		EM	75310	61121	70131	75235
302 267	(302 307)		NTSX		EM	75355	61223†	70223	75280†
302 272			NTSX		EM	75359	61131	70086	75245
302 273	(302 306)		NTSX		EM	75354	61222	70202	75279
302 280			NGEX		IL	75328	61196	70196	75253
302 283			NGEX		IL	75331	61199	70130	75256
302 284			NTSX		EM	75332	61200	70200	75257
302 285			NGEX		IL	75298	61201	70103	75258
302 286			NTSX		EM	75322	61202†	70190	75259†
302 292			NTSX		EM	75340	61208	70208	75265
302 296			NGEX		IL	75344	61116†	70116	75269
302 298			NTSX		EM	75346	61214	70214	75271

Spare vehicles:

61082	70108	70222	75212	75227	75247	75326†	75337†
61113	70117	75040	75216†	75229	75261	75334	75347
61190	70123	75074	75220†	75237	75312†	75336	75353
61221	70197	75078†	75222†				

Class 302/9. Parcels Units. BDTPMV–MPMV–DTPMV.

68100. BDTPMV. Dia. EF501. Lot No. 30435. York 1959.
68101–2. BDTPMV. Dia. EE501. Lot No. 30439. York 1959.
68020–1. MPMV. Dia. ED501. Lot No. 30434. York 1959.
68022. MPMV. Dia. ED501. Lot No. 30438. York 1960.
68207. DTPMV. Dia. EE501. Lot No. 30435. York 1959.
68208–9. DTPMV. Dia. EE501. Lot No. 30439. York 1959.

302 990	RPMA	IL	68100	68020	68207

302 991	RPMA	IL	68101	68021	68208
302 992	RPMA	IL	68102	68022	68209

Former numbers of converted vehicles:

68020 (61090)	68100 (75084)	68102 (75218)	68208 (75217)
68021 (61067)	68101 (75221)	68207 (75082)	68209 (75219)
68022 (61227)			

CLASS 303

DTSO–MBSO–BDTSO. Sliding doors.
System: 25 kV a.c. overhead.
Bogies: Gresley.
Gangways: Non-gangwayed [Gangwayed within sets only (r)].
Traction Motors: Four MV 155 kW.
Dimensions: 19.50 x 2.82 m (outer cars), 19.36 x 2.82 m (inner cars).
Maximum Speed: 75 mph.

DTSO. Dia. EE206. 83S. 34.4 t.
MBSO. Dia. ED201. 70S. 56.4 t.
BDTSO. Dia. EF202. 83S. 38.4 t.

r–Refurbished with 2+2 seating (some with hopper-type window vents).
DTSO. Dia. EE241. 56S. 34.4 t.
MBSO. Dia. ED220. 48S. 56.4 t.
BDTSO. Dia. EF217. 56S. 38.4 t.

75566–75599. **DTSO**. Lot No. 30579 Pressed Steel 1959–60.
75747–75801. **DTSO**. Lot No. 30629 Pressed Steel 1960–61.
61481–61514. **MBSO**. Lot No. 30580 Pressed Steel 1959–60.
61813–61867. **MBSO**. Lot No. 30630 Pressed Steel 1960–61.
75601–75634. **BDTSO**. Lot No. 30581 Pressed Steel 1959–60.
75803–75857. **BDTSO**. Lot No. 30631 Pressed Steel 1960–61.

303 001	**S**	PGWX	r	GW	75566	61481	75601
303 003	**S**	PGWX	r	GW	75568	61483	75603
303 004	**S**	PGWX	r	GW	75569	61484	75604
303 006	**S**	PGWX	r	GW	75571	61486	75606
303 008	**S**	PGWX	r	GW	75573	61488	75608
303 009	**S**	PGWX	r	GW	75574	61489	75609
303 010	**S**	PGWX	r	GW	75575	61490	75610
303 011	**S**	PGWX	r	GW	75576	61491	75611
303 012	**S**	PGWX	r	GW	75577	61492	75612
303 013	**S**	PGWX	r	GW	75578	61493	75613
303 014	**S**	PGWX	r	GW	75579	61494	75614
303 016	**S**	PGWX	r	GW	75581	61496	75616
303 019	**S**	PGWX	r	GW	75584	61499	75619
303 020	**S**	PGWX	r	GW	75585	61500	75620
303 021	**S**	PGWX	r	GW	75586	61501	75621
303 023	**S**	PGWX	r	GW	75588	61503	75623
303 024	**S**	PGWX	r	GW	75589	61504	75624
303 025	**S**	PGWX	r	GW	75590	61505	75625
303 027	**S**	PGWX	r	GW	75592	61507	75627
303 028	**S**	PGWX	r	GW	75600	61508	75635

303 029	**S**	PGWX		GW	75594	61509	75629
303 032	**S**	PGWX	r	GW	75597	61512	75632
303 033	**S**	PGWX	r	GW	75595	61860	75817
303 034	**S**	PGWX	r	GW	75599	61514	75634
303 037	**S**	PGWX	r	GW	75747	61813	75803
303 038	**S**	PGWX	r	GW	75748	61814	75804
303 039	**S**	PGWX		GW	75749	61815	75805
303 040	**S**	PGWX	r	GW	75750	61816	75806
303 043	**S**	PGWX	r	GW	75572	61819	75809
303 045	**S**	PGWX	r	GW	75755	61821	75811
303 046	**S**	PGWX	r	GW	75756	61822	75812
303 047	**S**	PGWX	r	GW	75757	61823	75813
303 048	**S**	PGWX		GW	75752	61824	75808
303 049	**G**	PLGX		LG	75759	61825	75815
303 052		PGWX		GW	75762	61828	75818
303 054	**S**	PGWX	r	GW	75764	61830	75820
303 055	**S**	PGWX	r	GW	75765	61831	75821
303 056	**S**	PGWX	r	GW	75766	61832	75822
303 058	**S**	PGWX	r	GW	75768	61834	75824
303 060	**G**	PLGX		LG	75770	61836	75826
303 061	**S**	PGWX	r	GW	75771	61837	75827
303 062		PGWX		GW	75772	61838	75828
303 063	**S**	PGWX		GW	75773	61839	75829
303 065	**S**	PGWX	r	GW	75775	61841	75831
303 066	**G**	PLGX		LG	75776	61842	75832
303 069	**S**	PGWX		GW	75779	61845	75607
303 070	**S**	PGWX	r	GW	75780	61846	75836
303 072	**S**	PGWX		GW	75782	61848	75838
303 073	**S**	PGWX		GW	75783	61849	75839
303 075	**S**	PGWX		GW	75785	61851	75841
303 076	**S**	PGWX		GW	75786	61852	75842
303 077	**S**	PGWX	r	GW	75787	61853	75843
303 079	**S**	PGWX	r	GW	75789	61855	75845
303 080	**S**	PGWX	r	GW	75790	61856	75846
303 082	**G**	PLGX		LG	75792	61858	75848
303 083	**S**	PGWX	r	GW	75793	61859	75849
303 085	**S**	PGWX	r	GW	75795	61861	75851
303 087	**S**	PGWX	r	GW	75797	61863	75853
303 088	**S**	PGWX	r	GW	75798	61864	75854
303 089	**S**	PGWX	r	GW	75799	61865	75855
303 090	**S**	PGWX	r	GW	75800	61866	75856
303 091	**S**	PGWX	r	GW	75801	61867	75857

CLASS 304

BDTSOL–MBSO–DTBSO. Originally 4 cars.
System: 25 kV a.c. overhead.
Bogies: Gresley.
Gangways: Non-gangwayed.
Traction Motors: Four BTH 155 kW.
Maximum Speed: 75 mph.
Dimensions: 19.53 x 2.82 m (outer cars), 19.36 x 2.82 m (inner cars).

Class 304/1. These cars have pairs of narrow windows instead of wide windows and the MBSOs were formerly MBS and were refurbished with new seats etc.
BDTSOL. Dia. EF203. Lot No. 30429 Wolverton 1960. 80S 2L. 36.8 t.
MBSO. Dia. ED215. Lot No. 30428 Wolverton 1960. 72S. 54.5 t.
DTBSO. Dia. EG202. Lot No. 30430 Wolverton 1960. 82S. 32.5 t.

304 001	PLGX	LG	75045	61045	75645
304 002	PLGX	LG	75046	61046	75646
304 003	PLGX	LG	75047	61047	75647
304 004	PLGX	LG	75048	61048	75648
304 005	PLGX	LG	75049	61049	75649
304 006	PLGX	LG	75050	61050	75650
304 007	PLGX	LG	75051	61051	75651
304 008	PLGX	LG	75052	61052	75652
304 009	PLGX	LG	75053	61053	75653
304 010	PLGX	LG	75054	61054	75654
304 011	PLGX	LG	75055	61055	75655
304 012	PLGX	LG	75056	61056	75656
304 013	PLGX	LG	75057	61057	75657
304 014	PLGX	LG	75058	61058	75658
304 015	PLGX	LG	75059	61059	75659

Class 304/2. Standard design with wide windows.
75680–75699. BDTSOL. Dia. EF204. Lot No. 30610 Wolverton 1960–61. 80S 2L. 36.8 t.
75868–75877. BDTSOL. Dia. EF204. Lot No. 30645 Wolverton 1961. 80S 2L. 36.8 t.
61628–61647. MBSO. Dia. ED203. Lot No. 30607 Wolverton 1960–61. 72S. 54.5 t.
61873–61882. MBSO. Dia. ED203. Lot No. 30642 Wolverton 1961. 72S. 54.5 t.
75660–75679. DTBSO. Dia. EG203. Lot No. 30608 Wolverton 1960–61. 82S. 32.5 t.
75858–75867. DTBSO. Dia. EG203. Lot No. 30643 Wolverton 1961. 82S. 32.5 t.

304 016	PLGX	LG	75680	61628	75660
304 017	PLGX	LG	75681	61629	75661
304 019	PLGX	LG	75683	61631	75663
304 020	PLGX	LG	75684	61632	75664
304 021	PLGX	LG	75685	61633	75665
304 023	PLGX	LG	75687	61635	75667
304 024	PLGX	LG	75688	61636	75668
304 027	PLGX	LG	75691	61639	75671
304 028	PLGX	LG	75692	61640	75672
304 029	PLGX	LG	75693	61641	75673
304 030	PLGX	LG	75694	61642	75674
304 032	PLGX	LG	75696	61644	75676
304 033	PLGX	LG	75697	61645	75677
304 034	PLGX	LG	75698	61646	75678
304 035	PLGX	LG	75699	61647	75679
304 036	PLGX	LG	75868	61873	75858
304 037	PLGX	LG	75869	61874	75859

304 039	PLGX	LG	75871	61876	75861
304 040	PLGX	LG	75872	61877	75862
304 041	PLGX	LG	75873	61878	75863
304 042	PLGX	LG	75874	61879	75864
304 043	PLGX	LG	75875	61880	75865
304 045	PLGX	LG	75877	61882	75867

CLASS 305/1

BDTSO–MBSO–DTSO. All facelifted with fluorescent lighting and PA.
System: 25 kV/6.25 kV a.c. overhead.
Bogies: Gresley.
Gangways: Non-gangwayed.
Traction Motors: Four GEC WT380 of 153 kW.
Dimensions: 19.53 x 2.82 m (outer cars), 19.34 x 2.82 m (inner cars).
Maximum Speed: 75 mph.

BDTSO. Dia. EF205. Lot No. 30570 York 1960. 34.9 t. 92S.
MBSO. Dia. ED204. Lot No. 30571 York 1960. 56.4 t. 84S.
DTSO. Dia. EE209. Lot No. 30572 York 1960. 31.5 t. 92S.

* Class 302 vehicle.

305 401		NNEX	IL	75462	61429	75514
305 402		NNEX	IL	75463	61430	75515
305 403	(305 445)	NNEX	IL	75506	61473	75558
305 404	(305 447)	NNEX	IL	75508	61475	75560
305 405		NNEX	IL	75466	61433	75518
305 406		NNEX	IL	75467	61434	75519
305 407		NNEX	IL	75468	61435	75520
305 408		NNEX	IL	75469	61436	75521
305 409		NNEX	IL	75470	61437	75522
305 410		NNEX	IL	75471	61438	75523
305 411	(305 448)	NNEX	IL	75509	61476	75561
305 412		NNEX	IL	75473	61440	75525
305 413		NNEX	IL	75474	61441	75526
305 414	(305 449)	NNEX	IL	75289*	61477	75562
305 415		NNEX	IL	75476	61443	75528
305 416		NNEX	IL	75477	61444	75529
305 417		NNEX	IL	75478	61445	75530
305 418		NNEX	IL	75479	61446	75531
305 419	(305 450)	NNEX	IL	75511	61478	75563
305 420		NNEX	IL	75481	61448	75533
305 421		NNEX	IL	75482	61449	75534
305 422	(305 452)	NNEX	IL	75513	61480	75565
Spare		NNEX	IL (S)	75510		

CLASS 305/2

BDTCOL–MBSO–TSOL–DTSO. All facelifted with fluorescent lighting, new seats and PA.
System: 25 kV/6.25 kV a.c. overhead.
Bogies: Gresley.

Gangways: Non-gangwayed.
Traction Motors: Four GEC WT380 of 153 kW.
Dimensions: 19.53 x 2.82 m (outer cars), 19.36 x 2.82 m (inner cars).
Maximum Speed: 75 mph.

BDTCOL. Dia. EF304. Lot No. 30566 York/Doncaster 1960.24F 52S 1L. 36.5 t.
MBSO. Dia. ED216. Lot No. 30567 York/Doncaster 1960. 76S. 56.5 t.
TSOL. Dia. EH223. Lot No. 30568 York/Doncaster 1960. 86S 1L. 31.5 t.
DTSO. Dia. EE220. Lot No. 30569 York/Doncaster 1960. 88S. 32.7 t.

305 502	**N**	NTSX	EM	75425	61421	70357	75444
305 503	**N**	NTSX	EM	75426	61412	70358	75445
305 506	**N**	NTSX	EM	75429	61415	70361	75448
305 508	**N**	NTSX	EM	75431	61417	70363	75450
305 511	**N**	NTSX	EM	75434	61420	70366	75453
305 513	**N**	NTSX	EM	75436	61422	70368	75455
305 515	**N**	NTSX	EM	75438	61424	70370	75457
305 516	**N**	NTSX	EM	75439	61425	70371	75458
305 517	**N**	NTSX	EM	75440	61426	70372	75459
305 518		NTSX	EM	75441	61427	70373	75460

CLASS 305/3

BDTSO–MBSO–TCsoL–DTSO. New batch formed by strengthening class 305/1 with Class 302 TCsoL. All facelifted with fluorescent lighting, new seats and PA.
System: 25 kV/6.25 kV a.c. overhead.
Bogies: Gresley.
Gangways: Non-gangwayed.
Traction Motors: Four GEC WT380 of 153 kW.
Dimensions: 19.53 x 2.82 m (outer cars), 19.36 x 2.82 m (inner cars).
Maximum Speed: 75 mph.

BDTSO. Dia. EF205. Lot No. 30570 York 1960. 34.9 t. 92S.
MBSO. Dia. ED204. Lot No. 30571 York 1960. 56.4 t. 84S.
70093. TCsoL. Lot No. 30437. York/Doncaster 1958–59.
70098–70221. TCsoL. Lot No. 30441. York 1959–61.
DTSO. Dia. EE209. Lot No. 30572 York 1960. 31.5 t. 92S.

305 521	(305 403)	NTSX	EM	75464	61431	70098	75516
305 522	(305 404)	NTSX	EM	75465	61432	70099	75517
305 523	(305 411)	NTSX	EM	75472	61439	70221	75524
305 524	(305 414)	NTSX	EM	75475	61442	70118	75527
305 525	(305 419)	NTSX	EM	75480	61447	70213	75532
305 526	(305 422)	NTSX	EM	75483	61450	70093	75535
305 527	(305 446)	NTSX	EM	75507	61474	70204	75559
305 528	(305 451)	NTSX	EM	75512	61479	70210	75564

CLASS 306

DMSO–TBSO–DTSO. Converted 1960–1 from 1500 V d.c. Kept for special workings. LNER design. Sliding doors. Screw couplings.
System: 25 kV/6.25 kV a.c. overhead.
Bogies: LNER bogies.

Gangways: Non-gangwayed.
Traction Motors: Four Crompton Parkinson of 155 kW.
Dimensions: 18.41 x 2.90 m (DMSO), 16.78 x 2.90 m (TBSO), 16.87 x 2.90 m (DTSO).
Maximum Speed: 65 mph.
Non-standard Livery: Original BR multiple unit green.

DMSO. Dia. EA217. Lot No. 363 Metro-Cammell. 1949. 51.7 t. 62S.
TBSO. Dia. EJ201. Lot No. 364 1949. Metro-Cammell. 26.4 t. 46S.
DTSO. Dia. EE211. Lot No. 365 1949. BRCW. 27.9 t. 60S.

306 017 **0**	NGEX	IL	65217	65417	65617

CLASS 307

BDTBSO–MSO–TSO–DTCOL. 25 kV a.c. overhead. Converted 1960–1 from 1500 V d.c. All refurbished with new seats, fluorescent lighting, PA.
System: 25 kV/6.25 kV a.c. overhead.
Bogies: Gresley (MSO), B4 (TSO and DTCO) B5 (BDTBSO).
Gangways: Originally non-gangwayed, but now gangwayed within unit.
Traction Motors: Four GEC WT344 of 130 kW.
Dimensions: 19.50 x 2.83 m.
Maximum Speed: 75 mph.

BDTBSO. Dia. EO202. Lot No. 30205 Ashford/Eastleigh 1954–6. 66S. 43 t.
MSO. Dia. EC204. Lot No. 30203 Ashford/Eastleigh 1954–6. 86S. 47.5 t.
TSOL. Dia. EH222. Lot No. 30204 Ashford/Eastleigh 1954–6. 86S 1L. 31 t.
DTCOL. Dia. EE307. Lot No. 30206 Afd./Elh 1954–6. 24F 52S 1L. 33 t.

Subject to withdrawal during 1990.

307 101	NGEX	IL (S)	75001	61001	70001	75101
307 102 **N**	NGEX	IL	75002	61002	70002	75102
307 103	NGEX	IL	75003	61003	70003	75103
307 104	NGEX	IL	75004	61004	70004	75104
307 105 **N**	NGEX	IL	75005	61005	70005	75105
307 106	NGEX	IL (S)	75006	61006	70006	75106
307 107	NGEX	IL	75007	61007	70007	75107
307 108	NGEX	IL	75008	61008	70008	75108
307 109	NGEX	IL	75009	61009	70009	75109
307 110	NGEX	IL	75010	61010	70010	75110
307 111 **N**	NGEX	IL	75011	61011	70011	75111
307 112	NGEX	IL	75012	61012	70012	75112
307 113	NGEX	IL	75013	61013	70013	75113
307 114	NGEX	IL	75014	61014	70014	75114
307 115	NGEX	IL	75015	61015	70015	75115
307 116	NGEX	IL	75016	61016	70016	75116
307 117 **N**	NGEX	IL	75017	61017	70017	75117
307 118	NGEX	IL	75018	61018	70018	75118
307 119	NGEX	IL	75019	61019	70019	75119
307 120 **N**	NGEX	IL	75020	61020	70020	75120
307 121	NGEX	IL	75021	61021	70021	75121
307 122 **N**	NGEX	IL	75022	61022	70022	75122
307 123	NGEX	IL	75023	61023	70023	75123

307 124	NGEX	IL	75024	61024	70024	75124
307 125 **N**	NGEX	IL	75025	61025	70025	75125
307 126 **N**	NGEX	IL	75026	61026	70026	75126
307 127	NGEX	IL	75027	61027	70027	75127
307 128	NGEX	IL	75028	61028	70028	75128
307 129	NGEX	IL	75029	61029	70029	75129
307 130 **N**	NGEX	IL	75030	61030	70030	75130
307 131	NGEX	IL	75031	61031	70031	75131
307 132 **N**	NGEX	IL	75032	61032	70032	75132

CLASS 308/1

BDTCOL–MBSO–TSOL–DTSO.
System: 25 kV a.c. overhead.
Bogies: Gresley.
Gangways: Originally non-gangwayed, but now gangwayed within unit.
Traction Motors: Four English Electric 536A of 143.5 kW.
Dimensions: 19.36 x 2.82 m (outer cars), 19.35 x 2.82 m (inner cars).
Maximum Speed: 75 mph.

75878–75886. BDTCOL. Dia. EF304. Lot No. 30652 Yk 1961. 24F 52S 1L. 36.3 t.
75896–75919. BDTCOL. Dia. EF304. Lot No. 30656 Yk 1961. 24F 52S 1L. 36.3 t.
61883–61891. MBSO. Dia. ED216. Lot No. 30653 York 1961. 76S. 55.0 t.
61892–61915. MBSO. Dia. ED216. Lot No. 30657 York 1961. 76S. 55.0 t.
70367. TSOL. Dia. EH223. Lot No. 30568 York/Doncaster 1960. 86S 1L. 31.4 t.
70611–70619. TSOL. Dia. EH223. Lot No. 30654 York 1961. 86S 1L. 31.4 t.
70620–70643. TSOL. Dia. EH223. Lot No. 30658 York 1961. 86S 1L. 31.4 t.
75887–75895. DTSO. Dia. EE220. Lot No. 30655 York 1961. 88S. 33 t.
75929–75952. DTSO. Dia. EE220. Lot No. 30659 York 1961. 88S. 33 t.

308 133	NTSX	EM	75878	61883	70611	75887
308 134	NTSX	EM	75879	61884	70612	75888
308 135 **N**	NTSX	EM	75880	61885	70613	75889
308 136 **N**	NTSX	EM	75881	61886	70614	75890
308 137	NTSX	EM	75882	61887	70615	75891
308 138	NTSX	EM	75883	61888	70367	75892
308 139	NTSX	EM	75884	61889	70617	75893
308 140 **N**	NTSX	EM	75885	61890	70618	75894
308 141	NTSX	EM	75886	61891	70619	75895
308 142	NTSX	EM	75896	61892	70620	75929
308 143	NTSX	EM	75897	61893	70621	75930
308 144	NTSX	EM	75898	61894	70622	75931
308 145	NTSX	EM	75899	61895	70623	75932
308 146	NTSX	EM	75900	61896	70624	75933
308 147	NTSX	EM	75901	61897	70625	75934
308 148	NTSX	EM	75902	61898	70626	75935
308 149 **N**	NTSX	EM	75903	61899	70627	75936
308 150	NTSX	EM	75904	61900	70628	75937
308 151	NTSX	EM	75905	61901	70629	75938
308 152 **N**	NTSX	EM	75906	61902	70630	75939
308 153	NTSX	EM	75907	61903	70631	75940
308 154	NTSX	EM	75908	61904	70632	75941
308 155	NTSX	EM	75909	61905	70633	75942

308 156 **N**	NTSX	EM	75910	61906	70634	75943
308 157	NTSX	EM	75911	61907	70635	75944
308 158	NTSX	EM	75912	61908	70636	75945
308 159	NTSX	EM	75913	61909	70637	75946
308 160	NTSX	EM	75914	61910	70638	75947
308 161	NTSX	EM	75915	61911	70639	75948
308 162	NTSX	EM	75916	61912	70640	75949
308 163	NTSX	EM	75917	61913	70641	75950
308 164 **N**	NTSX	EM	75918	61914	70642	75951
308 165	NTSX	EM	75919	61915	70643	75952

CLASS 309/1 ESSEX EXPRESS STOCK

DMBSO(T)–TSOL–TCsoL–BDTSOL. Built 1962–3 as 2 car units. Made up to four cars by the conversion of loco-hauled stock in 1973/80–1. All now refurbished with fluorescent lighting, hopper ventilators, new seating, PA.
System: 25 kV a.c. overhead.
Bogies: Commonwealth.
Gangways: Throughout.
Traction Motors: Four GEC of 210 kW.
Dimensions: 19.76 x 2.82 m (outer cars), 19.67 x 2.82 m (inner cars).
Maximum Speed: 100 mph.

DMBSO(T). Dia. EB206. Lot No. 30684 York 1962–63. 44S. 60 t.
71107–71110. TSOL. Dia. EH227. Lot No. 30871 W'ton 1973–74. 64S 2L. 35 t.
71569–71572. TSOL. Dia. EH227. Lot No. 30954 W'ton 1978–81. 64S 2L. 35 t.
71111–71114. TCsoL. Dia. EH309. Lot No. 30872 Wolverton 1973–74. 24F 28S 1L. 36 t.
71573–71576. TCsoL. Dia. EH309. Lot No. 30954 Wolverton 1978–81. 24F 28S 1L. 36 t.
BDTSOL. Dia. EF213. Lot No. 30683 York 1960–62. 60S 1L. 40 t.

Note: 61940/2 are DMBSO and seat 52S.

308 993	RPXX	IL	75926	68012	75960	
308 994	RPXX	IL	75927	68017	75959	
308 995	RPXX	IL	75347	68018	75961	
309 601 **N**	NGEX	CC	61940	71569	71573	75984
309 602 **N**	NGEX	CC	61941	71570	71574	75985
309 603 **N**	NGEX	CC	61942	71571	71575	75986
309 604 **N**	NGEX	CC	61943	71572	71576	75987
309 605 **N**	NGEX	CC	61944	71108	71113	75988
309 606 **N**	NGEX	CC	61945	71109	71112	75989
309 607 **N**	NGEX	CC	61946	71107	71111	75990
309 608 **N**	NGEX	CC	61947	71110	71114	75991

CLASS 309/2 ESSEX EXPRESS STOCK

BDTCsoL–MBSOL(T)–TSO–DTSOL. Built 1962–3 as 2 car units. Units 309 611–309 618 formerly contained griddle cars, but these were withdrawn and their place has been taken by the conversion of loco-hauled TSOs on refurbishment. All refurbished with fluorescent lighting, hopper ventilators, new seating, PA.

System: 25 kV a.c. overhead.
Bogies: Commonwealth.
Gangways: Throughout.
Traction Motors: Four GEC of 210 kW.
Dimensions: 19.76 x 2.82 m (outer cars), 19.67 x 2.82 m (inner cars).
Maximum Speed: 100 mph.

75637–44. BDTCsoL. Dia. EF301. Lot No. 30679 York 1962. 18F 32S 2L. 40 t.
75962–68. BDTCsoL. Dia. EF213. Lot No. 30675 York 1962. 18F 32S 2L. 40 t.
61925–31. MBSO(T). Dia. ED209. Lot No. 30676 York 1962. 44S 2L. 58 t.
61932–39. MBSO(T). Dia. ED209. Lot No. 30680 York 1962. 44S 2L. 58 t.
70253–59. TSO. Dia. EH229. Lot No. 30677 York 1962. 68S 35 t.
71754–61. TSO. Dia. EH228. Lot No. 31001 Wolverton 1984–87. 68S. 35 t.
75969 75. DTSOL. Dia. EF213. Lot No. 30678 York 1962. 56S 2L 37 t.
75976–83. DTSOL. Dia. EF213. Lot No. 30682 York 1962–1963. 56S 2L. 37 t.

61933/29/30/31 are MBSO and seat 48S 2L.
p MBSO(T) is fitted with public telephone. 42S 2L.

309 611	**N**	NGEX		CC	75637	61932	71754	75970
309 612	**N**	NGEX		CC	75638	61933	71755	75977
309 613	**N**	NGEX	p	CC	75639	61934	71756	75978
309 614	**N**	NGEX		CC	75640	61935	71757	75979
309 615	**N**	NGEX		CC	75641	61936	71758	75980
309 616	**N**	NGEX	p	CC	75642	61937	71759	75981
309 617	**N**	NGEX	p	CC	75643	61938	71760	75982
309 618	**N**	NGEX	p	CC	75966	61939	71761	75983
309 621	**N**	NGEX		CC	75962	61925	70253	75969
309 622	**N**	NGEX		CC	75963	61926	70254	75970
309 623	**N**	NGEX		CC	75964	61927	70255	75971
309 624	**N**	NGEX		CC	75965	61928	70256	75972
309 625	**N**	NGEX		CC (U)	75644	61929	70257	75973
309 626	**N**	NGEX		CC	75967	61930	70258	75974
309 627	**N**	NGEX		CC	75968	61931	70259	75975

Former numbers of converted hauled stock:

71107 (26203)	71113 (16244)	71573 (16264)	71756 (5068)
71108 (26189)	71114 (16252)	71574 (16257)	71757 (5069)
71109 (26196)	71569 (5047)	71575 (16242)	71758 (5058)
71110 (26204)	71570 (5050)	71576 (16259)	71759 (5062)
71111 (16246)	71571 (5059)	71754 (5065)	71760 (5056)
71112 (16249)	71572 (5061)	71755 (5051)	71761 (5066)

CLASS 310

Disc brakes. All facelifted. with new panels and PA.
System: 25 kV/6.25 kV a.c. overhead.
Bogies: B4.
Gangways: Gangwayed within unit.
Traction Motors: Four EE546 of 201.5 kW.

Dimensions: 19.86 x 2.82 m (outer cars), 19.93 x 2.82 m (inner cars).
Maximum Speed: 75 mph.

BDTSOL. Dia. EF211. Lot No. 30745 Derby 1965–67. 80S 2L. 37.3 t.
76228. BDTSOL. Formerly a DTCOL to Lot 39748. Dia. EF210. Seats 68S 2L.
76998. BDTSOL. Rebuilt from TSO 70756 to Lot 30747. Dia. EF214. Seats 75S 2L.
MBSO. Dia. ED219. Lot No. 30746 Derby 1965–67. 68S. 57.2 t.
TSO. Dia. EH232. Lot No. 30747 Derby 1965–67. 98S. 31.7 t.
DTCOL (310/0). Dia. EE306. Lot No. 30748 Derby 1965–67. 25F 43S 2L. 34.4 t.
DTSOL (310/1). Dia. EE237. Lot No. 30748 Derby 1965–67. 75S 2L. 34.4 t.

Class 310/0. BDTSOL–MBSO–TSO–DTCOL.

310 046	**N**	NNEX	EM	76130	62071	70731	76180
310 047	**N**	NGEX	EM	76131	62072	70732	76181
310 049	**N**	NGEX	EM	76133	62074	70734	76183
310 050	**N**	NGEX	EM	76134	62075	70735	76184
310 051	**N**	NNLX	BY	76135	62076	70736	76185
310 052	**N**	NNEX	EM	76136	62077	70737	76186
310 056	**N**	NGEX	EM	76140	62081	70741	76190
310 057	**N**	NGEX	EM	76141	62082	70742	76191
310 058	**N**	NGEX	EM	76142	62083	70743	76192
310 059		NGEX	EM	76143	62084	70744	76205
310 060		NGEX	EM	76144	62085	70745	76194
310 061	**N**	NGEX	EM	76145	62086	70746	76195
310 064	**N**	NNEX	EM	76148	62089	70749	76198
310 065	**N**	NNEX	EM	76149	62090	70750	76199
310 066	**N**	NNEX	EM	76228	62091	70751	76200
310 067	**N**	NNEX	EM	76151	62092	70752	76201
310 068	**N**	NGEX	EM	76152	62093	70753	76202
310 069	**N**	NGEX	EM	76153	62094	70754	76203
310 070	**N**	NGEX	EM	76154	62095	70755	76204
310 074		NGEX	EM	76158	62099	70759	76208
310 075	**N**	NNEX	EM	76159	62100	70760	76209
310 077	**N**	NNEX	EM	76161	62102	70762	76211
310 079	**N**	NNEX	EM	76163	62104	70764	76213
310 080	**N**	NNEX	EM	76164	62105	70765	76214
310 081		NGEX	EM	76165	62106	70766	76215
310 082	**N**	NNEX	EM	76166	62107	70767	76216
310 083	**N**	NGEX	EM	76167	62108	70768	76217
310 084	**N**	NNEX	EM	76168	62109	70769	76218
310 085	**N**	NNEX	EM	76169	62110	70770	76219
310 086		NNLX	BY	76170	62111	70771	76220
310 087	**N**	NNEX	EM	76171	62112	70772	76221
310 088	**N**	NNEX	EM	76172	62113	70773	76222
310 089	**N**	NNEX	EM	76173	62114	70774	76223
310 091	**N**	NNEX	CC	76175	62116	70776	76225
310 092	**N**	NGEX	EM	76176	62117	70777	76226
310 093	**N**	NGEX	CC	76177	62118	70778	76227
310 094	**N**	NNEX	CC	76998	62119	70780	76193
310 095	**N**	NNEX	CC	76179	62120	70779	76229

Class 310/1. BDTSOL–MBSO–TSO–DTSOL.

310 101	(310 073) **PM**	PBYX	BY	76157	62098	70758	76207
310 102	(310 055) **PM**	PBYX	BY	76139	62080	70740	76189
310 103	(310 076) **PM**	PBYX	BY	76160	62101	70761	76210
310 104	(310 078) **PM**	PBYX	BY	76162	62103	70763	76212
310 105	(310 090) **PM**	PBYX	BY	76174	62115	70775	76224
310 106	(310 072) **PM**	PBYX	BY	76156	62097	70757	76206
310 107	(310 062) **PM**	PBYX	BY	76146	62087	70747	76196
310 108	(310 048) **PM**	PBYX	BY	76132	62073	70733	76182
310 109	(310 053) **PM**	PBYX	BY	76137	62078	70738	76187
310 110	(310 054) **PM**	PBYX	BY	76138	62079	70739	76188
310 111	(310 063) **PM**	PBYX	BY	76147	62088	70748	76197

CLASS 311

DTSO–MBSO–BDTSO. Sliding doors.
System: 25 kV a.c. overhead.
Bogies: Gresley.
Gangways: Non-gangwayed.
Traction Motors: Four AEI 165 kW.
Dimensions: 19.50 x 2.82 m (outer cars), 19.36 x 2.82 m (inner cars).
Maximum Speed: 75 mph.

DTSO. Dia. EE214. Lot No. 30767 Cravens 1967. 83S. 34.4 t.
MBSO. Dia. ED211. Lot No. 30768 Cravens 1967. 70S. 56.4 t.
BDTSO. Dia. EF212. Lot No. 30769 Cravens 1967. 83S. 38.4 t.

311 098 **S**	PGWX	GW	76409	62169	76428
311 099	PGWX	GW	76410	62170	76429
311 102	PGWX	GW	76413	62173	76432
311 103	PGWX	GW	76414	62174	76433
311 104	PGWX	GW	76415	62175	76434
311 105 **S**	PGWX	GW	76416	62176	76435
311 107	PGWX	GW	76418	62178	76437
311 108	PGWX	GW	76419	62179	76438
311 109 **S**	PGWX	GW	76420	62180	76439
311 110	PGWX	GW	76421	62181	76440
Spare	PGWX	GW (S)	76426		

CLASS 312

BDTSOL–MBSO–TSO–DTCOL. Disc brakes PA.
System: 25 kV a.c. overhead.
Bogies: B4.
Gangways: Gangwayed within unit.
Traction Motors: Four EE546 of 201.5 kW.
Dimensions: 19.86 x 2.82 m (outer cars), 19.93 x 2.82 m (inner cars).
Maximum Speed: 90 mph.

Class 312/0. Standard design.

76994–97 BDTSOL. Dia. EF213. Lot No. 30891 York 1976. 84S 1L. 34.9 t.
62657–60 MBSO. Dia. ED214. Lot No. 30892 York 1976. 68S. 56 t.
71277–80 TSO. Dia. EH209. Lot No. 30893 York 1976. 98S. 30.5 t.
78045–48 DTCOL. Dia. EE305. Lot No. 30894 York 1976. 25F 47S 2L.
76949–74 BDTSOL. Dia. EF213. Lot No. 30863 York 1977–78. 84S 1L. 34.9 t.
62484–509 MBSO. Dia. ED212. Lot No. 30864 York 1977–78. 68S. 56 t.
71168–93 TSO. Dia. EH209. Lot No. 30865 York 1977–78. 98S. 30.5 t.
78000–25 DTCOL. Dia. EE305. Lot No. 30866 York 1977–78. 25F 47S 2L.

Notes: 312 727–730 were formerly numbered 312 201–204.

312 701		NGEX	CC		76949	62484	71168	78000
312 702	**N**	NGEX	CC		76950	62485	71169	78001
312 703	**N**	NGEX	CC		76951	62486	71170	78002
312 704	**N**	NGEX	CC		76952	62487	71171	78003
312 705	**N**	NGEX	CC		76989	62488	71172	78004
312 706	**N**	NGEX	CC		76954	62489	71173	78005
312 707	**N**	NGEX	CC		76955	62490	71174	78006
312 708	**N**	NGEX	CC		76956	62491	71175	78007
312 709	**N**	NGEX	CC		76957	62492	71176	78008
312 710	**N**	NGEX	CC		76974	62493	71193	78025
312 711	**N**	NGEX	CC		76959	62494	71178	78010
312 712	**N**	NGEX	CC		76960	62495	71179	78011
312 713	**N**	NGEX	CC		76961	62496	71180	78012
312 714	**N**	NGEX	CC		76962	62497	71181	78013
312 715	**N**	NGEX	CC		76963	62498	71182	78014
312 716	**N**	NGEX	CC		76964	62499	71183	78015
312 717	**N**	NGEX	CC	(U)	76958	62500	71177	78021
312 718	**N**	NGEX	CC		76966	62501	71185	78017
312 719	**N**	NGEX	CC		76967	62502	71186	78018
312 720	**N**	NGEX	CC		76968	62503	71187	78019
312 721	**N**	NGEX	CC		76969	62504	71188	78020
312 722	**N**	NGEX	CC		76970	62505	71189	78009
312 723	**N**	NGEX	CC		76953	62506	71208	78040
312 724	**N**	NGEX	CC		76972	62507	71191	78023
312 725	**N**	NGEX	CC		76973	62508	71192	78024
312 726	**N**	NGEX	CC		76965	62509	71184	78016
312 727	**N**	NGEX	CC		76994	62657	71277	78045
312 728	**N**	NGEX	CC		76995	62658	71278	78046
312 729	**N**	NGEX	CC		76996	62659	71279	78047
312 730	**N**	NGEX	CC		76997	62660	71280	78048

Class 312/1. Can also operate on 6.25 kV a.c. overhead.

BDTSOL. Dia. EF213. Lot No. 30867 York 1975–76. 84S 2L. 34.9 t.
MBSO. Dia. ED213. Lot No. 30868 York 1975–76. 68S. 56 t.
TSO. Dia. EH209. Lot No. 30869 York 1975–76. 98S. 30.5 t.
DTCOL. Dia. EE305. Lot No. 30870 York 1975–76. 25F 47S 2L.

312 781	**N**	NGEX	CC	76975	62510	71194	78026
312 782	**N**	NGEX	CC	76976	62511	71195	78027
312 783		NGEX	CC	76977	62512	71196	78028
312 784	**N**	NGEX	CC	76978	62513	71197	78029

312 785 **N**	NGEX	CC	76979	62514	71198	78030
312 786 **N**	NGEX	CC	76980	62515	71199	78031
312 787 **N**	NGEX	CC	76981	62516	71200	78032
312 788 **N**	NGEX	CC	76982	62517	71201	78033
312 789 **N**	NGEX	CC	76983	62518	71202	78034
312 790 **N**	NGEX	CC	76984	62519	71203	78035
312 791 **N**	NGEX	CC	76985	62520	71204	78036
312 792 **N**	NGEX	CC	76986	62521	71205	78037
312 793 **N**	NGEX	CC	76987	62522	71206	78038
312 794 **N**	NGEX	CC	76988	62523	71207	78039
312 795 **N**	NGEX	CC	76971	62524	71190	78022
312 796	NGEX	CC	76990	62525	71209	78041
312 797 **N**	NGEX	CC	76991	62526	71210	78042
312 798 **N**	NGEX	CC	76992	62527	71211	78043
312 799 **N**	NGEX	CC	76993	62528	71212	78044

CLASS 313

DMSO–TSO–DMSO. Tightlock couplers. Sliding doors. Disc and rheostatic brakes. PA. Cab to shore radio.
System: 25 kV a.c. overhead/750 V d.c. third rail.
Bogies: BX1.
Gangways: Gangwayed within unit. End doors.
Traction Motors: Four GEC G310AZ of 82.125 kW.
Dimensions: 19.80 x 2.82 m (outer cars), 19.92 x 2.82 m (inner cars).
Maximum Speed: 75 mph.

DMSO (A). Dia. EA204. Lot No. 30879 York 1976–77. 74S. 36.4 t.
TSO. Dia. EH210. Lot No. 30880 York 1976–77. 84S. 30.5 t.
DMSO (B). Dia. EI201. Lot No. 30885 York 1976–77. 74S. 37.6 t.
*–Extra shoegear for Euston–Watford line services.

313 001 **N**	NNLX	*	BY	62529	71213	62593
313 002 **N**	NNLX	*	BY	62530	71214	62594
313 003 **N**	NNLX	*	BY	62531	71215	62595
313 004 **N**	NNLX	*	BY	62532	71216	62596
313 005 **N**	NNLX	*	BY	62533	71217	62597
313 006 **N**	NNLX	*	BY	62534	71218	62598
313 007 **N**	NNLX	*	BY	62535	71219	62599
313 008 **N**	NNLX	*	BY	62536	71220	62600
313 009 **N**	NNLX	*	BY	62537	71221	62601
313 010 **N**	NNLX	*	BY	62538	71222	62602
313 011 **N**	NNLX	*	BY	62539	71223	62603
313 012 **N**	NNLX	*	BY	62540	71224	62604
313 013 **N**	NNLX	*	BY	62541	71225	62605
313 014 **N**	NNLX	*	BY	62542	71226	62606
313 015 **N**	NNLX	*	BY	62543	71227	62607
313 016 **N**	NNLX	*	BY	62544	71228	62608
313 017 **N**	NGNX	*	HE	62545	71229	62609
313 018 **N**	NGNX	*	HE	62546	71230	62610
313 019 **N**	NGNX	*	HE	62547	71231	62611
313 020 **N**	NGNX	*	HE	62548	71232	62612
313 021 **N**	NNLX	*	BY	62549	71233	62613

313 022	**N**	NNLX	*	BY	62550	71234	62614
313 023	**N**	NNLX	*	BY	62551	71235	62615
313 024	**N**	NGNX		HE	62552	71236	62616
313 025		NGNX		HE	62553	71237	62617
313 026		NGNX		HE	62554	71238	62618
313 027	**N**	NGNX		HE	62555	71239	62619
313 028	**N**	NGNX		HE	62556	71240	62620
313 029	**N**	NGNX		HE	62557	71241	62621
313 030		NGNX		HE	62558	71242	62622
313 031	**N**	NGNX		HE	62559	71243	62623
313 032	**N**	NGNX		HE	62560	71244	62624
313 033	**N**	NGNX		HE	62561	71245	62625
313 034		NNLX	*	BY	62562		62626
313 035	**N**	NGNX		HE	62563	71247	62627
313 036	**N**	NGNX		HE	62564	71248	62628
313 037	**N**	NGNX		HE	62565	71249	62629
313 038	**N**	NGNX		HE	62566	71250	62630
313 039	**N**	NGNX		HE	62567	71251	62631
313 040	**N**	NGNX		HE	62568	71252	62632
313 041	**N**	NGNX		HE	62569	71253	62633
313 042	**N**	NGNX		HE	62570	71254	62634
313 043		NGNX		HE	62571	71255	62635
313 044	**N**	NGNX		HE	62572	71256	62636
313 045	**N**	NGNX		HE	62573	71257	62637
313 046	**N**	NGNX		HE	62574	71258	62638
313 047	**N**	NGNX		HE	62575	71259	62639
313 048	**N**	NGNX		HE	62576	71260	62640
313 049		NGNX		HE	62577	71261	62641
313 050		NGNX		HE	62578	71262	62642
313 051	**N**	NGNX		HE	62579	71263	62643
313 052		NGNX		HE	62580	71264	62644
313 053		NGNX		HE	62581	71265	62645
313 054		NGNX		HE	62582	71266	62646
313 055		NGNX		HE	62583	71267	62647
313 056		NGNX		HE	62584	71268	62648
313 057	**N**	NGNX		HE	62585	71269	62649
313 058		NGNX		HE	62586	71270	62650
313 059		NGNX		HE	62587	71271	62651
313 060		NGNX		HE	62588	71272	62652
313 061	**N**	NGNX		HE	62589	71273	62653
313 062		NGNX		HE	62590	71274	62654
313 063	**N**	NGNX		HE	62591	71275	62655
313 064		NGNX		HE	62592	71276	62656

CLASS 314

DMSO–PTSO–DMSO. Thyristor control. Tightlock couplers. Sliding doors. Disc and rheostatic brakes. PA. Cab to shore radio.
System: 25 kV a.c. overhead.
Bogies: BX1.
Gangways: Gangwayed within unit. End doors.
Traction Motors: Four GEC G310AZ (Brush TM61-53*) of 82.125 kW.

Dimensions: 19.80 x 2.82 m (outer cars), 19.92 x 2.82 m (inner cars).
Maximum Speed: 75 mph.

DMSO. Dia. EA206. Lot No. 30912 York 1979. 68S. 34.5 t.
PTSO. Dia. EH211. Lot No. 30913 York 1979. 76S. 33.0 t.

314 201	**S**	PGWX	*	GW	64583	71450	64584
314 202	**S**	PGWX	*	GW	64585	71451	64586
314 203	**S**	PGWX	*	GW	64587	71452	64588
314 204	**S**	PGWX	*	GW	64589	71453	64590
314 205	**S**	PGWX	*	GW	64591	71454	64592
314 206	**S**	PGWX	*	GW	64593	71455	64594
314 207	**S**	PGWX		GW	64595	71456	64596
314 208	**S**	PGWX		GW	64597	71457	64598
314 209	**S**	PGWX		GW	64599	71458	64600
314 210	**S**	PGWX		GW	64601	71459	64602
314 211	**S**	PGWX		GW	64603	71460	64604
314 212	**S**	PGWX		GW	64605	71461	64606
314 213	**S**	PGWX		GW	64607	71462	64608
314 214	**S**	PGWX		GW	64609	71463	64610
314 215	**S**	PGWX		GW	64611	71464	64612
314 216	**S**	PGWX		GW	64613	71465	64614

CLASS 315

DMSO–TSO–PTSO–DMSO. Thyristor control. Tightlock couplers. Sliding doors. Disc and rheostatic brakes. PA.
System: 25 kV a.c. overhead.
Bogies: BX1.
Gangways: Gangwayed within unit. End doors.
Traction Motors: Four Brush TM61-53 (GEC G310AZ*) of 82.125 kW.
Dimensions: 19.80 x 2.82 m (outer cars), 19.92 x 2.82 m (inner cars).
Maximum Speed: 75 mph.

64461–64582. DMSO. Dia. EA207. Lot No. 30902 York 1980–81. 74S. 35 t.
71281–71341. TSO. Dia. EH216. Lot No. 30904 York 1980–81. 86S. 25.5 t.
71389–71449. PTSO. Dia. EH217. Lot No. 30903 York 1980–81. 84S. 32 t.

315 801	**N**	NGEX	IL	64461	71281	71389	64462
315 802	**N**	NGEX	IL	64463	71282	71390	64464
315 803	**N**	NGEX	IL	64465	71283	71391	64466
315 804		NGEX	IL	64467	71284	71392	64468
315 805		NGEX	IL	64469	71285	71393	64470
315 806	**N**	NGEX	IL	64471	71286	71394	64472
315 807		NGEX	IL	64473	71287	71395	64474
315 808		NGEX	IL	64475	71288	71396	64476
315 809		NGEX	IL	64477	71289	71397	64478
315 810		NGEX	IL	64479	71290	71398	64480
315 811		NGEX	IL	64481	71291	71399	64482
315 812	**N**	NGEX	IL	64483	71292	71400	64484
315 813		NGEX	IL	64485	71293	71401	64486
315 814		NGEX	IL	64487	71294	71402	64488
315 815		NGEX	IL	64489	71295	71403	64490
315 816	**N**	NGEX	IL	64491	71296	71404	64492

315 817	**N**	NGEX		IL	64493	71297	71405	64494
315 818	**N**	NGEX		IL	64495	71298	71406	64496
315 819	**N**	NGEX		IL	64497	71299	71407	64498
315 820	**N**	NGEX		IL	64499	71300	71408	64500
315 821	**N**	NGEX		IL	64501	71301	71409	64502
315 822		NGEX		IL	64503	71302	71410	64504
315 823	**N**	NGEX		IL	64505	71303	71411	64506
315 824	**N**	NGEX		IL	64507	71304	71412	64508
315 825	**N**	NGEX		IL	64509	71305	71413	64510
315 826	**N**	NGEX		IL	64511	71306	71414	64512
315 827	**N**	NGEX		IL	64513	71307	71415	64514
315 828	**N**	NGEX		IL	64515	71308	71416	64516
315 829	**N**	NGEX		IL	64517	71309	71417	64518
315 830	**N**	NGEX		IL	64519	71310	71418	64520
315 831	**N**	NGEX		IL	64521	71311	71419	64522
315 832	**N**	NGEX		IL	64523	71312	71420	64524
315 833	**N**	NGEX		IL	64525	71313	71421	64526
315 834	**N**	NGEX		IL	64527	71314	71422	64528
315 835	**N**	NGEX		IL	64529	71315	71423	64530
315 836	**N**	NGEX		IL	64531	71316	71424	64532
315 837	**N**	NGEX		IL	64533	71317	71425	64534
315 838	**N**	NGEX		IL	64535	71318	71426	64536
315 839	**N**	NGEX		IL	64537	71319	71427	64538
315 840	**N**	NGEX		IL	64539	71320	71428	64540
315 841	**N**	NGEX		IL	64541	71321	71429	64542
315 842	**N**	NGEX	*	IL	64543	71322	71430	64544
315 843		NGEX	*	IL	64545	71323	71431	64546
315 844	**N**	NGEX	*	IL	64547	71324	71432	64548
315 845	**N**	NGEX	*	IL	64549	71325	71433	64550
315 846	**N**	NGEX	*	IL	64551	71326	71434	64552
315 847	**N**	NGEX	*	IL	64553	71327	71435	64554
315 848	**N**	NGEX	*	IL	64555	71328	71436	64556
315 849	**N**	NGEX	*	IL	64557	71329	71437	64558
315 850	**N**	NGEX	*	IL	64559	71330	71438	64560
315 851	**N**	NGEX	*	IL	64561	71331	71439	64562
315 852	**N**	NGEX	*	IL	64563	71332	71440	64564
315 853	**N**	NGEX	*	IL	64565	71333	71441	64566
315 854	**N**	NGEX	*	IL	64567	71334	71442	64568
315 855	**N**	NGEX	*	IL	64569	71335	71443	64570
315 856	**N**	NGEX	*	IL	64571	71336	71444	64572
315 857	**N**	NGEX	*	IL	64573	71337	71445	64574
315 858	**N**	NGEX	*	IL	64575	71338	71446	64576
315 859	**N**	NGEX	*	IL	64577	71339	71447	64578
315 860	**N**	NGEX	*	IL	64579	71340	71448	64580
315 861	**N**	NGEX	*	IL	64581	71341	71449	64582

CLASS 316 NETWORKER PROTOTYPE

DMSO–TSO(A)–TSO(B)–DMSO. Sliding doors. Disc brakes. Fluorescent lighting. PA. The DMSOs and TSO(A) were converted from class 210 vehicles with 7000 being added to the former numbers. The power cars have three-phase motors.

System: 25 kV a.c. overhead.
Bogies:
Gangways: Through gangwayed.
Traction Motors:
Dimensions: 19.83 x 2.82 m. (outer cars), 19.92 x 2.82 m (inner cars).
Maximum Speed: 90 mph.

DMSO. Dia. EA209. Lot No. 30934 Derby 1981. 74S. 45 t.
TSO(A). Dia. EH236. Lot No. 30932 York 1981. 84S. 30 t.
TSO(B). Dia. EH210. Lot No. 30880 York 1976–77. 84S. 30.5 t.

316 999 **N**	NNEX	CC	67300	67401	71246	67301

CLASS 317

DTSO–MSO–TCOL–DTSO. Thyristor control. Tightlock couplers. Sliding doors. Disc brakes. PA.
System: 25 kV a.c. overhead.
Bogies: BP20 (MSO), BT13 (others).
Gangways: Throughout.
Traction Motors: Four GEC G315BZ of 247.5 kW.
Dimensions: 19.83 x 2.82 m (outer cars), 19.92 x 2.82 m (inner cars).
Maximum Speed: 100 mph.

Class 317/1. Pressure ventilated.

DTSO(A) Dia. EE216. Lot No. 30955 York 1981–82. 74S. 29.44 t.
MSO. Dia. EC202. Lot No. 30958 York 1981–82. 79S. 49.76 t.
TCOL. Dia. EH307. Lot No. 30957 Derby 1981–82. 22F 46S 2L. 28.80 t. Controlled emission toilets (but decommisioned except for units allocated to NMLX).
DTSO(B) Dia. EE235 (EE232†). Lot No. 30956 York 1981–82. 70S. (71S†). 29.28 t.

317 301 **N**	NNWX	BY	77024	62661	71577	77048
317 302 **N**	NNWX	BY	77001	62662	71578	77049
317 303 **N**	NNWX	BY	77002	62663	71579	77050
317 304 **N**	NGNX	HE	77003	62664	71580	77051
317 305 **N**	NGNX	HE	77004	62665	71581	77052
317 306 **N**	NGNX	HE	77005	62666	71582	77053
317 307 **N**	NGNX	HE	77006	62667	71583	77054
317 308 **N**	NGNX	HE	77007	62668	71584	77055
317 309 **N**	NNWX	BY	77008	62669	71585	77056
317 310 **N**	NGNX	HE	77009	62670	71586	77057
317 311 **N**	NGNX	HE	77010	62671	71587	77058
317 312 **N**	NGNX	HE	77011	62672	71588	77059
317 313 **N**	NGNX	HE	77012	62673	71589	77060
317 314 **N**	NNWX	BY	77013	62674	71590	77061
317 315 **N**	NGNX	HE	77014	62675	71591	77062
317 316 **N**	NGNX	HE	77015	62676	71592	77063
317 317 **N**	NGNX	HE	77016	62677	71593	77064
317 318 **N**	NMLX	HE	77017	62678	71594	77065
317 319 **N**	NMLX	HE	77018	62679	71595	77066
317 320 **N**	NMLX	HE	77019	62680	71596	77067
317 321 **N**	NMLX	HE	77020	62681	71597	77068

317 322	**N**	NMLX	HE	77021	62682	71598	77069
317 323	**N**	NMLX	HE	77022	62683	71599	77070
317 324	**N**	NMLX	HE	77023	62684	71600	77071
317 325	**N**	NMLX	HE	77000	62685	71601	77072
317 326	**N**	NMLX	HE	77025	62686	71602	77073
317 327	**N**	NMLX	HE	77026	62687	71603	77074
317 328	**N**	NMLX	HE	77027	62688	71604	77075
317 329	**N**	NMLX	HE	77028	62689	71605	77076
317 330	**N**	NMLX	HE	77029	62690	71606	77077
317 331	**N**	NGNX	HE	77030	62691	71607	77078
317 332	**N**	NGNX	HE	77031	62692	71608	77079
317 333	**N**	NGNX	HE	77032	62693	71609	77080
317 334	**N**	NGNX	HE	77033	62694	71610	77081
317 335	**N**	NGNX	HE	77034	62695	71611	77082
317 336	**N**	NGNX	HE	77035	62696	71612	77083
317 337	**N**	NGNX †	HE	77036	62697	71613	77084
317 338	**N**	NGNX †	HE	77037	62698	71614	77085
317 339	**N**	NGNX †	HE	77038	62699	71615	77086
317 340	**N**	NGNX †	HE	77039	62700	71616	77087
317 341	**N**	NGNX †	HE	77040	62701	71617	77088
317 342	**N**	NGNX †	HE	77041	62702	71618	77089
317 343	**N**	NGNX †	HE	77042	62703	71619	77090
317 344	**N**	NGNX †	HE	77043	62704	71620	77091
317 345	**N**	NGNX †	HE	77044	62705	71621	77092
317 346	**N**	NGNX †	HE	77045	62706	71622	77093
317 347	**N**	NGNX †	HE	77046	62707	71623	77094
317 348	**N**	NGNX †	HE	77047	62708	71624	77095

Class 317/2. Convection heating.

77200–19. DTSO(A). Dia. EE224. Lot No. 30994 York 1985–86. 74S. 29.31 t.
77280–83. DTSO(A). Dia. EE224. Lot No. 31007 York 1987. 74S. 29.31 t.
62846–65. MSO.Dia. EC205. Lot No. 30996 York 1985–86. 79S. 50.08 t.
62886–89. MSO. Dia. EC205. Lot No. 31009 York 1987. 79S. 50.08 t.
71734–53. TCOL. Dia. EH308. Lot No. 30997 Yk 1985–86. 22F 46S 2L. 28.28 t.
71762–65. TCOL. Dia. EH308. Lot No. 31010 York 1987. 22F 46S 2L. 28.28 t.
77220–39. DTSO(B). Dia. EE225. Lot No. 30995 York 1985–86. 29.28 t. 71S.
77284–87. DTSO(B). Dia. EE225. Lot No. 31008 York 1987. 29.28 t. 71S.

317 349	**N**	NGNX	HE	77200	62846	71734	77220
317 350		NGNX	HE	77201	62847	71735	77221
317 351	**N**	NGNX	HE	77202	62848	71736	77222
317 352		NGNX	HE	77203	62849	71739	77223
317 353	**N**	NGNX	HE	77204	62850	71738	77224
317 354	**N**	NGNX	HE	77205	62851	71737	77225
317 355	**N**	NGNX	HE	77206	62852	71740	77226
317 356	**N**	NGNX	HE	77207	62853	71742	77227
317 357		NGNX	HE	77208	62854	71741	77228
317 358	**N**	NGNX	HE	77209	62855	71743	77229
317 359	**N**	NGNX	HE	77210	62856	71744	77230
317 360	**N**	NGNX	HE	77211	62857	71745	77231
317 361	**N**	NGNX	HE	77212	62858	71746	77232
317 362	**N**	NGNX	HE	77213	62859	71747	77233

317 363	**N**	NGNX	HE	77214	62860	71748	77234
317 364	**N**	NGNX	HE	77215	62861	71749	77235
317 365	**N**	NGNX	HE	77216	62862	71750	77236
317 366	**N**	NGNX	HE	77217	62863	71752	77237
317 367	**N**	NGNX	HE	77218	62864	71751	77238
317 368	**N**	NGNX	HE	77219	62865	71753	77239
317 369	**N**	NGNX	HE	77280	62886	71762	77284
317 370	**N**	NGNX	HE	77281	62887	71763	77285
317 371	**N**	NGNX	HE	77282	62888	71764	77286
317 372	**N**	NGNX	HE	77283	62889	71765	77287

CLASS 318

DTSOL–MSO–DTSO. Thyristor control. Tightlock couplers. Sliding doors. Disc brakes. PA. Cab to shore radio.
System: 25 kV a.c. overhead.
Bogies: BP20 (MSO), BT13 (others).
Gangways: Throughout.
Traction Motors: Four Brush TM 2141 of 268 kW.
Dimensions: 19.83 x 2.82 m (outer cars), 19.92 x 2.82 m (inner cars).
Maximum Speed: 90 mph.

77240–59. DTSOL. Dia. EE227. Lot No. 30999 York 1985–86. 66S 1L. 30.01 t.
77288. DTSOL. Dia. EE227. Lot No. 31020 York 1986–87. 66S 1L. 30.01 t.
62866–85. MSO. Dia. EC207. Lot No. 30998 York 1985–86. 79S. 50.90 t.
62890. MSO. Dia. EC207. Lot No. 31019 York 1987. 79S. 50.90 t.
77260–79. DTSO. Dia. EE228. Lot No. 31000 York 1985–86. 71S. 26.60 t.
77289. DTSO. Dia. EE228. Lot No. 31021 York 1987. 71S. 26.60 t.

318 250	**S**	PGWX	GW	77260	62866	77240
318 251	**S**	PGWX	GW	77261	62867	77241
318 252	**S**	PGWX	GW	77262	62868	77242
318 253	**S**	PGWX	GW	77263	62869	77243
318 254	**S**	PGWX	GW	77264	62870	77244
318 255	**S**	PGWX	GW	77265	62871	77245
318 256	**S**	PGWX	GW	77266	62872	77246
318 257	**S**	PGWX	GW	77267	62873	77247
318 258	**S**	PGWX	GW	77268	62874	77248
318 259	**S**	PGWX	GW	77269	62875	77249
318 260	**S**	PGWX	GW	77270	62876	77250
318 261	**S**	PGWX	GW	77271	62877	77251
318 262	**S**	PGWX	GW	77272	62878	77252
318 263	**S**	PGWX	GW	77273	62879	77253
318 264	**S**	PGWX	GW	77274	62880	77254
318 265	**S**	PGWX	GW	77275	62881	77255
318 266	**S**	PGWX	GW	77276	62882	77256
318 267	**S**	PGWX	GW	77277	62883	77257
318 268	**S**	PGWX	GW	77278	62884	77258
318 269	**S**	PGWX	GW	77279	62885	77259
318 270	**S**	PGWX	GW	77289	62890	77288

Name: DTSOL No. 77240 of set 318 250 is named 'GEOFF SHAW'.

CLASS 319

Thyristor control. Tightlock couplers. Sliding doors. Disc brakes. PA. Cab to shore radio.
System: 25 kV a.c. overhead./750 V d.c. third rail.
Bogies: P7-4 (MSO), T3-7 (others).
Gangways: Throughout.
Traction Motors: Four GEC G315BZ of 247.5 kW.
Dimensions: 19.83 x 2.82 m (outer cars), 19.92 x 2.82 m (inner cars).
Maximum Speed: 100 mph.

Class 319/0. DTSO–MSO–TSOL–DTSO.

77291–381. DTSO. Dia. EE233. Lot No. 31022 (odd nos.) York 1987–8. 82S. 30 t.
77431–457. DTSO. Dia. EE233. Lot No. 31038 (odd nos.) York 1988. 82S. 30 t.
62891–936. MSO. Dia. EC209. Lot No. 31023 York 1987–8. 82S. 51 t.
62961–974. MSO. Dia. EC209. Lot No. 31039 York 1988. 82S. 51 t.
71772–817. TSOL. Dia. EH234. Lot No. 31024 York 1987–8. 77S 2L. 51 t.
71866–879. TSOL. Dia. EH234. Lot No. 31040 York 1988. 77S 2L. 51 t.
77290–380. DTSO. Dia. EE234. Lot No. 31025 (even nos.) York 1987–8. 78S. 30 t.
77430–456. DTSO. Dia. EE234. Lot No. 31041 (even nos.) York 1988. 78S. 30 t.

319 001	**N**	NMLX	SU	77291	62891	71772	77290
319 002	**N**	NSXX	SU	77293	62892	71773	77292
319 003	**N**	NSXX	SU	77295	62893	71774	77294
319 004	**N**	NSXX	SU	77297	62894	71775	77296
319 005	**N**	NSXX	SU	77299	62895	71776	77298
319 006	**N**	NSXX	SU	77301	62896	71777	77300
319 007	**N**	NSXX	SU	77303	62897	71778	77302
319 008	**N**	NSXX	SU	77305	62898	71779	77304
319 009	**N**	NSXX	SU	77307	62899	71780	77306
319 010	**N**	NSXX	SU	77309	62900	71781	77308
319 011	**N**	NSXX	SU	77311	62901	71782	77310
319 012	**N**	NMLX	SU	77313	62902	71783	77312
319 013	**N**	NMLX	SU	77315	62903	71784	77314
319 014	**N**	NMLX	SU	77317	62904	71785	77316
319 015	**N**	NMLX	SU	77319	62905	71786	77318
319 016	**N**	NMLX	SU	77321	62906	71787	77320
319 017	**N**	NMLX	SU	77323	62907	71788	77322
319 018	**N**	NMLX	SU	77325	62908	71789	77324
319 019	**N**	NMLX	SU	77327	62909	71790	77326
319 020	**N**	NMLX	SU	77329	62910	71791	77328
319 021	**N**	NMLX	SU	77331	62911	71792	77330
319 022	**N**	NMLX	SU	77333	62912	71793	77332
319 023	**N**	NMLX	SU	77335	62913	71794	77334
319 024	**N**	NMLX	SU	77337	62914	71795	77336
319 025	**N**	NMLX	SU	77339	62915	71796	77338
319 026	**N**	NMLX	SU	77341	62916	71797	77340

319 027	**N**	NMLX	SU	77343	62917	71798	77342
319 028	**N**	NMLX	SU	77345	62918	71799	77344
319 029	**N**	NMLX	SU	77347	62919	71800	77346
319 030	**N**	NMLX	SU	77349	62920	71801	77348
319 031	**N**	NMLX	SU	77351	62921	71802	77350
319 032	**N**	NMLX	SU	77353	62922	71803	77352
319 033	**N**	NMLX	SU	77355	62923	71804	77354
319 034	**N**	NMLX	SU	77357	62924	71805	77356
319 035	**N**	NMLX	SU	77359	62925	71806	77358
319 036	**N**	NMLX	SU	77361	62926	71807	77360
319 037	**N**	NMLX	SU	77363	62927	71808	77362
319 038	**N**	NMLX	SU	77365	62928	71809	77364
319 039	**N**	NMLX	SU	77367	62929	71810	77366
319 040	**N**	NMLX	SU	77369	62930	71811	77368
319 041	**N**	NMLX	SU	77371	62931	71812	77370
319 042	**N**	NMLX	SU	77373	62932	71813	77372
319 043	**N**	NMLX	SU	77375	62933	71814	77374
319 044	**N**	NMLX	SU	77377	62934	71815	77376
319 045	**N**	NMLX	SU	77379	62935	71816	77378
319 046	**N**	NMLX	SU	77381	62936	71817	77380
319 047	**N**	NMLX	SU	77431	62961	71866	77430
319 048	**N**	NMLX	SU	77433	62962	71867	77432
319 049	**N**	NMLX	SU	77435	62963	71868	77434
319 050	**N**	NMLX	SU	77437	62964	71869	77436
319 051	**N**	NMLX	SU	77439	62965	71870	77438
319 052	**N**	NMLX	SU	77441	62966	71871	77440
319 053	**N**	NMLX	SU	77443	62967	71872	77442
319 054	**N**	NMLX	SU	77445	62968	71873	77444
319 055	**N**	NMLX	SU	77447	62969	71874	77446
319 056	**N**	NMLX	SU	77449	62970	71875	77448
319 057	**N**	NMLX	SU	77451	62971	71876	77450
319 058	**N**	NMLX	SU	77453	62972	71877	77452
319 059	**N**	NMLX	SU	77455	62973	71878	77454
319 060	**N**	NMLX	SU	77457	62974	71879	77456

Class 319/1. DTCO–MSO–TSOL–DTSO.

DTCO. Dia. EE310. Lot No. 31063 York 1990. 16F 56S. 30 t.
MSO. Dia. EC214. Lot No. 31064 York 1990. 79S. 51 t.
TSOL. Dia. EH238. Lot No. 31065 York 1990. 74S 2L. 51 t.
DTSO. Dia. EE240. Lot No. 31066 York 1990. 78S. 30 t.

319 161	77459	63043	71929	77458
319 162	77461	63044	71930	77460
319 163	77463	63045	71931	77462
319 164	77465	63046	71932	77464
319 165	77467	63047	71933	77466
319 166	77469	63048	71934	77468
319 167	77471	63049	71935	77470
319 168	77473	63050	71936	77472
319 169	77475	63051	71937	77474
319 170	77477	63052	71938	77476
319 171	77479	63053	71939	77478

319 172	77481	63054	71940	77480
319 173	77483	63055	71941	77482
319 174	77485	63056	71942	77484
319 175	77487	63057	71943	77486
319 176	77489	63058	71944	77488
319 177	77491	63059	71945	77490
319 178	77493	63060	71946	77492
319 179	77495	63061	71947	77494
319 180	77497	63062	71948	77496
319 181	77973	63093	71979	77974
319 182	77975	63094	71980	77976
319 183	77977	63095	71981	77978
319 184	77979	63096	71982	77980
319 185	77981	63097	71983	77982
319 186	77983	63098	71984	77984

CLASS 320

DTSO–MSO–DTSO. Thyristor control. Tightlock couplers. Sliding doors. Disc brakes. PA. New units under construction for Sc.R.
System: 25 kV a.c. overhead.
Bogies: P7-4 (MSO), T3-7 (others).
Gangways: Gangwayed within unit.
Traction Motors: Brush TM2141B (268 kW).
Dimensions: 19.83 x 2.82 m (outer cars), 19.92 x 2.82 m (inner car).
Maximum Speed: 100 mph.

DTSO (A). Dia. EE238. Lot No. 31060 York 1990– . S. t.
MSO. Dia. EC212. Lot No. 31062 York 1990– . S. t.
DTSO (B). Dia. EE239. Lot No. 31061 York 1990– . S t.

320 301	77899	63021	77921
320 302	77900	63022	77922
320 303	77901	63023	77923
320 304	77902	63024	77924
320 305	77903	63025	77925
320 306	77904	63026	77926
320 307	77905	63027	77927
320 308	77906	63028	77928
320 309	77907	63029	77929
320 310	77908	63030	77930
320 311	77909	63031	77931
320 312	77910	63032	77932
320 313	77911	63033	77933
320 314	77912	63034	77934
320 315	77913	63035	77935
320 316	77914	63036	77936
320 317	77915	63037	77937
320 318	77916	63038	77938
320 319	77917	63039	77939
320 320	77918	63040	77940
320 321	77919	63041	77941
320 322	77920	63042	77942

CLASS 321

DTCO–MSO–TSOL–DTSO. Thyristor control. Tightlock couplers. Sliding doors. Disc brakes. PA.
System: 25 kV a.c. overhead.
Bogies: P7-4 (MSO), T3-7 (others).
Gangways: Gangwayed within unit.
Traction Motors: Brush TM2141B (268 kW).
Dimensions: 19.83 x 2.82 m (outer cars), 19.92 x 2.82 m (inner cars).
Maximum Speed: 100 mph.

Class 321/3. Anglia Region Units.
78049–78094. DTCO. Dia. EE308 (EE311*). Lot No. 31053 York 1988–9. 19F 58S (16F 56S*) 29.3 t.
78131–78150. DTCO. Dia. EE309. Lot No. 31075 York 1990. 20F 52S 29.3 t.
62975–63020. MSO. Dia. EC210. (EC213*) Lot No. 31054 York 1988–9. 81S. (79S*) 51.5 t.
63105–63124. MSO. Dia. EC213. Lot No. 31076 York 1990. 79S. 51.5 t.
71880–71925. TSOL. Dia. EH235 (EH237*). Lot No. 31055 York 1988–9. 76S 2L. (74S 2L*) 28 t.
71991–72010. TSOL. Dia. EH237. Lot No. 31077 York 1990. 74S 2L. 28 t.
77853–77898. DTSO. Dia. EE236. Lot No. 31056 York 1988–9. 78S. 29.1 t.
78280–78299. DTSO. Dia. EE236. Lot No. 31078 York 1990. 78S. 29.1 t.

321 301	**N**	NGEX		IL	78049	62975	71880	77853
321 302	**N**	NGEX		IL	78050	62976	71881	77854
321 303	**N**	NGEX		IL	78051	62977	71882	77855
321 304	**N**	NGEX		IL	78052	62978	71883	77856
321 305	**N**	NGEX		IL	78053	62979	71884	77857
321 306	**N**	NGEX		IL	78054	62980	71885	77858
321 307	**N**	NGEX		IL	78055	62981	71886	77859
321 308	**N**	NGEX		IL	78056	62982	71887	77860
321 309	**N**	NGEX		IL	78057	62983	71888	77861
321 310	**N**	NGEX		IL	78058	62984	71889	77862
321 311	**N**	NGEX	†	IL	78059	62985	71890	77863
321 312	**N**	NGEX	†	IL	78060	62986	71891	77864
321 313	**N**	NGEX	†	IL	78061	62987	71892	77865
321 314	**N**	NGEX	†	IL	78062	62988	71893	77866
321 315	**N**	NGEX	†	IL	78063	62989	71894	77867
321 316	**N**	NGEX	†	IL	78064	62990	71895	77868
321 317	**N**	NGEX	†	IL	78065	62991	71896	77869
321 318	**N**	NGEX	†	IL	78066	62992	71897	77870
321 319	**N**	NGEX	†	IL	78067	62993	71898	77871
321 320	**N**	NGEX	†	IL	78068	62994	71899	77872
321 321	**N**	NGEX		IL	78069	62995	71900	77873
321 322	**N**	NGEX		IL	78070	62996	71901	77874
321 323	**N**	NGEX		IL	78071	62997	71902	77875
321 324	**N**	NGEX	†	IL	78072	62998	71903	77876
321 325	**N**	NGEX	†	IL	78073	62999	71904	77877
321 326	**N**	NGEX	†	IL	78074	63000	71905	77878
321 327	**N**	NGEX		IL	78075	63001	71906	77879
321 328	**N**	NGEX		IL	78076	63002	71907	77880
321 329	**N**	NGEX		IL	78077	63003	71908	77881

321 330 **N**	NGEX	IL	78078	63004	71909	77882
321 331 **N**	NGEX	IL	78079	63005	71910	77883
321 332 **N**	NGEX	IL	78080	63006	71911	77884
321 333 **N**	NGEX	IL	78081	63007	71912	77885
321 334 **N**	NGEX	IL	78082	63008	71913	77886
321 335 **N**	NGEX	IL	78083	63009	71914	77887
321 336 **N**	NGEX	IL	78084	63010	71915	77888
321 337 **N**	NGEX †	IL	78085	63011	71916	77889
321 338 **N**	NGEX †	IL	78086	63012	71917	77890
321 339 **N**	NGEX	IL	78087	63013	71918	77891
321 340 **N**	NGEX	IL	78088	63014	71919	77892
321 341 **N**	NGEX †	IL	78089	63015	71920	77893
321 342 **N**	NGEX	IL	78090	63016	71921	77894
321 343 **N**	NGEX	IL	78091	63017	71922	77895
321 344 **N**	NGEX †	IL	78092	63018	71923	77896
321 345 **N**	NGEX	IL	78093	63019	71924	77897
321 346 **N**	NGEX	IL	78094	63020	71925	77898
321 347			78131	63105	71991	78280
321 348			78132	63106	71992	78281
321 349			78133	63107	71993	78282
321 350			78134	63108	71994	78283
321 351			78135	63109	71995	78284
321 352			78136	63110	71996	78285
321 353			78137	63111	71997	78286
321 354			78138	63112	71998	78287
321 355			78139	63113	71999	78288
321 356			78140	63114	72000	78289
321 357			78141	63115	72001	78290
321 358			78142	63116	72002	78291
321 359			78143	63117	72003	78292
321 360			78144	63118	72004	78293
321 361			78145	63119	72005	78294
321 362			78146	63120	72006	78295
321 363			78147	63121	72007	78296
321 364			78148	63122	72008	78297
321 365			78149	63123	72009	78298
321 366			78150	63124	72010	78299

Name: DTSOL No. 71891 of set 321 312 is named 'Southend-on-Sea'.

Class 321/4. London Midland Region Units.

78095–78124. DTCO. Dia. EE309. Lot No. 31067 York 1989. 28F 40S 29.3 t.
78125–78150. DTCO. Dia. EE309. Lot No. 31075 York 1990. 28F 40S 29.3 t.
78151–78162. DTCO. Dia. EE309. Lot No. 31079 York 1990. 28F 40S 29.3 t.
63063–63092. MSO. Dia. EC213. Lot No. 31068 York 1989. 79S. 51.5 t.
63099–63104. MSO. Dia. EC213. Lot No. 31076 York 1990. 79S. 51.5 t.
63125–63136. MSO. Dia. EC213. Lot No. 31080 York 1990. 79S. 51.5 t.
71749–71778. TSOL. Dia. EH237. Lot No. 31069 York 1989. 74S 2L. 28.8 t.
71985–72010. TSOL. Dia. EH237. Lot No. 31077 York 1990. 74S 2L. 28.8 t.
72011–72022. TSOL. Dia. EH237. Lot No. 31081 York 1990. 74S 2L. 28.8 t.
77943–77972. DTSO. Dia. EE236. Lot No. 31070 York 1989. 78S. 29.06 t.
78274–78299. DTSO. Dia. EE236. Lot No. 31078 York 1990. 78S. 29.06 t.
78300–78311. DTSO. Dia. EE236. Lot No. 31082 York 1990. 78S. 29.06 t.

321 401	**N**	NNWX	BY	78095	63063	71949	77943
321 402	**N**	NNWX	BY	78096	63064	71950	77944
321 403	**N**	NNWX	BY	78097	63065	71951	77945
321 404	**N**	NNWX	BY	78098	63066	71952	77946
321 405	**N**	NNWX	BY	78099	63067	71953	77947
321 406	**N**	NNWX	BY	78100	63068	71954	77948
321 407	**N**	NNWX	BY	78101	63069	71955	77949
321 408	**N**	NNWX	BY	78102	63070	71956	77950
321 409	**N**	NNWX	BY	78103	63071	71957	77951
321 410	**N**	NNWX	BY	78104	63072	71958	77952
321 411	**N**	NNWX	BY	78105	63073	71959	77953
321 412	**N**	NNWX	BY	78106	63074	71960	77954
321 413	**N**	NNWX	BY	78107	63075	71961	77955
321 414	**N**	NNWX	BY	78108	63076	71962	77956
321 415	**N**	NNWX	BY	78109	63077	71963	77957
321 416	**N**	NNWX	BY	78110	63078	71964	77958
321 417	**N**	NNWX	BY	78111	63079	71965	77959
321 418	**N**	NNWX	BY	78112	63080	71966	77960
321 419	**N**	NNWX	BY	78113	63081	71967	77961
321 420	**N**	NNWX	BY	78114	63082	71968	77962
321 421	**N**	NNWX	BY	78115	63083	71969	77963
321 422	**N**	NNWX	BY	78116	63084	71970	77964
321 423	**N**	NNWX	BY	78117	63085	71971	77965
321 424	**N**	NNWX	BY	78118	63086	71972	77966
321 425	**N**	NNWX	BY	78119	63087	71973	77967
321 426	**N**	NNWX	BY	78120	63088	71974	77968
321 427	**N**	NNWX	BY	78121	63089	71975	77969
321 428	**N**	NNWX	BY	78122	63090	71976	77970
321 429	**N**	NNWX	BY	78123	63091	71977	77971
321 430	**N**	NNWX	BY	78124	63092	71978	77972
321 431				78151	63125	72011	78300
321 432				78152	63126	72012	78301
321 433				78153	63127	72013	78302
321 434				78154	63128	72014	78303
321 435				78155	63129	72015	78304
321 436				78156	63130	72016	78305
321 437				78157	63131	72017	78306
321 438				78158	63132	72018	78307
321 439				78159	63133	72019	78308
321 440				78160	63134	72020	78309
321 441				78161	63135	72021	78310
321 442				78162	63136	72022	78311
321 443				78125	63099	71985	78274
321 444				78126	63100	71986	78275
321 445				78127	63101	71987	78276
321 446				78128	63102	71988	78277
321 447				78129	63103	71989	78278
321 448				78130	63104	71990	78279

4.2. SOUTHERN REGION 750 V d.c. EMUs

These classes are all allocated to the Southern Region and operate on the third rail system at 750–850 V d.c. Except where stated otherwise, all multiple units can run in multiple with one another. Buffet cars have electric cooking. In addition to the class number, the old SR designations e.g. 2 Hap are quoted together with the year of introduction of their type of control gear. Whilst outer couplings are buckeyes on all units, 1951 and 1957-type suburban units have centre buffers and three link couplings within a unit.

CLASS 421/1 — 4 Cig (PHASE 1)

DTCsoL (A)–MBSO–TSO–DTCsoL (B). Express stock. Fitted with electric parking brake.

Electrical Equipment: 1963-type.
Bogies: Two Mk. 4 motor bogies (MBSO). B5 (SR) bogies (trailer cars).
Gangways: Throughout.
Traction Motors: Four EE507 of 185 kW.
Dimensions: 19.75 x 2.82 m.
Maximum Speed: 90 mph.

DTCsoL(A). Dia. EE364. Lot No. 30741 York 1964–65. 18F 36S 2L. 35.5 t.
MBSO. Dia. ED260. Lot No. 30742 York 1964–65. 56S. 49 t.
70699–70721. TSO. Dia. EH275. Lot No. 30730 York 1964–65. 72S. 31.5 t.
71051. TSO. Dia. EH275. Lot No. 30817 York 1971. 72S. 31.5 t. (ex 421/2).
DTCsoL(B). Dia. EE363. Lot No. 30740 York 1964–65. 24F 28S 2L. (* 18F 36S 2L) 35 t.

Renumbered from 7300–7327.

1100	**N**	NSXX		BI	76125	62066	71051	76071
1105	**N**	NSXX		BI	76080	62021	70699	76039
1110	**N**	NSXX	*	BI	76085	62026	70704	76031
1111	**N**	NSXX		BI	76086	62027	70705	76032
1114	**N**	NSXX		BI	76089	62030	70708	76035
1116	**N**	NSXX		BI	76091	62032	70710	76037
1118	**N**	NSXX		BI	76026	62034	70712	76093
1123	**N**	NSXX		BI	76119	62060	70717	76065
1127	**N**	NSXX		BI	76102	62043	70721	76048
Spare					76108			

CLASS 421/2 — 4 Cig (PHASE 2)

DTCsoL (A)–MBSO–TSO–DTCsoL (B). Express stock.

Electrical Equipment: 1963-type.
Bogies: Two Mk. 6 motor bogies (MBSO). B5 (SR) bogies (trailer cars).
Gangways: Throughout.
Traction Motors: Four EE507 of 185 kW.
Dimensions: 19.75 x 2.82 m.
Maximum Speed: 90 mph.

76581–76610. DTCsoL(A). Dia. EE364. Lot No. 30806 York 1970. 35.5 t.
76717–76787. DTCsoL(A). Dia. EE364. Lot No. 30814 York 1970–72. 35.5 t.
62282. MBSO. Dia. ED260. Lot No. 30804 York 1970. 56S. 49t.
62287–62316. MBSO. Dia. ED260. Lot No. 30808 York 1970. 56S. 49t.
62355–62425. MBSO. Dia. ED260. Lot No. 30816 York 1970. 56S. 49t.
70967–70996. TSO. Dia. EH275. Lot No. 30809 York 1970–71. 72S. 31.5t.
71035–71105. TSO. Dia. EH275. Lot No. 30817 York 1970. 72S. 31.5t.
76611–76640. DTCsoL(B). Dia. EE363. Lot No. 30807 York 1970. 35 t.
76788–76858. DTCsoL(B). Dia. EE363. Lot No. 30815 York 1970–72. 35 t.

DTCsoL (A) seat 18F 36S 2L. DTCsoL (B) seat 24F 28S 2L. 76812 has hopper ventilators.

Renumbered from the series 7401–7426 and 7337–7400.

1201	**N**	NSXX	BI	76751	62389	71069	76822
1202	**N**	NSXX	BI	76752	62390	71070	76823
1203	**N**	NSXX	BI	76753	62391	71071	76804
1204	**N**	NSXX	BI	76754	62392	71072	76825
1205	**N**	NSXX	BI	76755	62393	71073	76826
1206	**N**	NSXX	BI	76756	62394	71074	76827
1208	**N**	NSSX	FR	76758	62396	71076	76829
1209	**N**	NSSX	FR	76759	62397	71077	76830
1210	**N**	NSSX	FR	76760	62398	71078	76831
1211	**N**	NSSX	FR	76761	62399	71079	76832
1212	**N**	NSSX	FR	76762	62400	71080	76833
1213	**N**	NSSX	FR	76763	62401	71081	76834
1214	**N**	NSSX	FR	76764	62402	71082	76835
1215	**N**	NSSX	FR	76765	62403	71083	76836
1216	**N**	NSSX	FR	76766	62404	71084	76837
1217	**N**	NSSX	FR	76767	62405	71085	76838
1218	**N**	NSSX	FR	76768	62406	71086	76839
1219	**N**	NSSX	FR	76769	62407	71087	76840
1220	**N**	NSSX	FR	76770	62408	71088	76841
1221	**N**	NSSX	FR	76771	62409	71089	76842
1222	**N**	NSSX	FR	76772	62410	71090	76843
1223	**N**	NSSX	FR	76773	62411	71091	76844
1224	**N**	NSSX	FR	76774	62412	71092	76845
1225	**N**	NSSX	FR	76775	62413	71093	76846
1226	**N**	NSSX	FR	76776	62414	71094	76847
1238		NSSX	FR	76582	62288	70968	76612
1241		NSSX	FR	76585	62291	70971	76615
1243		NSSX	FR	76587	62293	70973	76617
1244		NSSX	FR	76588	62294	70974	76618
1245	**N**	NSSX	FR	76589	62295	70975	76619
1246	**N**	NSSX	FR	76590	62296	70976	76620
1247	**N**	NSSX	FR	76591	62297	70977	76621
1249	**N**	NSSX	FR	76593	62299	70979	76623
1252		NSSX	FR	76596	62302	70982	76626
1253	**N**	NSSX	FR	76597	62303	70983	76627
1254		NSXX	BI	76598	62304	70984	76628
1255	**N**	NSXX	BI	76599	62305	70985	76629
1256	**N**	NSXX	BI	76600	62306	70986	76630

1257	**N**	NSXX	BI		76601	62307	70987	76631
1258	**N**	NSXX	BI		76566	62282	70988	76576
1259	**N**	NSXX	BI		76603	62309	70989	76633
1260	**N**	NSXX	BI		76604	62310	70990	76634
1261	**N**	NSXX	BI		76605	62311	70991	76635
1262	**N**	NSXX	BI		76606	62312	70992	76636
1263		NSXX	BI		76607	62313	70993	76637
1264		NSXX	BI		76608	62314	70994	76638
1266	**N**	NSXX	BI		76610	62316	70996	76640
1268	**N**	NSXX	BI		76718	62356	71036	76789
1269	**N**	NSXX	BI		76719	62357	71037	76790
1270	**N**	NSXX	BI		76720	62358	71038	76791
1271	**N**	NSXX	BI		76721	62359	71039	76792
1272		NSXX	BI		76722	62360	71040	76793
1274		NSXX	BI		76724	62362	71042	76795
1275	**N**	NSXX	BI		76725	62363	71043	76796
1277	**N**	NSXX	BI		76727	62365	71045	76798
1280	**N**	NSXX	BI	(U)	76730	62368	71048	76801
1281	**N**	NSXX	BI		76731	62369	71049	76802
1285	**N**	NSXX	BI		76735	62373	71053	76806
1286	**N**	NSXX	BI		76736	62374	71054	76807
1287	**N**	NSXX	BI		76737	62375	71055	76808
1288	**N**	NSXX	BI		76738	62376	71056	76809
1289	**N**	NSXX	BI		76739	62377	71057	76810
1291	**N**	NSXX	BI		76741	62379	71059	76812
1292	**N**	NSXX	BI		76742	62380	71060	76813
1293	**N**	NSXX	BI		76743	62381	71061	76814
1294	**N**	NSXX	BI		76744	62382	71062	76815
1295	**N**	NSXX	BI		76745	62383	71063	76816
Spare		NSXX	BI		76639		70995	76824

CLASS 411/3§, 411/4* & 411/5 REFURBISHED 4 Cep

DMSO (A)–TBCK–TSOL–DMSO (B). Kent Coast Express Stock. Refurbished and renumbered from the 71/72xx series. Fitted with hopper ventilators, Inter-City 70 seats, fluorescent lighting and PA.

Electrical Equipment: 1957-type (*1951-type).
Bogies: One Mk. 4 (Mk 3B†) motor bogie (DMSO). Commonwealth trailer bogies.
Gangways: Throughout.
Traction Motors: Two EE507 of 185 kW.
Dimensions: 19.75 x 2.82 m.
Maximum Speed: 90 mph.
§–Original refurbished unit (class 411/3). Does not have hopper ventilators, has a different interior colour scheme and has a TBFK with one compartment declassified.
61305 has four seats at inner vestibule end replaced by luggage racks.

DMSO (A). Dia. EA263. 64S. 44.15 t.
TBCK. Dia. EJ361. 24F 6S 2L. 36.17 t.
TSOL. Dia. EH282. 64S 2L. 33.78 t.
DMSO (B). Dia. EA264. 64S. 43.54 t.

Lot numbers are as follows, all cars being built at Ashford/Eastleigh:

61033–61040. Lot No. 30108 1956.
61041–61044. Lot No. 30111 1956.
61229–61240. Lot No. 30449 1958.
61304–61409. Lot No. 30454 1958–59.
61694–61811. Lot No. 30619 1960–61.
61868–61870. Lot No. 30638 1960–61.
61948–61961. Lot No. 30708 1963.
70033–70036. Lot No. 30109 1956.
70037–70040. Lot No. 30110 1956.
70041–70042. Lot No. 30112 1956.
70043–70044. Lot No. 30639 1961.
70229–70234. Lot No. 30450 1958.
70235–70240. Lot No. 30451 1958.
70241. Lot No. 30640 1961.
70260–70302. Lot No. 30455 1958–59.
70303–70355. Lot No. 30456 1958–59.
70503–70551. Lot No. 30620 1960–61.
70552–70610. Lot No. 30621 1960–61.

1500	**J**	NKCX	§	RE	61389	70345	70302	61388
1501	**N**	NKCX	*	RE	61041	70041	70034	61042
1502	**J**	NKCX	*	RE	61040	70040	70036	61039
1503	**J**	NKCX	*	RE	61033	70037	70033	61034
1504	**J**	NKCX	*	RE	61043	70042	71712	61037
1505	**N**	NKCX	*	RE	61044	70039	70035	61038
1506	**N**	NKCX		RE	61349	70325	70282	61348
1507	**N**	NKCX		RE	61363	70332	70289	61362
1508	**N**	NKCX		RE	61305	70303	70260	61304
1509	**N**	NKCX		RE	61335	70318	70275	61334
1510	**N**	NKCX		RE	61365	70333	70290	61364
1511	**N**	NKCX		RE	61367	70334	70291	61366
1512	**N**	NKCX		RE	61321	70311	70268	61320
1513	**N**	NKCX		RE	61796	70321	70278	61340
1514		NKCX		RE	61327	70314	70271	61326
1515	**J**	NKCX		RE	61345	70323	70280	61344
1516	**N**	NKCX		RE	61319	70310	70267	61318
1517	**J**	NKCX		RE	61317	70309	70266	61316
1518	**J**	NKCX		RE	61333	70317	70274	61332
1519	**J**	NKCX		RE	61403	70352	70516	61402
1520	**J**	NKCX		RE	61343	70327	70284	61380
1521	**J**	NKCX		RE	61353	70324	70281	61352
1522	**J**	NKCX		RE	61347	70341	70665	61346
1523	**J**	NKCX		RE	61383	70342	70299	61382
1524	**J**	NKCX		RE	61309	70305	70262	61308
1525	**J**	NKCX		RE	61235	70238	70232	61236
1526	**J**	NKCX		RE	61239	70240	70234	61240
1527	**J**	NKCX		RE	61237	70239	70233	61238
1528	**J**	NKCX		RE	61379	70340	70297	61378
1529	**J**	NKCX		RE	61355	70328	70285	61354
1530	**J**	NKCX		RE	61331	70316	70273	61330
1531	**J**	NKCX		RE	61233	70237	70231	61234
1532	**N**	NKCX		RE	61391	70346	71626	61390
1533	**J**	NKCX		RE	61393	70347	71627	61385
1534	**J**	NKCX		RE	61405	70353	71628	61404
1535	**J**	NKCX		RE	61397	70349	71629	61396
1536	**J**	NKCX		RE	61399	70350	71631	61398
1537	**J**	NKCX		RE	61229	70235	70229	61230
1538	**J**	NKCX		RE	61307	70304	70261	61306
1539	**J**	NKCX		RE	61401	70351	71632	61400
1540	**J**	NKCX		RE	61870	70343	70300	61384

1541	**J**	NKCX		RE	61409	70355	71633	61408
1542	**J**	NKCX		RE	61395	70348	71634	61394
1543	**J**	NKCX		RE	61323	70312	70269	61322
1544	**J**	NKCX		RE	61315	70308	70265	61314
1545	**J**	NKCX		RE	61359	70330	70287	61358
1546	**J**	NKCX		RE	61357	70329	70286	61356
1547	**J**	NKCX		RE	61329	70315	70272	61328
1548	**J**	NKCX		RE	61375	70338	70295	61374
1549	**J**	NKCX		RE	61339	70320	70277	61338
1550	**J**	NKCX		RE	61313	70307	70264	61312
1551	**J**	NKCX		RE	61325	70313	70270	61324
1552	**J**	NKCX		RE	61373	70337	70294	61372
1553	**J**	NKCX		RE	61351	70306	70263	61350
1554	**J**	NKCX		RE	61369	70335	70292	61368
1555	**J**	NKCX		RE	61311	70326	70283	61310
1556	**J**	NKCX		RE	61371	70336	70293	61370
1557	**J**	NKCX		RE	61337	70331	70288	61360
1558	**J**	NKCX		RE	61361	70319	70276	61336
1559	**J**	NKCX		RE	61377	70339	70296	61376
1560	**N**	NKCX		RE	61387	70344	70301	61386
1561	**N**	NKCX		RE	61231	70604	70230	61232
1562	**N**	NKCX	†	RE	61407	70236	70241	61406
1563	**J**	NKCX	†	RE	61740	70575	70526	61741
1564	**J**	NKCX	†	RE	61788	70599	70550	61789
1565	**J**	NKCX	†	RE	61762	70586	71711	61763
1566	**J**	NKCX	†	RE	61722	70566	70517	61723
1567	**J**	NKCX	†	RE	61786	70598	70549	61787
1568	**J**	NKCX	†	RE	61766	70588	70539	61767
1569	**J**	NKCX	†	RE	61782	70596	70547	61783
1570	**J**	NKCX	†	RE	61738	70574	70525	61739
1571	**J**	NKCX	†	RE	61806	70608	71636	61807
1572	**J**	NKCX	†	RE	61734	70572	70523	61735
1573	**J**	NKCX	†	RE	61726	70568	70519	61727
1574	**J**	NKCX	†	RE	61792	70601	71635	61793
1575	**J**	NKCX	†	RE	61768	70583	70540	61769
1576	**J**	NKCX	†	RE	61770	70590	70541	61771
1577	**J**	NKCX	†	RE	61718	70564	70515	61719
1578	**J**	NKCX	†	RE	61700	70555	70506	61701
1579	**J**	NKCX	†	RE	61772	70591	70542	61773
1580	**J**	NKCX	†	RE	61756	70589	70534	61757
1581	**J**	NKCX	†	RE	61784	70597	70548	61785
1582	**J**	NKCX	†	RE	61748	70603	71630	61797
1583	**J**	NKCX	†	RE	61746	70578	70529	61747
1584	**J**	NKCX	†	RE	61752	70581	70532	61753
1585	**J**	NKCX	†	RE	61710	70560	70511	61711
1586	**J**	NKCX	†	RE	61714	70562	70513	61715
1587	**J**	NKCX	†	RE	61764	70587	71625	61765
1588	**J**	NKCX	†	RE	61720	70044	70520	61721
1589	**J**	NKCX	†	RE	61742	70576	70527	61743
1590	**J**	NKCX	†	RE	61696	70553	70504	61697
1591	**J**	NKCX	†	RE	61790	70600	70551	61791
1592	**J**	NKCX	†	RE	61778	70594	70545	61779

1593	**J**	NKCX	†	RE	61730	70570	70521	61731
1594	**J**	NKCX	†	RE	61754	70582	70533	61755
1595	**J**	NKCX	†	RE	61704	70557	70508	61705
1596	**J**	NKCX	†	RE	61716	70563	70514	61717
1597	**J**	NKCX	†	RE	61708	70559	70510	61709
1598	**J**	NKCX	†	RE	61780	70595	70546	61781
1599	**J**	NKCX	†	RE	61706	70558	70509	61707
1600	**J**	NKCX	†	RE	61724	70567	70518	61725
1601	**J**	NKCX	†	RE	61776	70593	70544	61777
1602	**J**	NKCX	†	RE	61958	70565	70279	61959
1603		NKCX	†	RE	61728	70569	70298	61729
1604		NKCX	†	RE	61732	70571	70522	61733
1605	**N**	NKCX	†	RE	61712	70561	70512	61713
1606	**N**	NKCX	†	RE	61694	70552	70503	61695
1607	**N**	NKCX	†	RE	61698	70554	70505	61699
1608	**N**	NKCX	†	RE	61960	70659	70666	61961
1609	**J**	NKCX	†	RE	61744	70577	70528	61745
1610	**J**	NKCX	†	RE	61750	70580	70531	61751
1611	**J**	NKCX	†	RE	61758	70584	70537	61759
1612	**J**	NKCX	†	RE	61794	70602	70535	61795
1613	**J**	NKCX	†	RE	61760	70585	70536	61761
1614	**J**	NBTX	†	RE	61702	70556	70507	61703
1615	**J**	NBTX	†	RE	61956	70657	70664	61957
1616	**J**	NBTX	†	RE	61950	70654	70543	61951
1617	**J**	NBTX	†	RE	61800	70605	70661	61801
1618	**J**	NBTX	†	RE	61868	70043	70663	61869
1619	**J**	NBTX	†	RE	61952	70655	70662	61953
1620	**N**	NBTX	†	RE	61948	70653	70660	61949
1621	**N**	NBTX	†	RE	61810	70610	70524	61811
Spare		NKCX		SH	61035			61342

Former numbers of converted hauled stock:

71625 (4381)	71629 (3992)	71633 (4072)	71636 (4065)
71626 (3916)	71630 (3988)	71634 (4059)	71711 (3994)
71627 (3921)	71631 (4436)	71635 (3990)	71712 (4062)
71628 (3844)	71632 (4063)		

CLASS 421/3 FACELIFTED 4 Cig (PHASE 1)

DTCsoL–MBSO–TSO–DTCsoL. Express stock. Fitted with electric parking brake. Facelifted with new trim, fluorescent lighting in saloons, PA.

Electrical Equipment: 1963-type.
Bogies: Two Mk. 4 motor bogies (MBSO). B5 (SR) bogies (trailer cars).
Gangways: Throughout.
Traction Motors: Four EE507 of 185 kW.
Dimensions: 19.75 x 2.82 m.
Maximum Speed: 90 mph.

76022–75. DTCsoL. Dia. EE369. Lot 30740 Yk 1964–66. 18F 36S 2L. 35.5 t.
76076–129. DTCsoL. Dia. EE369. Lot. 30741 Yk 1964–66. 18F 36S 2L. 35 t.
62017–70. MBSO. Dia. ED264. Lot No. 30742 York 1964–66. 56S. 49 t.

70695–730. TSO. Dia. EH287. Lot No. 30743 York 1964–66. 72S. 31.5 t. **71044–97. TSO.** Dia. EH287. Lot No. 30817 York 1970. 72S. 31.5 t. **71766–70. TSO.** Dia. EH287. Lot No. 30744 York 1965–66. 72S. 31.5 t.

1701	(7312)	**N**	NSXX	BI	76087	62028	70706	76033
1702	(7326)	**N**	NSXX	BI	76101	62042	70720	76047
1703	(7322)	**N**	NSXX	BI	76097	62038	70716	76043
1704	(7317)	**N**	NSXX	BI	76092	62033	70711	76038
1705	(7301)	**N**	NSXX	BI	76076	62017	70695	76022
1706	(7319)	**N**	NSXX	BI	76094	62035	70713	76040
1707	(7309)	**N**	NSXX	BI	76084	62025	70703	76030
1708	(7335)	**N**	NSXX	BI	76110	62051	70729	76056
1709	(7328)	**N**	NSXX	BI	76103	62044	70722	76049
1710	(7303)	**N**	NSXX	BI	76078	62019	70697	76024
1711	(7033)	**N**	NSXX	BI	76114	62055	71766	76060
1712	(7304)	**N**	NSXX	BI	76079	62020	70698	76025
1713	(7047)	**N**	NSXX	BI	76128	62069	71767	76074
1714	(7302)		NSXX	BI	76077	62018	70696	76023
1715	(7307)	**N**	NSXX	BI	76082	62023	70701	76028
1716	(7325)	**N**	NSXX	BI	76100	62041	71768	76046
1717	(7308)	**N**	NSXX	BI	76083	62024	70702	76029
1718	(7306)	**N**	NSXX	BI	76081	62022	70700	76027
1719	(7035)	**N**	NSXX	BI	76116	62057	70719	76062
1720	(7038)	**N**	NSXX	BI	76098	62039	71769	76044
1721	(7315)	**N**	NSXX	BI	76090	62031	70709	76036
1722	(7331)	**N**	NSXX	BI	76106	62047	70725	76052
1723	(7332)	**N**	NSXX	BI	76107	62048	70726	76053
1724	(7039)	**N**	NSXX	BI	76120	62061	71770	76066
1725	(7313)	**N**	NSXX	BI	76088	62029	70707	76034
1726	(7334)	**N**	NSXX	BI	76109	62050	70728	76055
1727	(7336)	**N**	NSXX	BI	76111	62052	70730	76057
1728	(7324)	**N**	NSXX	BI	76099	62040	70718	76045
1729	(7329)	**N**	NSXX	BI	76104	62045	70723	76050
1730	(7330)	**N**	NSXX	BI	76105	62046	70724	76113
1731	(7320)	**N**	NSXX	BI	76095	62036	70714	76041
1732	(7321)	**N**	NSXX	BI	76096	62037	70715	76042

Former numbers of converted buffet cars:

71766 (69303)	71768 (69317)	71769 (69305)	71770 (69308)
71767 (69314)			

CLASS 421/6 — 4 Cig (PHASE 1/2)

DTCsoL–MBSO–TSO–DTCsoL. Express stock. Fitted with electric parking brake. Units reformed from Class 422 to enable all Class 422 power cars to have Mk. 6 motor bogies. Phase 1 units with phase 2 TSOs.

Electrical Equipment: 1963-type.
Bogies: Two Mk. 4 motor bogies (MBSO). B5 (SR) bogies (trailer cars).
Gangways: Throughout.
Traction Motors: Four EE507 of 185 kW.
Dimensions: 19.75 x 2.82 m.
Maximum Speed: 90 mph.

Lot Nos as for Class 421/3 except:
71044–97. TSO. Dia. EH287. Lot No. 30817 York 1970. 72S. 31.5 t.

f Facelifted vehicle.

1751	(2101)	**N**	NSXX	BI	76063	62054	71044	76059
1752	(2102)	**N**	NSXX	BI	76129	62064	71046	76069
1753	(2103)	**N**	NSXX	BI	76124	62065	71052	76070
1754	(2104)	**N**	NSXX	BI	76117	62058	71050	76051
1755	(2105)	**N**	NSXX	BI	76121	62062	71058	76067
1756	(2106)	**N**	NSXX	BI	76118	62059	71065	76064
1757	(2107)	**N**	NSXX	BI	76122	62063	71047	76068
1758	(2108)	**N**	NSXX	BI	76127	62068	71064	76073
1759	(2109)	**N**	NSXX	BI	76123	62070	71066	76075
1760	(2110)	**N**	NSXX	BI	76115	62056	71067	76061
1761	(2111)	**N**	NSXX	BI	76112	62053	71068	76058
1762	(2112)	**N**	NSXX	BI	76126	62067	71097	76072

CLASS 421/4 FACELIFTED 4 Cig (PHASE 2)

DTCsoL–MBSO–TSO–DTCsoL. Express stock. Facelifted with new trim, fluorescent lighting in saloons, PA.

Electrical Equipment: 1963-type.
Bogies: Two Mk. 6 motor bogies (MBSO). B5 (SR) bogies (trailer cars).
Gangways: Throughout.
Traction Motors: Four EE507 of 185 kW.
Dimensions: 19.75 x 2.82 m.
Maximum Speed: 90 mph.

DTCsoL. Dia. EE369. 18F 36S 2L. 35 t.
MBSO. Dia. ED264. 56S. 49t.
TSO. Dia. EH287. 72S. 31.5t.

For lot numbers see classes 421/2 or 422/2 except for:

76859. DTCsoL. Dia. EE369. Lot No. 30827 York 1972.
76860. DTCsoL. Dia. EE369 Lot No. 30828 York 1972.
62430. MBSO. Dia. ED264. Lot No. 30829 York 1972.
71106. TSO. Dia. EH287. Lot No. 30830 York 1972.

* "Greyhound" units. Fitted with an additional stage of field weakening to improve the maximum attainable speed.

1801	(7427)	**N**	NSXX	BI	76777	62415	71095	76848
1803	(7430)	**N**	NSXX	BI	76780	62418	71098	76851
1804	(7428)	**N**	NSXX	BI	76778	62416	71096	76849
1805	(7432)		NSXX	BI	76782	62420	71100	76853
1806	(7433)		NSXX	BI	76783	62421	71101	76854
1807	(7434)	**N**	NSXX	BI	76784	62422	71102	76855
1808	(7435)	**N**	NSXX	BI	76785	62423	71103	76856
1809	(7436)	**N**	NSXX	BI	76786	62424	71104	76857
1810	(7437)	**N**	NSXX	BI	76787	62425	71105	76858
1811	(7431)	**N**	NSXX	BI	76781	62419	71099	76852
1812	(7407)	**N**	NSXX	BI	76757	62395	71075	76828

1813	(7438)	**N**	NSXX		BI	76859	62430	71106	76860
1814	(1251)	**N**	NSSX		FR	76595	62301	70981	76625
1815	(1240)	**N**	NSSX		FR	76584	62290	70970	76614
1816	(1237)	**N**	NSSX		FR	76581	62287	70967	76611
1817	(1239)	**N**	NSSX		FR	76583	62289	70969	76613
1818	(1267)	**N**	NSSX		FR	76717	62355	71035	76788
1819	(1273)	**N**	NSSX	*	FR	76723	62361	71041	76794
1820	(1242)	**N**	NSSX	*	FR	76586	62292	70972	76616
1821	(1248)	**N**	NSSX	*	FR	76592	62298	70978	76622
1822	(1250)	**N**	NSSX	*	FR	76594	62300	70980	76624
1823	(2207)	**N**	NSSX	*	FR	76567	62283	71926	76577
1824	(2201)	**N**	NSSX	*	FR	76561	62277	71927	76571
1825	(2202)	**N**	NSSX	*	FR	76562	62278	71928	76572

Former numbers of converted buffet cars:

71926 (69315) | 71927 (69330) | 71928 (69331) |

Note: No new Lot Nos were issued for the above conversions.

CLASS 431 — MODIFIED 4 Rep.

DMSO–TCK–TBSK–DMSO. Express stock. Tractor units which work push & pull with class 438 (4 TC). The trailer cars were rebuilt from a loco-hauled FK and BSK and are from a withdrawn class 438 unit.

Electrical Equipment: 1966-type.
Bogies: Two Mk. 6 motor bogies (DMSO). B5 (SR) bogies (trailer cars).
Gangways: Throughout.
Traction Motors: Four EE546 of 300 kW.
Dimensions: 19.66 x 2.82 m.
Maximum Speed: 90 mph.

62141–62154. DMSO. Dia. EA260. Lot No. 30761 York 1967. 64S. 52.5 t.
62479. DMSO. Dia. EA260. Lot No. 30862 York 1974. 64S. 52.5 t.
TCK. Dia. EH362. Lot No. 30856 York 1974. 36F 6S 2L. 33.5 t.
TBSK. Dia. EJ260. Lot No. 30765 York 1966–67. 32S 1L. 35.5 t.

Note: These units are to be reformed as 6-car units with one power car per unit. The extra cars will be spare or reinstated vehicles (see Cl. 438).

1901	**N**	NSSX	BM	62142	71163	70830	62141
1902	**N**	NSSX	BM	62154	70845	70824	62153
1903	**N**	NSSX	BM	62145	70859	70827	62479
1904				62	7	7	
1905				62	7	7	
1906				62	7	7	

CLASS 422/2 — 4 Big (PHASE 2)

DTCsoL (A)–MBSO–TSRB–DTCsoL (B). Express stock.

Electrical Equipment: 1963-type.
Bogies: Two Mk. 6 motor bogies (MBSO). B5 (SR) bogies (trailer cars).
Gangways: Throughout.
Traction Motors: Four EE507 of 185 kW.

Dimensions: 19.75 x 2.82 m.
Maximum Speed: 90 mph.

76561–76570. DTCsoL(A). Dia. EE364. Lot No. 30802 York 1970. 18F 36S 2L. 35.5 t.
62279–62286. MBSO. Dia. ED260. Lot No. 30804 York 1970–71. 56S. 49 t.
62308. MBSO. Dia. ED260. Lot No. 30808 York 1971. 56S. 49 t.
69332–69339. TSRB. Dia. EN260. Lot No. 30805 York 1970. 40S. 35 t.
76573–76580. DTCsoL(B). Dia. EE363. Lot No. 30803 York 1970. 24F 28S 2L. 35.5 t.
76602. DTCsoL(A). Dia. EE363. Lot No. 30806 York 1970. 18F 36S 2L. 35.5 t.
76632. DTCsoL(B). Dia. EE363. Lot No. 30807 York 1970. 24F 28S 2L. 35.5 t.

2203	(7051)		NSXX	BI	76563	62279	69332	76573
2204	(7052)		NSXX	BI	76564	62280	69336	76574
2205	(7053)		NSXX	BI	76565	62281	69339	76575
2206	(7054)		NSXX	BI	76602	62308	69338	76632
2208	(7056)		NSXX	DI	76568	62284	69334	76578
2209	(7057)	N	NSXX	BI	76569	62285	69335	76579
2210	(7058)	N	NSXX	BI	76570	62286	69337	76580

CLASS 422/3 & 422/4(f) 4 Big (PHASE 2/1)

DTCsoL (A)–MBSO–TSRB–DTCsoL (B). Express stock. Units reformed from Class 421 to ensure that all Class 422 power cars have Mk. 6 motor bogies. Phase 2 units with phase 2 TSRBs (except for 69333 which is a phase 2 TSRB).

Electrical Equipment: 1963-type.
Bogies: Two Mk. 6 motor bogies (MBSO). B5 (SR) bogies (trailer cars).
Gangways: Throughout.
Traction Motors: Four EE507 of 185 kW.
Dimensions: 19.75 x 2.82 m.
Maximum Speed: 90 mph.
76726–76779. DTCsoL. Dia. EE364. Lot No. 30814 York 1970–72. 18F 36S 2L. 35.5 t.
62364–62388. MBSO. Dia. ED260. Lot No. 30816 York 1970. 56S. 49 t.
69301–69318. TSRB. Dia. EN260. Lot No. 30744 York 1966. 40S. 35 t.
69333. TSRB. Dia. EN260. Lot No. 30805 York 1970. 40S. 35 t.
76797–76850. DTCsoL. Dia. EE363. Lot No. 30815 York 1970–72. 24F 28S 2L. 35.5 t.
f Facelifted unit (except for 69333).

2251	(1276)	N	NSXX	BI	76726	62364	69302	76797
2252	(1278)	N	NSXX	BI	76728	62366	69312	76799
2253	(1284)	N	NSXX	BI	76734	62372	69313	76805
2254	(1282)	N	NSXX	BI	76732	62370	69306	76803
2255	(1290)	N	NSXX	BI	76740	62378	69310	76811
2256	(1297)	N	NSXX	BI	76747	62385	69307	76818
2257	(1279)	N	NSXX	BI	76729	62367	69311	76800
2258	(1296)	N	NSXX	BI	76746	62384	69316	76817
2259	(1298)	N	NSXX	BI	76748	62386	69318	76819
2260	(1299)	N	NSXX	BI	76749	62387	69304	76820
2261	(1300)	N	NSXX	BI	76750	62388	69301	76821

2262	(1802)	**N**	NSXX	BI (f)	76779	62417	69333	76850
Spare							69309	

CLASS 412 REFURBISHED 4 Bep

DMSO (A)–TBCK–TRB–DMSO (B). Kent Coast Express Stock. Refurbished and renumbered from the 70xx series. Fitted with hopper ventilators, Inter-City 70 seats, fluorescent lighting and PA.

Electrical Equipment: 1957-type.
Bogies: One Mk 3B motor bogie (DMSO). Commonwealth trailer bogies.
§ Rebogied with Mk 6 motor bogies and B5(SR) trailer bogies.
* Rebogied with B5(SR) trailer bogies.
Gangways: Throughout.
Traction Motors: Four EE507 of 185 kW.
Dimensions: 19.75 x 2.82 m.
Maximum Speed: 90 mph.

DMSO (A). Dia. EA263. 64S. 44.15 t.
TBCK. Dia. EJ361. 24F 6S 2L. 36.17 t.
TRSB. Dia. EN261. 24S 1L + 9 longitudinal buffet chairs. 35.5 t.
DMSO (B). Dia. EA264. 64S. 43.54 t.
Lot numbers are as follows, all cars being built at Ashford/Eastleigh:

61736–61809. Lot No. 30619 1960–61.
61954–61955. Lot No. 30708 1963.
69341–69347. Lot No. 30622 1961.
70354. Lot No. 30456 1959.
70573–70609. Lot No. 30621 1960–61.
70656. Lot No. 30709 1963.

2301	(7019)	**N**	NSSX	FR	61804	70607	69341	61805
2302	(7194)	**N**	NSSX	FR	61774	70592	69342	61775
2303	(7208)	**N**	NSSX	FR	61954	70656	69343	61955
2304	(7175)	**N**	NSSX	FR	61736	70573	69344	61737
2305		**N**	NSSX	FR	61798	70354	69345	61799
2306	(7021)	**N**	NSSX	FR	61808	70609	69346	61809
2307	(7018)	**N**	NSSX	FR	61802	70606	69347	61803

Former numbers of converted buffet cars:

69341 (69014)
69342 (69019)
69343 (69018)
69344 (69012)
69345 (69013)
69346 (69016)
69347 (69015)

CLASS 442 WESSEX EXPRESS STOCK

DTFsoL–TSOL(A)–MBRSM–TSOL(B)–DTSOL. New express stock for Waterloo–Bournemouth–Weymouth service. Air conditioned (heat pump system). Power-operated sliding doors. PA. Can be hauled and heated by any BR e.t.h. fitted locomotive. Multiple working with class 33/1 and 73 locomotives.

Electrical Equipment: 1986-type.
Bogies: Mk 6 motor bogies (MBRSM). T4 trailer bogies.
Gangways: Throughout.
Traction Motors: Four EE546 of 300 kW recovered from class 432.
Dimensions: 22.57 x 2.74 m.
Maximum Speed: 100 mph.

DTFsoL. Dia. EE160. Lot No. 31030 Derby 1988–89. 48F 1L. Public Telephone. 30.06 t.
TSOL (A). Dia. EH288. Lot No. 31032 Derby 1988–89. 80S 2L. 35.26 t.
MBRSM. Dia. ED265. Lot No. 31034 Derby 1988–89. 14S. 54.10 t.
TSOL (B). Dia. EH289. Lot No. 31033 Derby 1988–89. 76S 2L + wheelchair space. + 2 tip-up seats. 35.36 t.
DTSOL. Dia. EE273. Lot No. 31031 Derby 1988–89. 78S 1L. 39.06 t.

2401	**N**	NSSXBM	77382	71818	62937	BEAULIEU	71842	77406
2402	**N**	NSSXBM	77383	71819	62938	COUNTY OF HAMPSHIRE	71843	77407
2403	**N**	NSSXBM	77384	71820	62941		71844	77408
2404	**N**	NSSXBM	77385	71821	62939		71845	77409
2405	**N**	NSSXBM	77386	71822	62944		71846	77410
2406	**N**	NSSXBM	77387	71823	62941		71847	77411
2407	**N**	NSSXBM	77388	71824	62943		71848	77412
2408	**N**	NSSXBM	77389	71825	62945		71849	77413
2409	**N**	NSSXBM	77390	71826	62946		71850	77414
2410	**N**	NSSXBM	77391	71827	62948		71858	77415
2411	**N**	NSSXBM	77392	71828	62940		71852	77422
2412	**N**	NSSXBM	77393	71829	62947		71853	77417
2413	**N**	NSSXBM	77394	71830	62949		71854	77418
2414	**N**	NSSXBM	77395	71831	62950		71855	77419
2415	**N**	NSSXBM	77396	71832	62951		71856	77420
2416	**N**	NSSXBM	77397	71833	62952		71857	77421
2417	**N**	NSSXBM	77398	71834	62953		71852	77416
2418	**N**	NSSXBM	77399	71835	62954		71859	77423
2419	**N**	NSSXBM	77400	71836	62955	BBC SOUTH TODAY	71860	77424
2420	**N**	NSSXBM	77401	71837	62956		71861	77425
2421	**N**	NSSXBM	77402	71838	62957		71862	77426
2422	**N**	NSSXBM	77403	71839	62958		71863	77427
2423	**N**	NSSXBM	77404	71840	62959		71864	77428
2424	**N**	NSSXBM	77405	71841	62960		71865	77429

CLASS 423 — 4 Vep

DTCsoL–MBSO–TSO–DTCsoL. Outer suburban stock. Fitted with electric parking brake.

Electrical Equipment: 1963-type.
Bogies: Two Mk. 4 motor bogies (MBSO). B5 (SR) bogies (trailer cars).
Gangways: Throughout.
Traction Motors: Four EE507 of 185 kW.
Dimensions: 19.75 x 2.82 m.
Maximum Speed: 90 mph.

62121–40. MBSO. Dia. ED261. Lot No. 30760 Derby 1967. 58S. 49 t.
62182–216. MBSO. Dia. ED261. Lot No. 30773 York 1967–68. 58S. 49 t.
62217–66. MBSO. Dia. ED263. Lot No. 30794 York 1968–69. 58S. 49 t.
62267–76. MBSO. Dia. ED263. Lot No. 30800 York 1970. 58S. 49 t.
62317–54. MBSO. Dia. ED263. Lot No. 30813 York 1972–73. 58S. 49 t.
62435–75. MBSO. Dia. ED263. Lot No. 30851 York 1973–74. 58S. 49 t.
70781–800. TSO. Dia. EH276. Lot No. 30759 Derby 1967. 98S. 31.5 t.
70872–906. TSO. Dia. EH276. Lot No. 30772 York 1967–68. 98S. 31.5 t.

70907–56. TSO. Dia. EH283. Lot No. 30793 York 1968–69. 98S. 31.5 t.
70957–66. TSO. Dia. EH283. Lot No. 30801 York 1970. 98S. 31.5 t.
70997–71034. TSO. Dia. EH283. Lot No. 30812 York 1972–73. 98S. 31.5 t.
71115–55. TSO. Dia. EH283. Lot No. 30852 York 1973–74. 98S. 31.5 t.
76230–69. DTCsoL. Dia. EE365. Lot No. 30758 York 1967. 18F 46S 1L. 35 t.
76333–402. DTCsoL. Dia. EE365. Lot No. 30771 Yk 1967–68. 18F 46S 1L. 35 t.
76441–540. DTCsoL. Dia. EE367. Lot No. 30792 Yk 1968–69. 18F 46S 1L. 35 t.
76541–60. DTCsoL. Dia. EE367. Lot No. 30799 York 1970. 18F 46S 1L. 35 t.
76641–716. DTCsoL. Dia. EE367. Lot No. 30811 Yk 1972–73. 18F 46S 1L. 35 t.
76861–942. DTCsoL. Dia. EE368. Lot No. 30853 Yk 1973–74. 18F 46S 1L. 35 t.

Renumbered from 7700–7894 respectively with the last three digits unchanged.

3001	**N**	NSSX	BM	76230	62121	70781	76231
3002		NSSX	BM	76233	62122	70782	76232
3003	**N**	NSSX	BM	76234	62123	70783	76235
3006	**N**	NSSX	BM	76241	62126	70786	76240
3007	**N**	NSSX	BM	76243	62127	70787	76242
3008	**N**	NSSX	BM	76244	62128	70788	76245
3009	**N**	NSSX	BM	76246	62129	70789	76247
3010	**N**	NSSX	BM	76369	62130	70790	76249
3011	**N**	NSSX	BM	76251	62131	70791	76250
3012	**N**	NSSX	BM	76252	62132	70792	76253
3013	**N**	NSSX	BM	76255	62133	70793	76254
3014	**N**	NSSX	BM	76257	62134	70794	76248
3015	**N**	NSSX	BM	76258	62135	70795	76259
3016	**N**	NSSX	BM	76261	62136	70796	76260
3017	**N**	NSSX	BM	76262	62137	70797	76263
3018	**N**	NSSX	BM	76265	62138	70875	76264
3019	**N**	NSSX	BM	76267	62139	70799	76266
3020	**N**	NSSX	BM	76269	62140	70800	76268
3026	**N**	NSSX	BM	76344	62187	70877	76343
3030	**N**	NSSX	BM	76352	62191	70881	76351
3032	**N**	NSSX	BM	76356	62193	70883	76355
3034	**N**	NSSX	BM	76360	62195	70885	76359
3035	**N**	NSSX	BM	76362	62196	70890	76361
3037	**N**	NSSX	BM	76366	62198	70888	76365
3046	**N**	NSSX	BM	76384	62207	70897	76383
3068	**N**	NSSX	FR	76466	62229	70919	76370
3069	**N**	NSSX	FR	76468	62230	70920	76467
3070		NKCX	RE	76470	62231	70921	76469
3071	**N**	NSSX	FR	76472	62232	70922	76471
3075	**N**	NSSX	WD	76480	62236	70926	76479
3077	**N**	NSSX	WD	76484	62238	70928	76483
3078	**N**	NSSX	WD	76486	62239	70929	76485
3080	**N**	NSXX	BI	76490	62241	70931	76489
3088	**N**	NSXX	BI	76506	62249	70939	76505
3092	**N**	NSXX	BI	76514	62253	70943	76513
3093	**N**	NSXX	BI	76516	62254	70944	76515
3094	**N**	NSXX	BI	76518	62255	70945	76517
3095	**N**	NSXX	BI	76520	62256	70946	76519
3096	**N**	NSXX	BI	76522	62257	70947	76521

3097	**N**	NSXX	BI	76524	62258	70948	76523
3098	**N**	NSXX	BI	76364	62259	70949	76525
3099	**N**	NSXX	BI	76528	62260	70950	76527
3107	**N**	NSXX	BI	76544	62268	70958	76543
3112	**N**	NSXX	BI	76554	62273	70963	76553
3113	**N**	NSSX	WD	76556	62274	70964	76555
3114	**N**	NSXX	BI	76558	62275	70965	76557
3115	**N**	NSXX	BI	76560	62276	70966	76559
3116	**N**	NSSX	WD	76641	62317	70997	76642
3117	**N**	NSSX	WD	76643	62318	70998	76644
3118	**N**	NSSX	WD	76645	62319	70999	76646
3120	**N**	NSSX	WD	76649	62321	71001	76650
3121	**N**	NSSX	WD	76651	62322	71002	76652
3125	**N**	NSSX	WD	76659	62326	71006	76660
3134	**N**	NSXX	BI	76677	62335	71015	76678
3135	**N**	NSXX	BI	76679	62336	71016	76680
3136		NSXX	BI	76681	62337	71017	76682
3137	**N**	NSXX	BI	76683	62338	71018	76684
3138	**N**	NSXX	BI	76685	62339	71019	76686
3139	**N**	NSXX	BI	76687	62340	71020	76688
3140	**N**	NSXX	BI	76689	62341	70887	76363
3141	**N**	NSXX	BI	76691	62342	71022	76692
3142	**N**	NSBX	WD	76693	62343	71023	76694
3144	**N**	NSBX	WD	76697	62345	71025	76698
3147	**N**	NSXX	BI	76703	62348	71028	76704
3148	**N**	NSXX	BI	76705	62349	71029	76706
3149	**N**	NSXX	BI	76707	62350	71030	76708
3151		NSXX	BI	76711	62352	71032	76712
3152	**N**	NSXX	BI	76465	62353	71033	76714
3153	**N**	NSXX	RE	76715	62354	71034	76716
3154	**N**	NSBX	WD	76861	62435	71115	76862
3155	**N**	NSBX	WD	76863	62436	71116	76864
3156	**N**	NSBX	WD	76865	62437	71117	76866
3157	**N**	NSBX	WD	76867	62438	71118	76868
3158	**N**	NSBX	WD	76869	62439	71119	76870
3159	**N**	NSBX	WD	76871	62440	71120	76872
3160	**N**	NSBX	WD	76873	62441	71121	76874
3161	**N**	NKCX	RE	76875	62442	71122	76876
3162	**N**	NKCX	RE	76877	62443	71123	76878
3163	**N**	NKCX	RE	76879	62444	71124	76880
3164	**N**	NKCX	RE	76881	62445	71125	76882
3165	**N**	NKCX	RE	76883	62446	71126	76884
3166	**N**	NKCX	RE	76885	62447	71127	76886
3167	**N**	NKCX	RE	76887	62448	71128	76888
3169	**N**	NKCX	RE	76891	62450	71130	76892
3170	**N**	NKCX	RE	76893	62451	71131	76894
3171	**N**	NKCX	RE	76895	62452	71132	76896
3172	**N**	NKCX	RE	76897	62453	71133	76898
3173	**N**	NKCX	RE	76899	62454	71134	76900
3175	**N**	NKCX	RE	76903	62456	71136	76904
3176	**N**	NKCX	RE	76905	62457	71137	76906
3177	**N**	NKCX	RE	76907	62458	71138	76908

3178	**N**	NKCX	RE	76909	62463	71139	76910
3179	**N**	NKCX	RE	76911	62460	71140	76912
3180	**N**	NKCX	RE	76913	62461	71141	76914
3181	**N**	NKCX	RE	76915	62462	71142	76916
3182	**N**	NKCX	RE	76917	62459	71143	76918
3183	**N**	NKCX	RE	76919	62464	71144	76920
3184	**N**	NKCX	RE	76921	62465	71145	76922
3185	**N**	NKCX	RE	76923	62466	71146	76924
3186	**N**	NKCX	RE	76925	62467	71147	76926
3187	**N**	NKCX	RE	76927	62468	71148	76928
3188	**N**	NKCX	RE	76929	62469	71149	76930
3189	**N**	NKCX	RE	76931	62470	71150	76932
3190	**N**	NKCX	RE	76933	62471	71151	76934
3191	**N**	NKCX	RE	76935	62472	71152	76936
3192	**N**	NKCX	RE	76937	62473	71153	76938
3193	**N**	NKCX	RE	76939	62474	71154	76940
3194	**N**	NKCX	RE	76941	62475	71155	76942

CLASS 413/2 — 4 Cap

DTCsoL–MBSO–MLSO–DTSsoL. Formed 1982 by the combination of pairs of class 414 (2 Hap) units. Driving equipment removed from motor cars, plus one set of guard's equipment. Driving trailers all declassified.

Electrical Equipment: 1951-type.
Bogies: Mk. 4.
Gangways: Non-gangwayed.
Traction Motors: Two EE507 of 185 kW.
Dimensions: 19.49 x 2.82 m.
Maximum Speed: 90 mph.

All built at Eastleigh on frames laid at Ashford.

65393–65396. MBSO/MLSO. Dia. EB269. Lot No. 30314 1957. 84S. 42 t.
65397–65402. MBSO/MLSO. Dia. EB269. Lot No. 30319 1957. 84S. 42 t.
65404–65434. MBSO/MLSO. Dia. EB269. Lot No. 30388 1958. 84S. 42 t.
77115–77118. DTSsoL. Dia. EE221. Lot No. 30316 1957. 69S 2L. 32.5 t.
77119–77124. DTSsoL. Dia. EE221. Lot No. 30320 1957. 69S 2L. 32.5 t.
77126–77156. DTSsoL. Dia. EE221. Lot No. 30389 1958. 69S 2L. 32.5 t.

3201	**N**	NKCX	GI	77120	65398	65401	77123
3202	**N**	NKCX	GI	77118	65396	65412	77134
3203	**N**	NKCX	GI	77117	65395	65424	77146
3204	**N**	NKCX	GI	77132	65410	65420	77142
3205	**N**	NKCX	GI	77135	65413	65422	77144
3206	**N**	NKCX	GI	77141	65419	65423	77145
3207	**N**	NKCX	GI	77126	65404	65428	77150
3208	**N**	NKCX	GI	77147	65425	65429	77151
3209	**N**	NKCX	GI	77129	65407	65430	77152
3210		NKCX	GI	77124	65402	65434	77156
3211	**N**	NKCX	GI	77115	65393	65427	77149
3212		NKCX	GI	77128	65406	65433	77155
3213		NKCX	GI	77119	65397	65432	77154

CLASS 413/3 — 4 Cap

DTCsoL–MBSO–MLSO–DTCsoL. Formed 1982 by the combination of pairs of class 414 (2 Hap) units. Driving equipment removed from motor cars, plus one set of guard's equipment. Driving trailers all declassified.

Electrical Equipment: 1957-type.
Bogies: Mk. 4.
Gangways: Non-gangwayed.
Traction Motors: Two EE507 of 185 kW.
Dimensions: 19.49 x 2.82 m.
Maximum Speed: 90 mph.

MBSO/MLSO. Dia. EB270. Lot No. 30452 Afd/Elh 1958–59. 84S. 42 t.
DTSsoL. Dia. EE222. Lot No. 30453 Afd/Elh 1958–59. 69S 2L. 32.5 t.

3301	**N**	NKCX	GI	75373	61253	61255	75375
3302	**N**	NKCX	GI	75361	61241	61244	75364
3303		NKCX	GI	75370	61250	61252	75372
3304		NKCX	GI	75402	61282	61283	75403
3305		NKCX	GI	75399	61279	61302	75422
3306		NKCX	GI	75374	61254	61256	75376
3307		NKCX	GI	75378	61258	61271	75391
3308		NKCX	GI	75366	61246	61264	75304
3309		NKCX	GI	75371	61251	61257	75377
3310		NKCX	GI	75363	61243	61259	75386
3311	**N**	NKCX	GI	75411	61291	61297	75417

CLASS 423/1 — Facelifted 4 Vep.

DTCsoL–MBSO–TSO–DTCsoL. For details see 3000 series. Facelifted with fluorescent lighting, PA. The MBSO has been modified to seat 76S. New design codes not yet recored by BR Rolling Stock Library.

3405	(3005)	**N**	NSSX	BM	76239	62271	70785	76238
3421	(3168)	**N**	NKCX	RE	76889	62449	71129	76890
3422	(3040)	**N**	NKCX	RE	76372	62201	70891	76371
3423	(3061)	**N**	NKCX	RE	76452	62222	70912	76451
3424	(3031)	**N**	NKCX	RE	76354	62185	70882	76353
3425	(3023)	**N**	NSSX	BM	76338	62192	70874	76358
3426	(3047)	**N**	NSSX	BM	76386	62208	70898	76385
3427	(3041)	**N**	NSSX	BM	76374	62184	70892	76373
3428	(3062)	**N**	NSSX	BM	76454	62223	70913	76453
3429	(3021)	**N**	NSSX	BM	76334	62202	70872	76333
3430	(3028)	**N**	NSSX	WD	76348	62189	70879	76347
3431	(3064)	**N**	NSSX	WD	76458	62182	70915	76457
3432	(3054)	**N**	NSSX	WD	76400	62225	70905	76399
3433	(3057)	**N**	NSSX	WD	76444	62215	70908	76443
3434	(3066)	**N**	NSSX	WD	76462	62218	70917	76461
3435	(3025)	**N**	NSXX	BI	76342	62228	70876	76341
3436	(3029)	**N**	NSXX	BI	76350	62190	70880	76349
3437	(3027)	**N**	NSXX	BI	76346	62186	70878	76345
3438	(3100)	**N**	NSXX	BI	76530	62262	70951	76529

3439	(3055)	**N**	NSXX	BI	76402	62227	70906	76401
3440	(3102)	**N**	NSXX	BI	76534	62188	70953	76533
3441	(3043)	**N**	NSXX	BI	76378	62261	70894	76377
3442	(3081)	**N**	NSXX	BI	76492	62216	70932	76491
3443	(3082)	**N**	NSXX	BI	76494	62263	70933	76493
3444	(3038)	**N**	NSXX	BI	76368	62204	70889	76367
3445	(3060)	**N**	NKCX	RE	76450	62242	70911	76449
3446	(3101)	**N**	NKCX	RE	76532	62243	70952	76531
3447	(3044)	**N**	NKCX	RE	76380	62199	70895	76379
3448	(3042)	**N**	NKCX	RE	76376	62221	70886	76375
3449	(3022)	**N**	NKCX	RE	76336	62205	70873	76335
3450	(3060)	**N**	NKCX	RE	76460	62203	70916	76459
3451	(3079)	**N**	NKCX	RE	76488	62240	70930	76487
3452		**N**	NKCX	RE	76340	62183	71021	76690
3453	(3045)	**N**	NKCX	RE	76382	62226	70896	76381
3454		**N**	NKCX	RE	76390	62200	70798	76389
3455	(3048)	**N**	NSSX	FR	76388	62206	70899	76387
3456	(3063)	**N**	NSSX	FR	76456	62210	70914	76455
3457	(3050)	**N**	NSSX	FR	76392	62197	70901	76391
3458	(3051)	**N**	NSSX	FR	76394	62209	70902	76393
3459	(3052)	**N**	NSSX	FR	76396	62224	70903	76395
3460	(3105)	**N**	NSXX	BI	76540	62211	70956	76539
3461	(3104)	**N**	NSXX	BI	76538	62212	70955	76537
3462	(3103)	**N**	NSXX	BI	76536	62213	70954	76535
3463	(3053)	**N**	NSXX	BI	76398	62266	70904	76397
3464	(3056)	**N**	NSXX	BI	76442	62265	70907	76441
3465	(3106)	**N**	NSXX	BI	76542	62264	70957	76541
3466	(3067)	**N**	NSBX	WD	76464	62214	70918	76463
3467	(3058)	**N**	NSBX	WD	76446	62217	70909	76445
3468	(3059)	**N**	NSBX	WD	76448	62267	70910	76447
3469	(3108)	**N**	NSBX	WD	76546	62219	70959	76545
3470	(3083)	**N**	NSBX	WD	76496	62220	70934	76495
3471	(3084)	**N**	NKCX	RE	76498	62269	70935	76497
3472	(3085)	**N**	NKCX	RE	76500	62244	70936	76499
3473	(3086)	**N**	NKCX	RE	76502	62245	70937	76501
3474	(3087)	**N**	NKCX	RE	76504	62246	70938	76503
3475	(3111)	**N**	NKCX	RE	76552	62270	70962	76551
3476	(3109)	**N**	NSXX	BI	76548	62247	70960	76547
3477	(3110)	**N**	NSXX	BI	76550	62248	70961	76549
3478	(3122)	**N**	NSXX	BI	76653	62125	71003	76654
3479	(3123)	**N**	NSBX	WD	76655	62272	71004	76656
3480	(3072)	**N**	NSBX	WD	76474	62323	70923	76473
3481	(3119)	**N**	NSBX	WD	76647	62324	70900	76648
3482	(3124)	**N**	NSBX	WD	76657	62320	71005	76658
3483	(3126)	**N**	NSBX	WD	76661	62233	71007	76662
3484	(3073)	**N**	NSBX	WD	76476	62325	70924	76475
3485	(3089)	**N**	NSBX	WD	76508	62327	70940	76507
3486	(3074)	**N**	NSBX	WD	76478	62234	70925	76477
3487	(3090)	**N**	NSBX	WD	76510	62250	70941	76509
3488	(3127)	**N**	NSBX	WD	76663	62235	71008	76664
3489	(3128)	**N**	NSBX	WD	76665	62251	71009	76666
3490	(3143)	**N**	NSBX	WD	76695	62328	71024	76696

3491	(3076)	**N**	NKCX	RE	76482	62329	70927	76481
3492	(3129)	**N**	NKCX	RE	76667	62344	71010	76668
3493	(3130)	**N**	NKCX	RE	76669	62237	71011	76670
3494	(3133)	**N**	NKCX	RE	76675	62330	71014	76676
3495	(3145)	**N**	NKCX	RE	76699	62331	71026	76700
3496	(3132)	**N**	NKCX	RE	76673	62334	71013	76674
3497	(3131)	**N**	NKCX	RE	76671	62346	71012	76672
3498	(3146)	**N**	NKCX	RE	76701	62333	71027	76702
3499	(3174)	**N**	NKCX	RE	76901	62347	71135	76902
3500	(3070)	**N**	NKCX	RE				
3501	(3091)	**N**	NKCX	RE	76512	62332	70942	76511
3502	(3150)	**N**	NKCX	RE	76709	62252	71031	76710
3503	(3136)	**N**						
3504	(3151)	**N**						
3505	(3071)	**N**						
3506								
3507								
3508								
3509								
3510								
3511								
3512								
3513								
3514								
3515								
3516								
3517								
3518								
3519								
3520								
3521								
3522								
3523								
3524								
3525								
3526								
3527								
3528								
3529								
3530								
3531								
3532								
3533								
3534								
3535								
3536								
3537								
3538								
3539								
3540								
3541								
3542								

3543			
3544			
3545			
3546			
3547			
3548			
3549			
3550			
3551			
Spare	76236	62194		76339
Spare		62351	62455	

CLASS 414/2 — 2 Hap

DMBSO–DTCsoL.

Electrical Equipment: 1951-type.
Bogies: Mk. 4.
Gangways: Non-gangwayed.
Traction Motors: Two EE507 of 185 kW.
Dimensions: 19.49 x 2.82 m.
Maximum Speed: 90 mph.

DMBSO. Dia. EB269. Lot No. 30388 Ashford/Eastleigh 1958. 82S. 42 t.
DTCsoL. Dia. EE361. Lot No. 30389 Ashford/Eastleigh 1958. 19F 60S 1L. 32.5 t.

4201 NSBX WD 65405 77127

CLASS 414/3 — 2 Hap

DMBSO–DTCsoL.

Electrical Equipment: 1957-type.
Bogies: Mk. 4.
Gangways: Non-gangwayed.
Traction Motors: Two EE507 of 185 kW.
Dimensions: 19.49 x 2.82 m.
Maximum Speed: 90 mph.

61249–61303. DMBSO. Dia. EB270. Lot No. 30452 Afd/Elh 1959. 84S. 42 t.
61654–61683. DMBSO. Dia. EB270. Lot No. 30617 Eastleigh 1961. 84S. 42 t.
75369–75423. DTCsoL. Dia. EE362. Lot No. 30453 Ashford/Eastleigh 1959. 19F 60S 1L. 32.5 t.
75706–75735. DTCsoL. Dia. EE362. Lot No. 30618 Eastleigh 1961. 19F 60S 1L. 32.5 t.

4301		NSBX	WD	61249	75369
4302	**N**	NSBX	WD	61260	75379
4303	**N**	NSBX	WD	61261	75381
4304		NSBX	WD	61262	75382
4305		NSBX	WD	61268	75388
4306		NSBX	WD	61270	75390
4307		NSBX	WD	61273	75393
4308	**N**	NSBX	WD	61275	75395
4309	**N**	NSBX	WD	61276	75396
4310	**N**	NSBX	WD	61278	75398
4311		NSBX	WD	61287	75407
4312		NSBX	WD	61288	75408
4313	**N**	NSBX	WD	61290	75410
4314		NSBX	WD	61294	75414
4315		NSBX	WD	61295	75415
4316		NSBX	WD	61296	75416

Converted Class 302 unit No. 302 990 is seen at Chelmsford Yard on its first day in service operating the 11.24 postal to Liverpool Street on 29th April 1989.
Dr. Iain Scotchman

Unit No. 303 060 is seen at Wilmslow, Cheshire on 14th January 1989. Painted in Greater Manchester PTE livery, it is working the 12.50 Altrincham to Alderley Edge service.
J.C. Hillmer

Class 305 No. 305 935 has been converted by InterCity into a mobile classroon to host classes for operating staff during the East Coast Main Line electrificatior project. Consisting of cars 75496, 61463 and 75548, the unit is pictured at Leed City station on 27th April 1989. *B.J. Nicoll*

After calling at Chelmsford, Class 309 No. 309 605 heads for Clacton with th 08.28 from Liverpool Street. *Dr. Iain Scotchmar*

Class 310 No. 310 106, painted in Provincial Midline livery, pauses at Sandwell and Dudley whilst operating the 12.05 Wolverhampton–Birmingham New Street Midline service. *Paul A. Biggs*

Class 312 No. 312 730 in Network SouthEast livery stands at Ilford on 19th May 1989. *Colin J. Marsden*

Strathclyde PTE-liveried Class 314 No. 314 202 is seen at Hyndland on 13th Ma
1989. *John Augustsor*

After calling at Broxbourne, 315 821 accelerates towards Liverpool Street witl
the 11.06 from Hartford East on 23rd February 1989. *Chris Wilsor*

·till in blue/grey livery, Class 317/2 No. 317 362 passes Hadley Wood with the 2.43 Peterborough–King's Cross service on 12th November 1988. *Chris Wilson*

·trathclyde PTE-liveried Class 318 No. 318 260 forms the 16.00 Glasgow Central–\yr near Falkland on 11th September 1989. *Brian Denton*

319 028 departs from Denmark Hill with the 12.11 Cricklewood–Orpington on 21st March 1989. Chris Wilson

One of the striking Class 321/4 units, No. 321 403 arrives at Euston with a service from Northampton on 7th August 1989.
John Augustson

Class 411 ('4 Cep') EMU No. 1618 in the popular but non-standard 'jaffa cake livery is seen outside Ashford station with the 06.53 Ramsgate–Cannon Stree service on 25th August 1988. *Chris Wilsor*

Southern Region Class 412 ('4 Bep') EMU No. 2306, in Network SouthEast livery passes Vauxhall with the 11.50 Waterloo–Portsmouth Harbour Network Expres service on 14th August 1989. This unit has recently been fitted with Mark 6 (motor) and B5S (trailer) bogies, replacing the Mark 4 and Commonwealth bogie previously fitted. *David Browr*

Class 414/3 ('2 Hap') EMUs Nos. 4315 and 4304, both still in blue/grey livery pass through Clapham Cutting with the 16.25 Waterloo–Alton service on 19th August 1989. In the background is an InterCity sponsored Gatwick Express.

David Brown

Inner-suburban Class 415/4 unit No. 5420 stands at BRML Eastleigh after works repaint in NSE livery. The date is 12th October 1989. *Brian Denton*

'Portsmouth Express' Class 421/4 ('4 Cig') No. 1822 passes Wimbledon Park flyover operating the 16.50 Waterloo–Portsmouth Harbour service. These units have had slight electrical modifications to give them better acceleration, and may be recognised by a rather ugly black trim on the yellow cab ends. The date is 8th August 1989. *David Brown*

Class 423/4 ('4 Vep') No. 3485 passes under the A27 trunk road at Portcreek with the 12.20 Portsmouth Harbour–Waterloo semi-fast on 20th September 1989.

Chris Wilson

Wimbledon-based Class 455 unit No. 5862 passes through Clapham Cutting with the 15.52 Waterloo–Effingham Junction service on 19th August 1989. The lines on the left were the scene of the Clapham crash on 12th December 1988, the results of the public inquiry having been published on 7th November 1989.

Alex Dasi-Sutton

NSE-liveried Class 438 ('4 TC') No. 8023 is seen on the Berks and Hants line near Bruton with the 13.30 Bristol Temple Meads–Weymouth service. Class 33 No. 33116 is propelling the formation on 12th August 1989. *John Chalcraft*

One of the attractive Class 442 'Wessex Electric' units heads out of Eastleigh on a Waterloo–Bournemouth working on 3rd April 1989.
Les Nixon

The Isle of Wight's 'new' stock – dating from 1938 – looks surprisingly smart. One of the ex-London Underground Class 483 units Nos. 483 002 stands at Ryde St. John's shed on 24th October 1989. The even more ancient Class 485 units they replace will be scrapped. *Brian Denton*

Built in 1940, the Class 487 Waterloo and City trains are due for replacement soon. Motored vehicle No. 58 is in the foreground in this view at Bank on 29th April 1989. Replacement stock has already been designated Class 482.

Chris Wilson

Class 504 No. 504 447 arrives at Besses o' th' Barn on a Bury–Manchester Victoria working on 22nd November 1988. The life of these units is drawing to a close as the route will be converted to a Metrolink light rail line. The unique 1200 V dc protected side-contact third rail, which has proved very reliable over the years, will be repaced by overhead catenary. *Hugh Ballantyne*

Class 507 No. 507 006 leads a 6-car Southport–Hunts Cross service into Formby station on 28th January 1989. Although dedicated to Merseyrail, these units are painted in standard BR blue and grey livery. The semaphore signals on this system are at long last to be replaced. *Les Nixon*

4317		NSBX	WD	61298 75418
4318	**N**	NSBX	WD	61300 75420
4319		NSBX	WD	61303 75423
4320		NSBX	WD	61654 75706
4321	**N**	NSBX	WD	61668 75720
4322	**N**	NSBX	WD	61683 75735

CLASS 405 — 1936 type 4 Sub

DMBSO–TS–TSO–DMBSO. This unit is kept for special workings and is not compatible electrically with other SR EMUs. It has automatic air brakes, but no electro-pneumatic brakes.

Electrical Equipment: 1936-type.
Bogies: Central 43" motor bogies and SR standard trailer bogies.
Gangways: Non-gangwayed.
Traction Motors: Two EE507 of 185 kW.
Dimensions: 19.05 x 2.82 m. (outer cars), 18.90 x 2.82 m (inner cars).
Maximum Speed: 75 mph.

DMBSO. Dia. EB265. Lot No. 3638 Eastleigh 1951. 82S. 42 t.
TS. Dia. EH262. Lot No. 3351 Eastleigh 1946. 120S. 27 t.
TSO. Dia. EH266. Lot No. 3384 Eastleigh 1948. 102S. 26 t.

Non-standard livery: SR malachite green.

4732 **0** NSXX BI 12795 10239 12354 12796

CLASS 415/1 — SR DESIGN 4 EPB

DMBSO–2TSO–DMBSO. Originally formed with a TS and a TSO, but all now have been reformed with 2TSO.

Electrical Equipment: 1951-type.
Bogies: Central 40" motor bogies and SR standard trailer bogies.
Gangways: Non-gangwayed.
Traction Motors: Two EE507 of 185 kW.
Dimensions: 19.05 x 2.82 m. (outer cars), 18.90 x 2.82 m (inner cars).
Maximum Speed: 75 mph (90 mph e).

Built to various SR lots as shown, all at Lancing/Eastleigh.

14001–14002. DMBSO. Dia. EB266. Lot No. 3638 1952. 82S. 40 t.
14039–14053. DMBSO. Dia. EB266. Lot No. 3756 1953. 82S. 40 t.
14069–14208. DMBSO. Dia. EB266. Lot No. 3757 1953–4. 82S. 40 t.
14213–14310. DMBSO. Dia. EB266. Lot No. 4016 1954–5. 82S. 40 t.
14311–14405. DMBSO. Dia. EB266. Lot No. 4099 1955–6. 82S. 40 t.
14417–14426. DMBSO. Dia. EB266. Lot No. 4172 1956. 82S. 40 t.
14433–14510. DMBSO. Dia. EB266. Lot No. 4173 1956–7. 82S. 40 t.
14522–14523. DMBSO. Dia. EB267. Lot No. 4281 1957. 84S. 40 t. Former class 418/0 vehicles.

15101–15113. TSO. Dia. EH270. Lot No. 3638 1951–2. 102S. 27 t.
15116–15133. TSO. Dia. EH270. Lot No. 3756 1953. 102S. 27 t.
15135–15158. TSO. Dia. EH270. Lot No. 3757 1953–4. 102S. 27 t.
15234–15277/279–283. TSO. Dia. EH270. Lot No. 4016 1954–5. 102S. 27 t.
15278. TSO. Dia. EH270. Lot No. 3504 1949. 102S. 27 t. (ex class 405 12392).
15288–15382. TSO. Dia. EH270. Lot No. 4099 1955–6. 102S. 27 t.
15395–15402. TSO. Dia. EH270. Lot No. 4172 1956. 102S. 27 t.

15405–15444. TSO. Dia. EH270. Lot No. 4173 1956–7. 102S. 27 t.

BR design DMBSOs (ex class 416/2):

65300–65310. DMBSO. Dia. EB269. Lot No. 30114 Ashford/Eastleigh 1954. 84S. 42 t.

65383. DMBSO. Dia. EB269. Lot No. 30314 Ashford/Eastleigh 1956. 84S. 42 t.

e–Express gear ratio.

5104		NKSX	SG	14207	15409	15157	14208
5107		NKSX	SG	14213	15153	15235	14214
5113		NKSX	SG	14225	15376	15241	14226
5114		NKSX	SG	14227	15426	15242	14228
5115		NKSX	SG	14230	15350	15243	14229
5121		NKSX	SG	14242	15247	15249	14241
5124		NKSX	SG	14248	15248	15252	14247
5126		NKSX	SG	14252	15244	15254	14376
5131		NKSX	SG	14261	15412	15259	14262
5133		NKSX	SG	14265	15132	15261	14266
5134		NKSX	SG	65383	15444	15262	14268
5138		NKSX	SG	14275	15423	15266	14276
5139		NKSX	SG	14277	15442	15267	14278
5145		NKSX	SG	14290	15158	15273	14289
5146		NKSX	SG	14291	15133	15274	14292
5148		NKSX	SG	14296	15343	15276	14295
5150		NKSX	SG	14099	15370	15278	14299
5153		NKSX	SG	14305	15256	15281	14306
5154		NKSX	SG	14308	15279	15282	14307
5155		NKSX	SG	14309	15107	15283	14310
5156		NKSX	SG	14311	15272	15334	14312
5157		NKSX	SG	14313	15381	15335	14314
5159		NKSX	SG	14317	15105	15337	14318
5160		NKSX	SG	14319	15268	15338	14320
5163		NKSX	SG	14325	15260	15341	14326
5166		NKSX	SG	14331	15253	15344	14332
5168		NKSX	SG	14335	15340	15346	14336
5169		NKSX	SG	65303	15109	15347	14338
5170		NKSX	SG	14339	15245	15348	14340
5173		NKSX	SG	14345	15116	15120	14346
5174		NKCX	SG	14348	15138	15352	14347
5176		NKSX	SG	14352	15396	15354	14351
5177		NKSX	SG	14354	15257	15355	14353
5182		NKSX	SG	14364	15357	15360	14363
5185		NKSX	SG	14369	15277	15363	14370
5189		NKSX	SG	14377	15339	15367	14378
5190	**N**	NKSX	SG	14380	15361	15368	14379
5191		NKSX	SG	14381	15108	15369	14382
5194		NKSX	SG	14388	15275	15372	14387
5195		NKSX	SG	14389	15239	15373	14390
5196		NKSX	SG	14392	15399	15374	14391
5201		NKSX	SG	14402	15126	15379	14401
5202		NKSX	SG	14404	15362	15380	14403
5209		NKSX	SG	14418	15395	15397	14417

5210		NKSX		SG	14420	15263	15398	14419
5213		NKSX		SG	14425	15353	15401	14426
5217		NKSX		SG	14434	15366	15405	14433
5220		NKSX		SG	14439	15349	15408	14260
5222		NKSX		SG	14490	15421	15410	14444
5223		NKSX		SG	14446	15427	15411	14445
5226		NKSX		SG	14451	15280	15414	14452
5228		NKSX		SG	14455	15415	15416	14456
5229		NKSX		SG	14333	15435	15417	14457
5230		NKSX		SG	14460	15110	15418	14459
5231		NKSX		SG	14461	15264	15419	14462
5232		NKSX		SG	14464	15422	15420	14463
5240		NKSX		SG	14103	15425	15428	14480
5242		NKSX		SG	14483	15402	15430	14484
5243		NKSX		SG	14405	15375	15356	14485
5248		NKSX		SG	14492	15443	15436	14510
5261		NKSX		SG	65300	15154	15413	65310
5264		NKSX	e	SG	14522	15258	15147	14523
5265	(5001)	NKCX		SG	14001	15429	15101	14002
5266	(5020)	NKCX		SG	14039	15234	15382	14040
5267	(5027)	NKCX		SG	14053	15123	15127	14042
5268	(5035)	NKCX		SG	14069	15113	15135	14070
5269	(5039)	NKCX		SG	14078	15118	15139	14094
5270	(5040)	NKCX		SG	14080	15122	15140	14079
5271	(5042)	NKCX		SG	14083	15130	15142	14084
5272	(5044)	NKCX		SG	14087	15345	15144	14088
5273	(5045)	NKCX		SG	14090	15129	15145	14089
5274	(5046)	NKCX		SG	14091	15364	15146	14092
5275	(5049)	NKCX		SG	14098	15271	15149	14097
5276	(5051)	NKCX		SG	14101	15143	15151	14102
5277	(5052)	NKCX		SG	14104	15103	15152	14495

CLASS 415/4 FACELIFTED SR DESIGN 4 EPB

DMBSO–2TSO–DMBSO. Facelifted with new trim, fluorescent lighting, PA.

Electrical Equipment: 1951-type.
Bogies: Central 40" motor bogies and SR standard trailer bogies.
Gangways: Non-gangwayed.
Traction Motors: Two EE507 of 185 kW.
Dimensions: 19.05 x 2.82 m. (outer cars), 18.90 x 2.82 m (inner cars).
Maximum Speed: 75 mph. (90 mph e).

Built to various SR lots as shown, all at Lancing/Eastleigh.
† Converted from class 405 (4 Sub) TS. Vehicles in the 154xx series were converted on facelifting.
§ Converted from class 415 TS.

14003–14030. DMBSO. Dia. EB277. Lot No. 3638 1952. 82S. 40 t.
14037–14066. DMBSO. Dia. EB277. Lot No. 3756 1953. 82S. 40 t.
14071–14206. DMBSO. Dia. EB277. Lot No. 3757 1953–4. 82S. 40 t.
14211–14304. DMBSO. Dia. EB277. Lot No. 4016 1954–5. 82S. 40 t.
14315–14410. DMBSO. Dia. EB277. Lot No. 4099 1955–6. 82S. 40 t.
14411–14430. DMBSO. Dia. EB277. Lot No. 4172 1956. 82S. 40 t.

14431–14520. DMBSO. Dia. EB277. Lot No. 4173 1956–7. 82S. 40 t.
14521–14570. DMBSO. Dia. EB278. Lot No. 4281 1957. 82S. 40 t. Former class 418/0 vehicle or 416/3 vehicles.

Note: All position 2 TSOs are Dia. EH280 (Non-smoking), whereas all position 3 TSOs are Dia. EH281 (Smoking).

15001/10. 15466/67/71/74/75. TSO†. Dia. EH270. Lot No. 1094 1946. 102S. 27 t.
15033/34/36/45–47/53/55/59/60/67. TSO†. Lot No. 3231 1947. 102S. 27 t.
15002/3/19/23/27/30/39–42/48/50. 15450. TSO†. Lot No. 3351 1947. 102S. 27 t.
15449/65/79/80. TSO†. Lot No. 3384 1948. 102S. 27 t.
15453/63. TSO†. Lot No. 3385 1948. 102S. 27 t.
15012/54/74. TSO†. Lot No. 3386 1948. 102S. 27 t.
15004/9/11/13/20/29/31/37/44/51/52/56–58/61–63/65/66/68–71/73/75–78. TSO†. Lot No. 3463 1948. 102S. 27 t.
15454/56/57/58/60/62/68/70/72/76/78. TSO†. Lot No. 3464 1949. 102S. 27 t.
15451. TSO†. Lot No. 3617 1948. 102S. 27 t.
15452/59/61/64/69/73. TSO†. Lot No. 3504 1949. 102S. 27 t.
15481. TSO†. Lot No. 3505 1950. 102S. 27 t.
15455/77. TSO†. Lot No. 3506 1950. 102S. 27 t.
15104–15115. TSO. Lot No. 3638 1951–2. 102S. 27 t.
15119–15128. TSO. Lot No. 3756 1953. 102S. 27 t.
15136–15182. TSO§. Lot No. 3757 1953–4. 102S. 27 t.
15184–15227/230–265.§ TSO. Lot No. 4016 1954–5. 102S. 27 t.
15285–15383. TSO§. Lot No. 4099 1955–6. 102S. 27 t.
15384–15403. TSO§. Lot No. 4172 1956. 102S. 27 t.
15404–15448. TSO. Lot No. 4173 1956–7. 102S. 27 t.

e–Express gear ratio.

5401		NSLX	e	SU	14556	15449	15450	14521
5402		NSLX		SU	14449	15464	15465	14407
5403		NSLX		SU	14286	15174	15221	14285
5404		NSLX		SU	14435	15036	15406	14436
5405	**N**	NSLX		SU	14470	15053	15216	14469
5406		NSLX		SU	14303	15230	15285	14356
5407	**N**	NSLX		SU	14396	15392	15060	14427
5408		NSLX		SU	14297	15227	15313	14298
5409	**N**	NSLX		SU	14494	15065	15047	14206
5410		NSLX	e	SU	14540	15386	15304	14528
5411		NSLX		SU	14475	15056	15192	14476
5412		NSLX		SU	14415	15451	15452	14304
5413		NSLX		SU	14428	15326	15191	14395
5414		NSLX		SU	14441	15039	15182	14442
5415		NSLX		SU	14465	15040	15051	14466
5416		NSLX		SU	14421	15389	15324	14422
5417		NSLX		SU	14478	15057	15041	14477
5418	**N**	NSLX		SU	14443	15063	15219	14282
5419	**N**	NSLX		SU	14473	15055	15058	14474
5420	**N**	NSLX		SU	14240	15198	15202	14239
5421		NSLX		SU	14500	15068	15073	14499
5422	**N**	NSLX		SU	14467	15209	15050	14468

5423		NSLX		SU	14511	15074	15212	14512
5424		NSLX		SU	14447	15042	15052	14448
5425		NSLX	e	SU	14538	15453	15454	14570
5426		NSLX		SU	14517	15077	15066	14518
5427		NSLX		SU	14453	15045	15046	14454
5428		NSLX		SU	14430	15393	15061	14429
5429		NSLX		SU	14410	15333	15383	14409
5430		NSLX		SU	14496	15438	15437	14509
5431		NSLX		SU	14423	15390	15400	14424
5432		NSLX		SU	14416	15044	15447	14486
5433		NSLX		SU	14472	15054	15424	14471
5434		NSLX		SU	14497	15067	15455	14498
5435		NSLX		SU	14491	15403	15431	14267
5436		NSLX		SU	14534	15456	15457	14543
5437		NSLX		SU	14411	15384	15394	14412
5438		NSLX		SU	14530	15459	15458	14547
5439		NSLX		SU	14487	15062	15432	14488
5440		NSLX		SU	14514	15075	15445	14513
5441		NSLX		SU	14397	15327	15377	14398
5442		NSLX		SU	14537	15460	15461	14554
5443		NSLX		SU	14504	15070	15440	14503
5444		NSLX		SU	14535	15463	15462	14553
5445		NSLX		SU	14527	15020	15351	14529
5446		NSLX		SU	14531	15466	15467	14541
5447		NSLX		SU	14532	15468	15469	14550
5448		NSLX		SU	14539	15470	15471	14548
5449		NKSX		SG	14438	15037	15407	14437
5450		NKSX		SG	14533	15472	15473	14552
5451		NKSX	e	SG	14526	15474	15475	14551
5452	**N**	NKSX	e	SG	14536	15477	15476	14563
5453		NKSX	e	SG	14545	15479	15478	14525
5454	**N**	NKSX	e	SG	14555	15481	15480	14524
5455		NKSX		SG	14062	15023	15027	14054
5456		NKSX		SG	14482	15059	15001	14481
5457		NKSX		SG	14502	15069	15439	14501
5458		NKSX		SG	14431	15034	15404	14432
5459		NKSX		SG	14505	15071	15441	14506
5460		NKSX		SG	14515	15076	15446	14516
5461		NKSX		SG	14519	15078	15448	14520
5462		NKSX		SG	14021	15011	15111	14022
5463		NKSX		SG	14037	15019	15119	14038
5464		NKSX		SG	14081	15166	15159	14082
5465		NKSX		SG	14236	15196	15208	14350
5466		NKSX		SG	14004	15002	15220	14003
5467		NKSX		SG	14006	15003	15177	14005
5468		NKSX		SG	14212	15184	15332	14211
5469		NKSX		SG	14007	15004	15201	14008
5470		NKSX		SG	14315	15286	15215	14316
5471		NKSX		SG	14255	15206	15231	14256
5472		NKSX		SG	14508	15295	15169	14458
5473		NKSX		SG	14493	15029	15170	14058
5474		NKSX	e	SG	14065	15033	15224	14066

5475		NKSX	SG	14019	15010	15048	14020
5476		NKSX	SG	14016	15316	15319	14015
5477		NKSX	SG	14013	15180	15233	14057
5478		NKSX	SG	14059	15030	15167	14060
5479		NKSX	SG	14322	15289	15317	14321
5480		NKSX	SG	14264	15210	15291	14263
5481		NKSX	SG	14330	15293	15226	14329
5482		NKSX	SG	14017	15009	15297	14018
5483		NKSX	SG	14052	15175	15329	14051
5484		NKSX	SG	14061	15031	15308	14046
5485	**N**	NKSX	SG	14105	15178	15185	14106
5486	**N**	NKSX	SG	14071	15161	15125	14072
5487	**N**	NKSX	SG	14361	15309	15359	14362
5488	**N**	NKSX	SG	14386	15321	15371	14385
5489	**N**	NKSX	SG	14024	15012	15112	14023
5490	**N**	NKSX	SG	14056	15433	15128	14205
5491	**N**	NKSX	SG	14029	15136	15115	14030
5492	**N**	NKSX	SG	14223	15246	15240	14224
5493	**N**	NKSX	SG	14028	15237	15114	14027
5494		NKSX	SG	14073	15236	15137	14074
5495		NKSX	SG	14095	15265	15148	14096
5496		NKSX	SG	14400	15336	15378	14399
5497		NKSX	SG	14246	15104	15251	14245

CLASS 415/5 SR DESIGN 4 EPB

DMBSO–2TS–DMBSO. Originally formed with a TS and a TSO. Peak hour units which have been formed by concentrating all the compartment trailers.

Electrical Equipment: 1951-type.
Bogies: Central 40" motor bogies and SR standard trailer bogies.
Gangways: Non-gangwayed.
Traction Motors: Two EE507 of 185 kW.
Dimensions: 19.05 x 2.82 m. (outer cars), 18.90 x 2.82 m (inner cars).
Maximum Speed: 75 mph.

Built to various SR lots as shown, all at Lancing/Eastleigh.

† Converted from class 405 (4 Sub) TS.
§ Converted from class 405 (4 Sub) 9 compartment TS. These vehicles were intended to be composites.

14009–14026. DMBSO. Dia. EB266. Lot No. 3638 1952. 82S. 40 t.
14031–14064. DMBSO. Dia. EB266. Lot No. 3756 1953. 82S. 40 t.
14075–14210. DMBSO. Dia. EB266. Lot No. 3757 1953–4. 82S. 40 t.
14221–14302. DMBSO. Dia. EB266. Lot No. 4016 1954–5. 82S. 40 t.
14323–14394. DMBSO. Dia. EB266. Lot No. 4099 1955–6. 82S. 40 t.
14413–14414. DMBSO. Dia. EB266. Lot No. 4172 1956. 82S. 40 t.
15015/16/18/32. TS†. Dia. EH268. Lot No. 3351 1947. 120S. 28 t.
15022/49/81. TS†. Dia. EH268. Lot No. 3463 1948. 120S. 28 t.
15005/38. TS§. Lot No. 3231 1947. 108S. 28 t.
15084. TS§. Lot No. 1060 1946. 108S. 28 t.
15160–15183. TS. Dia. EH268. Lot No. 3757 1953–4. 120S. 28 t.
15189–15225. 15229/32. TS. Dia. EH268. Lot No. 4016 1954–55. 120S. 28 t.

15284–15330. TS. Dia. EH268. Lot No. 4099 1955–6. 120S. 28 t.
15385–15392. TS. Dia. EH268. Lot No. 4172 1956. 120S. 28 t.

5501	NKCX	SG	14250	15015	15294	14249
5502	NKCX	SG	14366	15311	15318	14365
5503	NKCX	SG	14359	15218	15288	14279
5504	NKCX	SG	14076	15163	15302	14075
5505	NKCX	SG	14323	15290	15296	14324
5506	NKCX	SG	14302	15229	15232	14301
5507	NKCX	SG	14341	15299	15038	14342
5508	NKCX	SG	14036	15018	15164	14035
5509	NKCX	SG	14238	15197	15199	14237
5510	NKCX	SG	14344	15193	15084	14343
5511	NKCX	SG	14031	15016	15301	14032
5512	NKCX	SG	14210	15183	15223	14209
5513	NKCX	SG	14357	15307	15310	14358
5514	NKCX	SG	14367	15312	15330	14368
5515	NKCX	SG	14413	15385	15387	14414
5516	NKCX	SG	14063	15032	15211	14064
5517	NKCX	SG	14043	15022	15165	14044
5518	NKCX	SG	14294	15225	15322	14293
5519	NKCX	SG	14274	15320	15217	14383
5520	NKCX	SG	14221	15189	15323	14222
5521	NKCX	SG	14050	15213	15308	14270
5522	NKCX	SG	14394	15325	15306	14393
5523	NKCX	SG	14232	15194	15204	14231
5524	NKCX	SG	14287	15222	15284	14288
5525	NKCX	SG	14272	15214	15049	14271
5526	NKCX	SG	14373	15315	15391	14349
5527	NKCX	SG	14234	15195	15298	14233
5528	NKCX	SG	14257	15207	15305	14258
5529	NKCX	SG	14085	15168	15176	14086
5530	NKCX	SG	14201	15179	15081	14202
5531	NKCX	SG	14010	15005	15287	14009
5532	NKCX	SG	14025	15013	15160	14026

Old numbers of Class 415 cars converted from class 405:

	0	*1*	*2*	*3*	*4*	*5*	*6*	*7*	*8*	*9*
15000–9		10398	10345	10333	10203	11454	10253	10228	10335	10201
15010–9	10391	10174	10169	10189	10184	10277	10251	10292	10238	10332
15020–9	10177	10254	10175	10309	10266	10456	10298	10281	10272	10202
15030–9	10278	10192	10280	10460	10470	10229	10459	10222	11451	10246
15040–9	10285	10322	10283	10275	10224	10452	10466	10468	10299	10176
15050–9	10260	10211	10223	10471	10167	10451	10190	10193	10188	10450
15060–9	10467	10226	10178	10194	10469	10212	10186	10453	10183	10225
15070–9	10180	10191	10179	10185	10168	10187	10227	10204	10182	10195
15080–4	11456	10208	10207	10395	11485					
15450–9	10337	8980	12361	12360	10139	12404	10133	10143	10132	12381
15460–9	10125	12393	10131	12359	12389	12355	10444	10448	10128	12390
15470–9	10124	10439	12391	10137	10446	10445	10129	12402	10141	12353
15480–1	12358	12398								
15228:	10394									
15449:	10480									

CLASSES 415/6 & 415/7 BR DESIGN 4 EPB

DMBSO(A)–2TSO–DMBSO(B). All remaining units have been facelifted with new trim, fluorescent lighting and PA. Units with express gear ratio (e) are classified Class 415/7.

Electrical Equipment: 1951-type.
Bogies: Mark 3C (Mk 3D*) bogies. † Mk 3C motor bogie and Mk 3D trailer bogies.
Gangways: Non-gangwayed.
Traction Motors: Two EE507 of 185 kW.
Dimensions: 19.50 x 2.82 m. (outer cars), 19.35 x 2.82 m (inner cars).
Maximum Speed: 75 mph (90 mph e).

DMBSO(A). Dia. EB271. Lot No. 30582 Eastleigh 1959–61. 84S. 41 t.
DMBSO(B). Dia. EB272. Lot No. 30582 Eastleigh 1959–61. 84S. 41 t.
TSO. Dia. EH271. Lot No. 30583 Eastleigh 1959–61. 112S. 29.5 t.

e Express gear ratio.

5601		NKCX	*	SG	61550	70409	70410	61551
5602	**N**	NKSX		SG	61582	70441	70442	61583
5603		NKSX	*	SG	61538	70397	70398	61539
5604		NKSX		SG	61588	70447	70448	61589
5605		NKSX	*	SG	61540	70399	70400	61541
5606	**N**	NKSX	*	SG	61536	70395	70396	61537
5610	**N**	NKSX		SG	61566	70425	70426	61567
5611	**N**	NKSX		SG	61570	70429	70430	61571
5612	**N**	NKSX	*	SG	61542	70401	70402	61543
5613	**N**	NKSX	*	SG	61532	70391	70392	61533
5614	**N**	NKSX	*	SG	61546	70405	70406	61547
5615	**N**	NKSX		SG	61612	70471	70472	61613
5616	**N**	NKSX		SG	61576	70435	70436	61577
5617	**N**	NKSX		SG	61592	70451	70452	61593
5618	**N**	NKSX	†	SG	61584	70443	70444	61585
5619	**N**	NKCX		SG	61562	70421	70422	61563
5620	**N**	NKCX		SG	61602	70461	70462	61603
5621	**N**	NKCX	*	SG	61520	70380	70379	61521
5622	**N**	NKCX		SG	61560	70419	70420	61561
5623	**N**	NKCX	e	RE	61578	70437	70438	61579
5624	**N**	NKCX	e	RE	61572	70431	70432	61573
5625	**N**	NKCX	e	RE	61608	70455	70468	61609
5626	**N**	NKCX	e	RE	61590	70449	70450	61591
5627	**N**	NKCX	e	RE	61600	70460	70459	61601
5628	**N**	NKCX	e	RE	61604	70464	70463	61605

CLASS 455/7

DTSO–MSO–TSO–DTSO. Sliding doors. Disc brakes. Fluorescent lighting. PA. Second series with TSOs originally in class 508. Pressure ventilation.

Bogies: BT13 (DTSO), BP27 (MSO), BX1 (TSO).
Gangways: Through gangwayed.
Traction Motors: Four EE507 of 185 kW.
Dimensions: 19.83 x 2.82 m. (outer cars), 19.92 x 2.82 m (inner cars).
Maximum Speed: 75 mph.

DTSO. Dia. EE218. Lot No. 30976 York 1984–85. 74S. 29.5 t.
MSO. Dia. EC203. Lot No. 30975 York 1984–85. 84S. 45 t.
TSO. Dia. EH219. Lot No. 30944 York 1977–80. 86S. 25.48 t.

5701	**N**	NSBX	WD		77727	62783	71545	77728
5702	**N**	NSBX	WD		77729	62784	71547	77730
5703	**N**	NSBX	WD		77731	62785	71540	77732
5704	**N**	NSBX	WD		77733	62786	71548	77734
5705	**N**	NSBX	WD		77735	62787	71565	77736
5706	**N**	NSBX	WD		77737	62788	71534	77738
5707	**N**	NSBX	WD		77739	62789	71536	77740
5708	**N**	NSBX	WD		77741	62790	71560	77742
5709	**N**	NSBX	WD		77743	62791	71532	77744
5710	**N**	NSBX	WD		77745	62792	71566	77746
5711	**N**	NSBX	WD		77747	62793	71542	77748
5712	**N**	NSBX	WD		77749	62794	71546	77750
5713	**N**	NSBX	WD		77751	62795	71567	77752
5714	**N**	NSBX	WD		77753	62796	71539	77754
5715	**N**	NSBX	WD		77755	62797	71535	77756
5716	**N**	NSBX	WD		77757	62798	71564	77758
5717		NSBX	WD		77759	62799	71528	77760
5718	**N**	NSBX	WD		77761	62800	71557	77762
5719	**N**	NSBX	WD		77763	62801	71558	77764
5720		NSBX	WD		77765	62802	71568	77766
5721		NSBX	WD		77767	62803	71553	77768
5722		NSBX	WD		77769	62804	71533	77770
5723		NSBX	WD		77771	62805	71526	77772
5724		NSBX	WD		77773	62806	71561	77774
5725		NSBX	WD		77775	62807	71541	77776
5726		NSBX	WD		77777	62808	71556	77778
5727		NSBX	WD		77779	62809	71562	77780
5728		NSBX	WD	(S)	77781	62810	71527	77782
5729		NSBX	WD		77783	62811	71550	77784
5730	**N**	NSBX	WD		77785	62812	71551	77786
5731		NSBX	WD		77787	62813	71555	77788
5732		NSBX	WD		77789	62814	71552	77790
5733		NSBX	WD		77791	62815	71549	77792
5734		NSBX	WD		77793	62816	71531	77794
5735		NSBX	WD		77795	62817	71563	77796
5736		NSBX	WD		77797	62818	71554	77798
5737		NSBX	WD		77799	62819	71544	77800
5738		NSBX	WD		77801	62820	71529	77802
5739		NSBX	WD		77803	62821	71537	77804
5740		NSBX	WD		77805	62822	71530	77806
5741		NSBX	WD		77807	62823	71559	77808
5742		NSBX	WD		77809	62824	71543	77810
5743		NSBX	WD		77811	62825	71538	77812

CLASS 455/8

DTSO–MSO–TSO–DTSO. Sliding doors. Disc brakes. Fluorescent lighting. PA. First series. Pressure ventilation.

Bogies: BP20 (MSO), BT13 (trailer cars).
Gangways: Through gangwayed.
Traction Motors: Four EE507 of 185 kW.
Dimensions: 19.83 x 2.82 m. (outer cars), 19.92 x 2.82 m (inner cars).
Maximum Speed: 75 mph.

DTSO. Dia. EE218. Lot No. 30972 York 1982–84. 74S. 29.5 t.
MSO. Dia. EC203. Lot No. 30973 York 1982–84. 84S. 45.6 t.
TSO. Dia. EH221. Lot No. 30974 York 1982–84. 84S. 27.1 t.

5801	**N**	NSLX	SU	77579	62709	71637	77580
5802	**N**	NSLX	SU	77581	62710	71638	77582
5803	**N**	NSLX	SU	77583	62711	71639	77584
5804	**N**	NSLX	SU	77585	62712	71640	77586
5805	**N**	NSLX	SU	77587	62713	71641	77588
5806	**N**	NSLX	SU	77589	62714	71642	77590
5807	**N**	NSLX	SU	77591	62715	71643	77592
5808	**N**	NSLX	SU	77593	62716	71644	77594
5809	**N**	NSLX	SU	77595	62717	71645	77596
5810	**N**	NSLX	SU	77597	62718	71646	77598
5811	**N**	NSLX	SU	77599	62719	71647	77600
5812	**N**	NSLX	SU	77601	62720	71648	77602
5813	**N**	NSLX	SU	77603	62721	71649	77604
5814	**N**	NSLX	SU	77605	62722	71650	77606
5815	**N**	NSLX	SU	77607	62723	71651	77608
5816	**N**	NSLX	SU	77609	62724	71652	77610
5817	**N**	NSLX	SU	77611	62725	71653	77612
5818	**N**	NSLX	SU	77613	62726	71654	77614
5819	**N**	NSLX	SU	77615	62727	71655	77616
5820	**N**	NSLX	SU	77617	62728	71656	77618
5821	**N**	NSLX	SU	77619	62729	71657	77620
5822	**N**	NSLX	SU	77621	62730	71658	77622
5823	**N**	NSLX	SU	77623	62731	71659	77624
5824	**N**	NSLX	SU	77637	62732	71660	77626
5825	**N**	NSLX	SU	77627	62733	71661	77628
5826	**N**	NSLX	SU (S)	77629	62734	71662	77680
5827	**N**	NSLX	SU	77631	62735	71663	77632
5828	**N**	NSLX	SU	77633	62736	71664	77634
5829	**N**	NSLX	SU	77635	62737	71665	77636
5830	**N**	NSLX	SU	77625	62743	71666	77638
5831	**N**	NSLX	SU	77639	62739	71667	77640
5832	**N**	NSLX	SU	77641	62740	71668	77642
5833	**N**	NSLX	SU	77643	62741	71669	77644
5834	**N**	NSLX	SU	77645	62742	71670	77646
5835	**N**	NSLX	SU	77647	62738	71671	77648
5836	**N**	NSLX	SU	77649	62744	71672	77650
5837	**N**	NSBX	WD	77651	62745	71673	77652
5838	**N**	NSBX	WD	77653	62746	71674	77654

5839	**N**	NSBX	WD	77655	62747	71675	77656
5840	**N**	NSBX	WD	77657	62748	71676	77658
5841	**N**	NSBX	WD	77659	62749	71677	77660
5842	**N**	NSBX	WD	77661	62750	71678	77662
5843	**N**	NSBX	WD	77663	62751	71679	77664
5844	**N**	NSBX	WD	77665	62752	71680	77666
5845	**N**	NSBX	WD	77667	62753	71681	77668
5846	**N**	NSBX	WD	77669	62754	71682	77670
5847	**N**	NSBX	WD	77671	62755	71683	77672
5848	**N**	NSBX	WD	77673	62756	71684	77674
5849	**N**	NSBX	WD	77675	62757	71685	77676
5850	**N**	NSBX	WD	77677	62758	71686	77678
5851	**N**	NSBX	WD	77679	62759	71687	77630
5852	**N**	NSBX	WD	77681	62760	71688	77682
5853	**N**	NSBX	WD	77683	62761	71689	77684
5854	**N**	NSBX	WD	77685	62762	71690	77686
5855	**N**	NSBX	WD	77687	62763	71691	77688
5856	**N**	NSBX	WD	77689	62764	71692	77690
5857	**N**	NSBX	WD	77691	62765	71693	77692
5858	**N**	NSBX	WD	77693	62766	71694	77694
5859	**N**	NSBX	WD	77695	62767	71695	77696
5860	**N**	NSBX	WD	77697	62768	71696	77698
5861	**N**	NSBX	WD	77699	62769	71697	77700
5862	**N**	NSBX	WD	77701	62770	71698	77702
5863	**N**	NSBX	WD	77703	62771	71699	77704
5864	**N**	NSBX	WD	77705	62772	71700	77706
5865	**N**	NSBX	WD	77707	62773	71701	77708
5866	**N**	NSBX	WD	77709	62774	71702	77710
5867	**N**	NSBX	WD	77711	62775	71703	77712
5868	**N**	NSBX	WD	77713	62776	71704	77714
5869	**N**	NSBX	WD	77715	62777	71705	77716
5870	**N**	NSBX	WD	77717	62778	71706	77718
5871	**N**	NSBX	WD	77719	62779	71707	77720
5872	**N**	NSBX	WD	77721	62780	71708	77722
5873	**N**	NSBX	WD	77723	62781	71709	77724
5874	**N**	NSBX	WD	77725	62782	71710	77726

CLASS 455/9

DTSO–MSO–TSO–DTSO. Sliding doors. Disc brakes. Fluorescent lighting. PA. Third series. Convection heating.

Bogies: BP20 (MSO), BT13 (trailer cars).
Gangways: Through gangwayed.
Traction Motors: Four EE507 of 185 kW.
Dimensions: 19.83 x 2.82 m. (outer cars), 19.92 x 2.82 m (inner cars).
Maximum Speed: 75 mph.

DTSO. Dia. EE226. Lot No. 30991 York 1985. 74S. 29.5 t.
MSO. Dia. EC206. Lot No. 30992 York 1985. 84S. 45.6 t.
TSO. Dia. EH224. Lot No. 30993 York 1985. 84S. 27.1 t.
TSO n. Dia. EH224. Lot No. 30932 Derby 1981. 84S. 27.1 t.

*–Chopper control. §–Tread brakes.
n Formerly class 210 No. 60400.

5901		NSBX		WD	77813	62826	71714	77814
5902		NSBX		WD	77815	62827	71715	77816
5903	**N**	NSBX		WD	77817	62828	71716	77818
5904		NSBX		WD	77819	62829	71717	77820
5905	**N**	NSBX		WD	77821	62830	71718	77822
5906		NSBX		WD	77823	62831	71719	77824
5907	**N**	NSBX		WD	77825	62832	71720	77826
5908	**N**	NSBX		WD	77827	62833	71721	77828
5909	**N**	NSBX		WD	77829	62834	71722	77830
5910	**N**	NSBX		WD	77831	62835	71723	77832
5911	**N**	NSBX		WD	77833	62836	71724	77834
5912	**N**	NSBX	*	WD	77835	62837	71725	77836
5913	**N**	NSBX	§	WD	77837	62838	71726	77838
5914	**N**	NSBX	§	WD	77839	62839	71727	77840
5915	**N**	NSBX	§	WD	77841	62840	71728	77842
5916	**N**	NSBX	*	WD	77843	62841	71729	77844
5917		NSBX	*	WD	77845	62842	71730	77846
5918		NSBX	*	WD	77847	62843	71731	77848
5919	**N**	NSBX	*	WD	77849	62844	71732	77850
5920	**N**	NSBX	*	WD	77851	62845	71733	77852
Spare	**N**	NSBX		WD (S)			67400	

CLASS 456

DMSO–DTSO. Sliding doors. Disc brakes. Fluorescent lighting. PA.

Bogies: P7 (motor) and T3 trailer.
Gangways: Through gangwayed.
Traction Motors: Two EE507 of 185 kW.
Dimensions: 19.83 x 2.82 m.
Maximum Speed: 75 mph.

DMSO. Dia. EA267. Lot No. 31073 York 1990– . 79S. t.
DTSO. Dia. EA218. Lot No. 31074 York 1990– . 73S. t.

6001 **N**	64735	78250
6002 **N**	64736	78251
6003 **N**	64737	78252
6004 **N**	64738	78253
6005 **N**	64739	78254
6006 **N**	64740	78255
6007 **N**	64741	78256
6008 **N**	64742	78257
6009 **N**	64743	78258
6010 **N**	64744	78259
6011 **N**	64745	78260
6012 **N**	64746	78261
6013 **N**	64747	78262
6014 **N**	64748	78263
6015 **N**	64749	78264
6016 **N**	64750	78265
6017 **N**	64751	78266
6018 **N**	64752	78267
6019 **N**	64753	78268
6020 **N**	64754	78269
6021 **N**	64755	78270
6022 **N**	64756	78271
6023 **N**	64757	78272
6024 **N**	64758	78273

CLASS 416/2 — BR DESIGN 2 EPB

DMBSO–DTSso.

Electrical Equipment: 1951-type.
Bogies: Mk. 3D.
Gangways: Non-gangwayed.
Traction Motors: Two EE507 of 185 kW.
Dimensions: 19.49 x 2.82 m.
Maximum Speed: 75 mph.

65301–65307. DMBSO. Dia. EB269. Lot No. 30114 Afd/Elh 1954. 84S. 42 t.
65326–65341. DMBSO. Dia. EB269. Lot No. 30119 Afd/Elh 1954. 84S. 42 t.
65343–65365. DMBSO. Dia. EB269. Lot No. 30167 Afd/Elh 1955. 84S. 42 t.
65367–65392. DMBSO. Dia. EB269. Lot No. 30314 Afd/Elh 1956–58. 84S. 42 t.
77501–77506. DTSso. Dia. EE264. Lot No. 30115 Afd/Elh 1954. 102S. 30.5 t.
77511–77526. DTSso. Dia. EE264. Lot No. 30120 Afd/Elh 1954. 102S. 30.5 t.
77528–77550. DTSso. Dia. EE264. Lot No. 30168 Afd/Elh 1955. 102S. 30.5 t.
77552–77577. DTSso. Dia. EE264. Lot No. 30315 Afd/Elh 1956–58. 102S. 30.5 t.

Note: 77537/9 are DTSO and seat 92S. (Dia. EE275).

6202	NKSX	SG	65301	77501
6203	NKSX	SG	65302	77502
6205	NKSX	SG	65304	77504
6207	NKSX	SG	65306	77506
6212	NKSX	SG	65326	77511
6213	NKSX	SG	65327	77512
6217	NKSX	SG	65331	77516
6218	NKSX	SG	65332	77517
6221	NKSX	SG	65335	77520
6222	NKSX	SG	65336	77521
6223	NKSX	SG	65337	77522
6224	NKSX	SG	65338	77523
6225	NKSX	SG	65339	77524
6226	NKSX	SG	65340	77525
6227	NKSX	SG	65341	77526
6229 **N**	NKSX	SG	65343	77528
6230	NKSX	SG	65344	77529
6231	NKSX	SG	65345	77530
6235	NKSX	SG	65349	77534
6236	NKSX	SG	65350	77535
6237	NKSX	SG	65351	77536
6238	NKSX	SG	65352	77537
6239	NKSX	SG	65353	77538
6240	NKSX	SG	65354	77539
6241	NKSX	SG	65355	77540
6243	NKSX	SG	65357	77542
6244	NKSX	SG	65358	77543
6245	NKSX	SG	65359	77544
6247	NKSX	SG	65361	77546
6249	NKSX	SG	65363	77548
6251	NKSX	SG	65365	77550
6253	NKSX	SG	65367	77552
6255	NKSX	SG	65369	77554
6256	NKSX	SG	65370	77555
6257	NKSX	SG	65371	77556
6259	NKSX	SG	65373	77558
6260	NKSX	SG	65374	77559
6261	NKSX	SG	65375	77560
6262	NKSX	SG	65376	77561
6263	NKSX	SG	65377	77562
6264	NKSX	SG	65378	77563
6265	NKSX	SG	65379	77564
6267	NKSX	SG	65381	77566
6268	NKSX	SG	65382	77567
6269	NKSX	SG	65307	77568
6270	NKSX	SG	65384	77569
6271	NKSX	SG	65385	77570
6272	NKSX	SG	65386	77571
6273	NKSX	SG	65387	77572
6274	NKSX	SG	65388	77573
6275	NKSX	SG	65389	77574
6276	NKSX	SG	65390	77575
6277	NKSX	SG	65391	77576
6278	NKSX	SG	65392	77577

CLASS 416/3 SR DESIGN 2 EPB

DMBSO–DTSso. These units were made by producing new bodies for former 2 Nol underframes. All have now been facelifted with new seat trim, fluorescent lighting and PA.

Electrical Equipment: 1951-type.
Bogies: Central 40" motor bogies and SR standard trailer bogies.
Gangways: Non-gangwayed.
Traction Motors: Two EE507 of 185 kW.
Dimensions: 19.05 x 2.82 m.
Maximum Speed: 75 mph.

14283. DMBSO. Dia. EB269. Lot No. 4016 Eastleigh 1954–5. 82S. 40 t.
14542–14590. DMBSO. Dia. EB269. Lot No. 4281 Eastleigh 1957–9. 82S. 40 t.
DTSO. Dia. EE269. Lot No. 4281 Eastleigh 1957–9. 92S. 30 t.

6301		NSLX	SU	14577	16121
6302		NSLX	SU	14580	16124
6303		NSLX	SU	14576	16120
6304		NSLX	SU	14589	16133
6305		NSLX	SU	14587	16131
6306		NSLX	SU	14571	16115
6307		NSLX	SU	14573	16117
6308	**N**	NSLX	SU	14564	16108
6309	**N**	NSLX	SU	14562	16106
6310	**N**	NSLX	SU	14558	16118
6311	**N**	NSLX	SU	14565	16109
6312	**N**	NSLX	SU	14579	16123
6313	**N**	NSLX	SU	14574	16102
6314	**N**	NSLX	SU	14586	16130
6315		NSLX	SU	14590	16134
6316		NSLX	SU	14559	16103
6317		NSLX	SU	14578	16122
6318		NSLX	SU	14566	16110
6319		NSLX	SU	14568	16112
6320		NSLX	SU	14561	16105
6321		NSLX	SU	14283	16128
6323		NSLX	SU	14581	16125
6324		NSLX	SU	14560	16104
6325		NSLX	SU	14567	16111
6326		NSLX	SU	14585	16129
6327		NSLX	SU	14572	16116
6328		NSLX	SU	14582	16126
6329		NSLX	SU	14542	16114
6330		NSLX	SU	14588	16132
6331		NSLX	SU	14583	16127
6332		NSLX	SU	14569	16113
6333		NSLX	SU	14557	16101
6334		NSLX	SU	14546	16107
Spare			SU(S)		16119

CLASS 416/4 FACELIFTED BR DESIGN 2 EPB

DMBSO–DTSO. Facelifted units with new trim, fluorescent lighting and PA.

Electrical Equipment: 1951-type.
Bogies: Mk. 3D.
Gangways: Non-gangwayed.
Traction Motors: Two EE507 of 185 kW.
Dimensions: 19.49 x 2.82 m.
Maximum Speed: 75 mph.

65305–65309. DMBSO. Dia. EB281. Lot No. 30114 Afd/Elh 1954. 82S. 42 t.
65328–65334. DMBSO. Dia. EB281. Lot No. 30119 Afd/Elh 1954. 82S. 42 t.
65342–65366. DMBSO. Dia. EB281. Lot No. 30167 Afd/Elh 1955. 82S. 42 t.
65368–65372. DMBSO. Dia. EB281. Lot No. 30314 Afd/Elh 1956–58. 79S. 42 t.
77113. DTSO. Dia. EE279. Lot No. 30117 Eastleigh 1955. Former South Tyneside trailer. 92s. 30.5 t.
77505. DTSO. Dia. EE274. Lot No. 30115 Afd/Elh 1954. 92S. 30.5 t.

77513–77519. DTSO. Dia. EE274. Lot No. 30120 Afd/Elh 1954. 92S. 30.5 t.
77527–77551. DTSO. Dia. EE274. Lot No. 30168 Afd/Elh 1955. 92S. 30.5 t.
77553–77557. DTSO. Dia. EE274. Lot No. 30315 Afd/Elh 1956–58. 92S. 30.5 t.

*–Modified for working Maidstone service. Doors fitted between saloon and guards compartment in motor car and between saloon and cab in trailer. DMBSO seats 79S (dia. EB280) and DTSO seats 90S (dia. EE271).

6401		NKSX*	SG	65346	77531
6402	**N**	NKSX*	SG	65362	77547
6403	**N**	NKSX*	SG	65356	77541
6404	**N**	NKSX*	SG	65329	77514
6405	**N**	NKSX*	SG	65347	77532
6406	**N**	NKSX*	SG	65305	77505
6407	**N**	NKSX*	SG	65330	77515
6408	**N**	NKSX*	SG	65342	77527
6409	**N**	NKSX*	SG	65309	77113
6410	**N**	NKSX	SG	65334	77519
6411	**N**	NKSX	SG	65333	77518
6412	**N**	NKSX	SG	65364	77549
6413	**N**	NKSX	SG	65372	77557
6414		NKSX	SG	65368	77553
6415		NKSX	SG	65348	77533
6416		NKSX	SG	65328	77513
6417		NKSX	SG	65366	77551
6418		NKSX	SG	65360	77545

CLASS 438 — 4 TC

DTSO–TFK–TBSK–DTSO. (Class 438/0), DTSO–TFK(T)–TBSK–DTSO (Class 438/1). Converted from loco-hauled stock. Unpowered units which work push & pull with class 431/2 tractor units and class 33/1 and 73 locos. Express stock.

Electrical Equipment: 1966-type.
Bogies: B5 (SR) bogies.
Gangways: Throughout.
Dimensions: 19.66 x 2.82 m.
Maximum Speed: 90 mph.

76270–76332. DTSO. Dia. EE266. Lot No. 30764 York 1966–67. 64S. 32 t.
76943–76948. DTSO. Dia. EE266. Lot No. 30857 York 1974. 64S. 32 t.
70844–70871. TFK. Dia. EH160. Lot No. 30766 York 1966–67. 42F 2L. 33.5 t.
71162–71167. TFK. Dia. EH160. Lot No. 30856 York 1974. 42F 2L. 33.5 t.
70812–70843. TBSK. Dia. EJ260. Lot No. 30765 York 1966–67. 32S 1L. 35.5 t.
71156–71161. TBSK. Dia. EJ260. Lot No. 30855 York 1974. 32S 1L. 35.5 t.

Notes:

(1) Most TFKs (including all those in pa fitted units are now declassified to TCK Dia. EH362 seating 36F 8S 2L.
(2) Because of the possibilty of reinstatement of certain cars, the details above refer to the complete series. 76943–8 are all withdrawn.

Renumbered from 401–428. formerly class 491.

8001	**N**	NSSX	pa	BM	76270	70844	70821	76332
8004	**N**	NSSX	pa	BM	76275	70847	70815	76276
8006	**N**	NSSX	pa	BM	76280	70849	70817	76279
8010	**N**	NSSX	pa	BM	76288	70853	70812	76287
8012	**N**	NSSX	pa	BM	76291	70858	70828	76292
8014		NSSX		BM	76295	70869	70825	76296
8015	**N**	NSSX	pa	BM	76297	70855	70823	76298
8017	**N**	NSSX	pa	BM	76302	70860	70826	76301

8018	**N**	NSSX	pa	BM	76303	70861	70829	76304
8021		NSSX		BM (S)	76309	70864	70832	76310
8023	**N**	NSSX	pa	BM	76313	70866	70834	76314
8027	**N**	NSSX	pa	BM	76321	70870	70838	76322
8028	**N**	NSSX	pa	BM	76323	70871	70839	76324

Spare Class 431 and 438 Vehicles:

70807 | 70818 | 70822 | 70837 | 70842 | 71160 | 76284 | 76327 | 76331

In addition, the following withdrawn vehicles have not been scrapped and certain of them will be reinstated in order to make Class 431 up to six 6-car units:
All withdrawn class 432 TBFKs.
All withdrawn class 438 TBSKs except for 70816.
All withdrawn class 438 TFKs except for 70848.
All withdrawn class 438 DTSOs except for 76277.

Former numbers of converted Class 431 and 438 vehicles converted from hauled stock:

70807 (15890)
70812 (34987)
70815 (34958)
70817 (34960)
70818 (34961)
70821 (34985)
70822 (34986)
70823 (34970)
70824 (34984)
70825 (34996)
70826 (34986)
70827 (34997)
70828 (34998)
70829 (34981)
70830 (34976)
70832 (34967)
70834 (34972)
70837 (34975)
70838 (34979)
70839 (34978)
70842 (34947)
70844 (13022)
70845 (13008)
70847 (13016)
70849 (13025)
70853 (13030)
70855 (13018)
70858 (13063)
70859 (13040)
70860 (13019)
70861 (13096)
70864 (13038)
70866 (13037)
70869 (13002)
70870 (13032)
70871 (13017)
71160 (34637)
71163 (13097)
76270 (4043)
76275 (3929)
76276 (4003)
76279 (3926)
76280 (4014)
76284 (4008)
76287 (4379)
76288 (4391)
76291 (3931)
76292 (3946)
76295 (3927)
76296 (3945)
76297 (3938)
76298 (4004)
76301 (4375)
76302 (4382)
76303 (4017)
76304 (4384)
76309 (4006)
76310 (4027)
76313 (3941)
76314 (3943)
76321 (3930)
76322 (3936)
76323 (3940)
76324 (4009)
76327 (4018)
76331 (4388)
76332 (3937)
76947 (4034)
76948 (4073)

CLASS 488 VICTORIA–GATWICK TRAILER SETS

TFOLH–TSOL (Class 488/3 only)–TSOLH. Converted 1983–84 from loco-hauled Mk. 2F FOs and TSOs for Victoria–Gatwick service. Express stock. Air conditioned. Fluorescent lighting. PA. Conversion consisted of a modified seating layout and the removal of one toilet to provide additional luggage space.

Bogies: B4.
Gangways: Throughout.
Dimensions: 20.12 x 2.82 m.
Maximum Speed: 90 mph.

72500–72509. TFOLH. Dia. EP101. Lot No. 30859 Derby 1973–74. 41F 1L. 35 t.
72602–72647. TSOLH. Dia. EP201. Lot No. 30860 Derby 1973–74. 48S 1L. 35 t.
72701–72718. TSOL. Dia. EH285. Lot No. 30860 Derby 1973–74. 48S 1L. 35 t.

CLASS 488/2. Note: TFOLH fitted with public telephone.

8201	I	IVGX	SL	72500 (3413)	72638 (6068)
8202	I	IVGX	SL	72501 (3382)	72617 (6086)
8203	I	IVGX	SL	72502 (3321)	72640 (6097)
8204	I	IVGX	SL	72503 (3407)	72641 (6079)
8205	I	IVGX	SL	72504 (3406)	72628 (6058)
8206	I	IVGX	SL	72505 (3415)	72629 (6048)
8207	I	IVGX	SL	72506 (3335)	72642 (6076)
8208	I	IVGX	SL	72507 (3412)	72643 (6040)
8209	I	IVGX	SL	72508 (3409)	72644 (6039)
8210	I	IVGX	SL	72509 (3398)	72635 (6128)

CLASS 488/3. TSOLH–TSOL–TSOLH.

8302	I	IVGX	SL	72602 (6130)	72701 (6088)	72604 (6087)
8303	I	IVGX	SL	72603 (6093)	72702 (6099)	72608 (6077)
8304	I	IVGX	SL	72606 (6084)	72703 (6075)	72611 (6083)
8305	I	IVGX	SL	72605 (6082)	72704 (6132)	72609 (6080)
8306	I	IVGX	SL	72607 (6020)	72705 (6032)	72610 (6074)
8307	I	IVGX	SL	72612 (6156)	72706 (6143)	72613 (6126)
8308	I	IVGX	SL	72614 (6090)	72707 (6127)	72615 (5938)
8309	I	IVGX	SL	72616 (6007)	72708 (6095)	72639 (6070)
8310	I	IVGX	SL	72618 (6044)	72709 (5982)	72619 (5909)
8311	I	IVGX	SL	72620 (6140)	72710 (6003)	72621 (6108)
8312	I	IVGX	SL	72622 (6004)	72711 (6109)	72623 (6118)
8313	I	IVGX	SL	72624 (5972)	72712 (6091)	72625 (6085)
8314	I	IVGX	SL	72626 (6017)	72713 (6023)	72627 (5974)
8315	I	IVGX	SL	72636 (6071)	72714 (6092)	72645 (5942)
8316	I	IVGX	SL	72630 (6094)	72715 (6019)	72631 (6096)
8317	I	IVGX	SL	72632 (6072)	72716 (6114)	72633 (6129)
8318	I	IVGX	SL	72634 (6089)	72717 (6069)	72637 (6098)
8319	I	IVGX	SL	72646 (6078)	72718 (5979)	72647 (6081)
Spare	I	IVGX	SL	72601	72700	

Note: 72601 and 72700 are officially withdrawn, but are retained for contingencies.

CLASS 419 — 1957 type MLV

Built 1959–61. Dual braked.

Electrical Equipment: 1957-type.
Bogies: Mk 3B.
Gangways: Non-gangwayed.
Traction Motors: Two EE507 of 185 kW.
Dimensions: 19.64 x 2.82 m.
Maximum Speed: 90 mph.

68001–2. DMLV. Dia. EX560. Lot No. 30458 Afd./Elh. 1959. 45.5 t.
68003–10. DMLV. Dia. EX560. Lot No. 30623 Afd./Elh. 1960–61. 45.5 t.

9001	**N**	REMD	RE	68001
9002	**N**	REMD	RE	68002
9003	**J**	REMD	RE	68003
9004	**N**	REMD	RE	68004
9005	**J**	REMD	RE	68005
9006	**J**	NBTX	RE	68006
9007	**J**	REMD	RE	68007
9008	**J**	REMD	RE	68008
9009	**R**	REMD	RE	68009
9010	**J**	NBTX	RE	68010

CLASS 489 VICTORIA–GATWICK GLV

Converted 1983–84 from class 414/3 (2 Hap) DMBSOs to work with class 488.

Bogies: Mk 4.
Gangways: Gangwayed at inner end only.
Traction Motors: Two EE507 of 185 kW.
Dimensions: 19.49 x 2.82 m.
Maximum Speed: 90 mph.

DMLV. Dia. EX561. Lot No. 30452 Ashford/Eastleigh 1959. 40.5 t.

9101	**I**	IVGX	SL	68500 (61269)
9102	**I**	IVGX	SL	68501 (61281)
9103	**I**	IVGX	SL	68502 (61274)
9104	**I**	IVGX	SL	68503 (61277)
9105	**I**	IVGX	SL	68504 (61286)
9106	**I**	IVGX	SL	68505 (61299)
9107	**I**	IVGX	SL	68506 (61292)
9108	**I**	IVGX	SL	68507 (61267)
9109	**I**	IVGX	SL	68508 (61272)
9110	**I**	IVGX	SL	68509 (61280)

4.3. SOUTHERN REGION TUBE STOCK

CLASS 483 'NEW' ISLE OF WIGHT STOCK

DMBSO(A)–DMBSO(B). Tube stock built 19 for LTE. Converted 1989–90 for Isle of Wight Line. End doors. dg. pa. Former LT numbers are shown in parentheses.

System: 660 V d.c. third rail.
Gangways: Non-gangwayed.
Traction Motors: Two of 130 kW.
Dimensions: 15.95 x 2.69 m.
Maximum Speed: 45 mph.

DMSO (A). Dia. EA265. 42S. 27.5 t.
DMSO (B). Dia. EA266. 42S. 27.5 t.

483 001	**N**	NSSX	RY	121 (10184)	221 (11184)
483 002	**N**	NSSX	RY	122 (10221)	222 (11221)
483 003	**N**	NSSX	RY	123 (10116)	223 (11116)
483 004				124 (10205)	224 (11205)
483 005				125 (10142)	225 (11142)
483 006				126 (10297)	226 (11297)
483 007				127 (10291)	227 (11291)
483 008				128 (10255)	228 (11255)
483 009				129 (10229)	229 (11229)
483 010				130 (10139)	230 (11172)

CLASSES 485 & 486 ISLE OF WIGHT STOCK

Tube stock built 1923–31 for London Electric Railway (later LT). Converted 1967 for Isle of Wight Line. End doors. Former LT numbers are shown in parentheses. Fixed formations not maintained.

System: 660 V d.c. third rail.
Gangways: Non-gangwayed.
Traction Motors: Two EE507 of 178 kW.
Dimensions: 15.16 x 2.64 m.
Maximum Speed: 45 mph.

All are **N** livery, sector NSSX and allocated to RY.
* Stored at FR.

DMBSO. Dia. EB261. Metro-Cammell 1932–5. 26S. 32 t.

1 (3703)	3 (3251)*	5 (3185)	7 (3209)	9 (3223)	11 (3705)
2 (3706)	4 (3702)*	6 (3084)	8 (3074)	10 (3686)	

DTSO. Dia. EE260. Metro-Cammell 1926–9. 38S. 17 t.

26 (5294)	28 (5304)	31 (5283)	31 (5283)	33 (5291)*	34 (5302)
27 (5279)	29 (5293)*	32 (5290)*			

TSO. Dia. EH261. Cammell-Laird 1924. 42S. 19 t.

44 (7281)	49 (7296)	93 (7282)	94 (7287)	95 (7292)

CLASS 487 WATERLOO & CITY

Built 1940 for Waterloo & City Line (known colloquially as the "drain". Do not run in permanent sets. Run as single motor cars or as pairs of motors with up to three trailers in between.

System: 630 V d.c. third rail.
Gangways: Non-gangwayed. End doors.
Traction Motors: Two EE500 of 140 kW.
Dimensions: 14.33 x 2.64 m.
Maximum Speed: 35 mph.

DMBSO. Dia. EB260. English Electric 1940. 40S. 29 t.
TSO. Dia. EH260. English Electric 1940. 52S. 19 t.

DMBSO

51 **N** NKSX WC
53 **N** NKSX WC
54 **N** NKSX WC
59 **N** NKSX WC(S)
56 **N** NKSX WC
57 **N** NKSX WC
58 **N** NKSX WC
60 **N** NKSX WC
61 **N** NKSX WC
62 **N** NKSX WC

TSO

72 **N** NKSX WC
73 **N** NKSX WC
74 **N** NKSX WC
75 **N** NKSX WC
76 **N** NKSX WC
77 **N** NKSX WC
78 **N** NKSX WC
80 **N** NKSX WC
81 **N** NKSX WC
83 **N** NKSX WC
84 **N** NKSX WC
85 **N** NKSX WC
86 **N** NKSX WC

4.4. LONDON MIDLAND REGION d.c. EMUs

CLASS 504 MANCHESTER–BURY

DMBSO–DTSO. These unique units will be phased out in a few years time when the route is converted to light rail.

System: 1200 V d.c. protected side-contact third rail.
Bogies: Gresley.
Traction Motors: Four EE of 90 kW.
Gangways: Non-gangwayed.
Dimensions: 19.53 x 2.82 m.
Maximum Speed: 90 mph.
Max Speed: 65 mph.

DMBSO. Dia. EB204. Lot No. 30477 Wolverton 1959. 84S. 50 t.
DTSO. Dia. EE215. Lot No. 30478 Wolverton 1959. 94S. 33 t.

Unit numbers not carried.

504 444		PBQX	BQ	65444	77165
504 445	**G**	PBQX	BQ	65445	77166
504 446	**G**	PBQX	BQ	65446	77167
504 447		PBQX	BQ	65447	77168
504 449	**G**	PBQX	BQ	65449	77170
504 450	**G**	PBQX	BQ	65450	77171
504 451	**G**	PBQX	BQ	65451	77172
504 452		PBQX	BQ	65452	77173
504 453	**G**	PBQX	BQ	65453	77174
504 454	**G**	PBQX	BQ	65454	77175
504 455	**G**	PBQX	BQ	65455	77176
504 456	**G**	PBQX	BQ	65456	77177
504 457	**G**	PBQX	BQ	65457	77178
504 458	**G**	PBQX	BQ	65458	77179
504 459	**G**	PBQX	BQ	65459	77180
504 460	**G**	PBQX	BQ	65460	77181
504 461	**G**	PBQX	BQ	65461	77182

CLASS 507 MERSEYRAIL

BDMSO–TSO–DMSO. Tightlock couplers. Sliding doors. Disc and rheostatic brakes. PA.

System: 750 V d.c. third rail.
Bogies: BX1.
Gangways: Gangwayed within unit. End doors.
Traction Motors: Four GEC G310AZ of 82.125 kW.
Dimensions: 19.80 x 2.82 m (outer cars), 19.92 x 2.82 m (inner cars).
Maximum Speed: 75 mph.

BDMSO. Dia. EI202. Lot No. 30906 York 1978–80. 74S. 37.06 t.
TSO. Dia. EH205. Lot No. 30907 York 1978–80. 82S. 25.60 t.
DMSO. Dia. EA201. Lot No. 30908 York 1978–80. 74S. 35.62 t.

507 001	PHRX	HR	64367	71342	64405
507 002	PHRX	HR	64368	71343	64406
507 003	PHRX	HR	64369	71344	64407
507 004	PHRX	HR	64370	71345	64408
507 005	PHRX	HR	64371	71346	64409
507 006	PHRX	HR	64372	71347	64410
507 007	PHRX	HR	64373	71348	64411
507 008	PHRX	HR	64374	71349	64412
507 009	PHRX	HR	64375	71350	64413
507 010	PHRX	HR	64376	71351	64414
507 011	PHRX	HR	64377	71352	64415
507 012	PHRX	HR (S)	64378	71353	64416
507 013	PHRX	HR	64379	71354	64417
507 014	PHRX	HR	64380	71355	64418
507 015	PHRX	HR	64381	71356	64419
507 016	PHRX	HR	64382	71357	64420
507 017	PHRX	HR	64383	71358	64421
507 018	PHRX	HR	64384	71359	64422
507 019	PHRX	HR	64385	71360	64423
507 020	PHRX	HR	64386	71361	64424
507 021	PHRX	HR	64387	71362	64425
507 022	PHRX	HR	64388	71363	64426
507 023	PHRX	HR	64389	71364	64427
507 024	PHRX	HR	64390	71365	64428
507 025	PHRX	HR	64391	71366	64429
507 026	PHRX	HR	64392	71367	64430
507 027	PHRX	HR	64393	71368	64431
507 028	PHRX	HR	64394	71369	64432
507 029	PHRX	HR	64395	71370	64433
507 030	PHRX	HR	64396	71371	64434
507 031	PHRX	HR	64397	71372	64435
507 032	PHRX	HR	64398	71373	64436
507 033	PHRX	HR	64399	71374	64437

CLASS 508 MERSEYRAIL

DMSO–TSO–BDMSO. Tightlock couplers. Sliding doors. Disc and rheostatic brakes. PA. Originally built as four car units and numbered 508 001–043. One trailer removed and used for class 455/7 on transfer from the SR.

System: 750 V d.c. third rail.
Bogies: BX1.
Gangways: Gangwayed within unit. End doors.
Traction Motors: Four GEC G310AZ of 82.125 kW.
Dimensions: 19.80 x 2.82 m (outer cars), 19.92 x 2.82 m (inner cars).
Maximum Speed: 75 mph.

64649–64687. DMSO. Dia. EA208. Lot No. 30942 York 1979–80. 74S. 36.15 t.
64688–64691. DMSO. Dia. EA208. Lot No. 30979 York 1980. 74S. 36.15 t.
71483–71520. TSO. Dia. EH218. Lot No. 30943 York 1979–80. 82S. 26.72 t.
71521–71525. TSO. Dia. EH218. Lot No. 30980 York 1980. 82S. 26.72 t.
64692–64729. BDMSO. Dia. EI203. Lot No. 30945 York 1979–80. 74S. 36.61 t.
64730–64734. BDMSO. Dia. EI203. Lot No. 30981 York 1980. 74S. 36.61 t.

508 101	PBDX	BD	64649	71483	64692
508 102	PBDX	BD	64650	71484	64693
508 103	PBDX	BD	64651	71485	64694
508 104	PBDX	BD	64652	71486	64695
508 105	PBDX	BD	64653	71487	64696
508 106	PBDX	BD	64654	71488	64697
508 107	PBDX	BD	64655	71489	64698
508 108	PBDX	BD	64656	71490	64699
508 109	PBDX	BD	64657	71491	64700
508 110	PBDX	BD	64658	71492	64701
508 111	PBDX	BD	64659	71493	64702
508 112	PBDX	BD	64660	71494	64703
508 113	PBDX	BD	64661	71495	64704
508 114	PBDX	BD	64662	71496	64705
508 115	PBDX	BD	64663	71497	64706
508 116	PBDX	BD	64664	71498	64707
508 117	PBDX	BD	64665	71499	64708
508 118	PBDX	BD	64666	71500	64709
508 119	PBDX	BD	64667	71501	64710
508 120	PBDX	BD	64668	71502	64711
508 121	PBDX	BD	64669	71503	64712
508 122	PBDX	BD	64670	71504	64713
508 123	PBDX	BD	64671	71505	64714
508 124	PBDX	BD	64672	71506	64715
508 125	PBDX	BD	64673	71507	64716
508 126	PBDX	BD	64674	71508	64717
508 127	PBDX	BD	64675	71509	64718
508 128	PBDX	BD	64676	71510	64719
508 129	PBDX	BD	64677	71511	64720
508 130	PBDX	BD	64678	71512	64721
508 131	PBDX	BD	64679	71513	64722
508 132	PBDX	BD	64680	71514	64723
508 133	PBDX	BD	64681	71515	64724
508 134	PBDX	BD	64682	71516	64725
508 135	PBDX	BD	64683	71517	64726
508 136	PBDX	BD	64684	71518	64727
508 137	PBDX	BD	64685	71519	64728
508 138	PBDX	BD	64686	71520	64729
508 139	PHRX	HR	64687	71521	64730
508 140	PHRX	HR	64688	71522	64731
508 141	PHRX	HR	64689	71523	64732
508 142	PHRX	HR	64690	71524	64733
508 143	PHRX	HR	64691	71525	64734

4.5. DEPARTMENTAL EMUs

For information in this section we wish to thank Roger Butcher.

Individual vehicles.

ADB 975032	(M 75165)	SH	SR class 932 experimental stock. 'Mars'.
DB 977335	(S 76277)	RTC	MTA Pool Generator coach for DB999550.
DB 977336	(S 76278)	RTC	MTA Pool Driving trailer for DB999550.
ADB 977362	(S 10392)	BI	SR class 930 deicing trailer.
ADB 977363	(S 10399)	FR	SR Class 930 deicing trailer.
ADB 977364	(S 10400)	RE	SR class 930 deicing trailer.
DB 977394	(S 65316)		MTA Pool Traction unit for DB999550 (pending conversion).
ADB 977578	(S 77101)	SU	Sandite vehicle (work with Class 317/9).
ADB 977579	(S 77109)	SU	Sandite vehicle (work with Class 317/9).
DB 999602	(S 62483)	RTC	Ultrasonic test train instrumentation coach. MTA pool.
DB 999603	(S 62482)	Cathays	Ultrasonic test train coach (pending conversion).
ADB 977609	(S 65414)		Southern Region deicing and sandite unit under conversion).
ADB 977610	(S 77136)		Southern Region deicing and sandite unit under conversion).

Complete Units

Southern Region Class 930 Deicing and Sandite units.

001	SU	ADB 977365	(S 10726)	ADB 977368	(S 10499)
002	SU	ADB 977366	(S 10500)	ADB 977367	(S 10497)
003	SU	ADB 975594	(S 12658)	ADB 975595	(S 10994)
004	EH	ADB 975586	(S 10907)	ADB 975587	(S 10908)
005	WD	ADB 975588	(S 10981)	ADB 975589	(S 10982)
006	WD	ADB 975590	(S 10833)	ADB 975591	(S 10834)
007	GI	ADB 975592	(S 10993)	ADB 975593	(S 12659)
008	AF	ADB 975596	(S 10844)	ADB 975597	(S 10987)
009	BI	ADB 975598	(S 10989)	ADB 975599	(S 10990)
010	BI	ADB 975600	(S 10988)	ADB 975601	(S 10843)
011	RE	ADB 975602	(S 10991)	ADB 975603	(S 10992)
012	FR	ADB 975604	(S 10939)	ADB 975605	(S 10940)
013	RE	ADB 975896	(S 11387)	ADB 975897	(S 11388)
015	WD	ADB 977531	(S 14047)	ADB 977532	(S 14048)
016	WD	ADB 977533	(S 14273)	ADB 977534	(S 14384)
017	BM	ADB 977566	(S 65312)	ADB 977567	(S 65314)

Southern Region Class 931 Stores Units:

018	FR	ADB 977290	(S 65318)	ADB 977291	(S 65324)
019	RE	ADB 977068	(S 14549)	ADB 977069	(S 16029)

Southern Region Class 930 Carriage Cleaning Fluid Unit:

026	SG	ADB 977559	(S 65313)	ADB 977560	(S 65320)

Southern Region Tractor Units:

021	WD	ADB 977304	(S 65317)	ADB 977305	(S 65322)
050	SH	ADB 977296	(S 65319)	ADB 977297	(S 77108)
053	BI	ADB 977211	(S 61684)	ADB 977212	(S 75736)
054	SH	ADB 977207	(S 61658)	ADB 977208	(S 75710)

Southern Region Test Units:

1053	SH	ADB 977505	(S 65321)	ADB 977507	(S 77110)
1054	SH	ADB 977506	(S 65323)	ADB 977508	(S 77112)

Class 935 Prototype High Density Stock:

057	CJ	ADB 975844	(S 64305)	ADB 975849	(S 62426)
		ADB 975850	(S 62429)	ADB 975851	(S 64304)

Anglia Region class 302 three-car Sandite Units:

996	CC	ADB 977598	(E 75080)	ADB 977599	(E 61073)
		ADB 977600	(E 75001)		
997	IL	ADB 977601	(E 75211)	ADB 977602	(E 61220)
		ADB 977603	(E 75035)		
998	EM	ADB 977604	(E 75077)	ADB 977605	(E 61052)
		ADB 977606	(E 75070)		

Eastern Region Instruction Unit:

305 935		ADB 977639	(E 75548)	ADB 977640	(E 61463)
		ADB 977641	(E 75214)		

London Midland Region Sandite Units:

	BD	ADB 977345	(M 61178)	ADB 977346	(M 75178)
	BD	ADB 977347	(M 61180)	ADB 977348	(M 75180)
	BD	ADB 977349	(M 61183)	ADB 977350	(M 75183)
	WN	ADB 977385	(M 61148)	ADB 977386	(M 75189)

5.3. LOCO–HAULED NON-PASSENGER STOCK

AK51 (RK) KITCHEN CAR

Dia. AK503. Mark 1. Gas cooking. Converted from RBR. Fluorescent lighting. xd*. ETH 2 (2X*).

Note: Kitchen cars have traditionally been numbered in the NPCCS series, but have passenger coach diagram numbers!

Lot No. 30628 Pressed Steel 1960–61. 39 t.

80041 (1690) **I** ICHV BN **I**

NN COURIER VEHICLE

Dia. NN504. Converted 1986–7 from Mark 1 BSKs. One compartment retained for courier use. Roller shutter doors.

80200–2/4–6/8/11–14/16–17/21–23/25. Lot No. 30699 Wolverton 1962. Commonwealth bogies. vd. ETH 2. 37t.
80203/9–10/15/20/24. Lot No. 30573 Gloucester 1960. B4 bogies. vd. ETH2. 35t.
80207. Lot No. 30721 Wolverton 1963. Commonwealth bogies. vd. ETH2. 37t.
80218. Lot No. 30427 Wolverton 1958–9. B4 bogies. vd. ETH 2. 35t.
80219. Lot No. 31027 Wolverton 1958. B4 bogies. vd. ETH 2. 35t.

80200	(35303)	**B**	RPXG	v	OM
80201	(35311)		RPXG	v	DY
80202	(35462)	**R**	RPXG	v	OM
80203	(35287)		RPXG	v	DY
80204	(35297)	**R**	RPXG	x	EN
80205	(35312)		RPXG	v	HT
80206	(35304)	**R**	RPXG	x	EN
80207	(35466)		RPXG	x	EN
80208	(35341)		RPXG	x	EN
80209	(35279)		RPXG	v	HT
80210	(35282)		RPXG	v	HT
80211	(35296)		RPXG	x	HT
80212	(35307)		RPXG	x	CA
80213	(35316)		RPXG	x	EN
80214	(35323)		RPXG	x	HT
80215	(35292)		RPXG	x	CA
80216	(35295)		RPXG	x	EN
80217	(35299)		RPXG	x	CA
80218	(35197)		RPXG	x	MA
80219	(35201)		RPXG	x	CA
80220	(35276)		RPXG	x	HT
80221	(35328)		RPXG	x	EN
80222	(35315)		RPXG	x	EN
80223	(35331)		RPXG	x	CA
80224	(35291)		RPXG	x	OM
80225	(35327)		RPXG	x	

NS (POS) POST OFFICE SORTING VAN

Used in travelling post office (TPO) trains. Mark 1. Various diagrams.

80300–80305. Lot No. 30486 Wolverton 1959. Dia. NS501. Originally built with nets for collecting mail bags in motion. Equipment now removed. BR2 bogies. vd. ETH 3X. 36t. (*–B5 bogies. xd. 34t).
80306–80308. Lot No. 30487 Wolverton 1959. Dia. NS502. BR2 bogies. vd. ETH 3. 36t.

80309–80314. Lot No. 30661 Wolverton 1961. Dia. NS501. BR2 bogies. vd. ETH 3. 37t.
80315–80316. Lot No. 30662 Wolverton 1961. Dia. NS501. BR2 bogies. vd. ETH 3X. 36t.
80318. Lot No. 30663 Wolverton 1961. Dia. NS501. BR2 bogies. vd. ETH 3X. 36t.

No.		Type		Depot
80300		RPOM		MA
80301	**R**	RPOM		MA
80302	**R**	RPOM		MA
80303		RPOA	*	CA
80304	**R**	RPOA		CA
80305	**R**	RPOA		CA
80306	**R**	RPOA		CA
80307	**R**	RPOA		CA
80308	**R**	RPOA	*	NC
80309	**R**	RPOA	*	CA
80310	**R**	RPOE		HT
80311	**R**	RPOE		HT
80312	**R**	RPOE		HT
80313	**R**	RPOE		HT
80314	**R**	RPOE	*	HT
80315	**R**	RPOE		HT
80316	**R**	RPOA	*	NC
80318	**R**	RPOA	*	NC

The following lots are pressure ventilated and have B5 bogies.

80319–80327. Dia. NS504. Lot No. 30778 York 1968–9. xd. ETH 4. 35t.
80328–80338. Dia. NS505. Lot No. 30779 York 1968–9. xd. ETH 4. 35t.
80339–80355. Dia. NS506. Lot No. 30780 York 1968–9. xd. ETH 4. 35t.

No.		Type	Depot
80319	**R**	RPOA	CA
80320	**R**	RPOE	HT
80321	**R**	RPOZ	EC
80322	**R**	RPOE	HT
80323	**R**	RPOE	HT
80324	**R**	RPOE	HT
80325	**R**	RPOM	DY
80326	**R**	RPOM	DY
80327	**R**	RPOM	DY
80328	**R**	RPOZ	EC
80329		RPOM	MA
80330		RPOE	HT
80331	**R**	RPOE	HT
80332	**R**	RPOE	HT
80333	**R**	RPOE	HT
80334		RPOM	MA
80335	**R**	RPOE	HT
80336	**R**	RPOE	HT
80337	**R**	RPOE	HT
80338	**R**	RPOM	EN
80339	**R**	RPOM	EN
80340	**R**	RPOM	MA
80341	**R**	RPOM	EN
80342	**R**	RPOM	EN
80343		RPOM	MA
80344		RPOM	EN
80345	**R**	RPOW	BJ
80346	**R**	RPOW	BJ
80347	**R**	RPOM	EN
80348		RPOM	EN
80349	**R**	RPOM	DY
80350		RPOA	NC
80351	**R**	RPOM	DY
80352	**R**	RPOA	NC
80353	**R**	RPOM	EN
80354	**R**	RPOM	EN
80355	**R**	RPOM	MA

80356–80380. Lot No. 30839 York 1972–3. Dia. NS501. Pressure ventilated. Fluorescent lighting. B5 bogies. xd. ETH 4X. 37t.

No.		Type	Depot
80356		RPOM	EN
80357	**R**	RPOM	EN
80358		RPOW	BJ
80359	**R**	RPOW	BJ
80360	**R**	RPOW	BJ
80361	**R**	RPOW	BJ
80362	**R**	RPOE	HT
80363	**R**	RPOE	HT
80364	**R**	RPOE	HT
80365	**R**	RPOE	HT
80366	**R**	RPOM	EN
80367	**R**	RPOM	EN
80368		RPOM	EN
80369		RPOM	EN
80370	**R**	RPOM	EN
80371		RPOM	EN
80372		RPOM	EN
80373	**R**	RPOM	EN
80374		RPOM	MA
80375	**R**	RPOM	MA
80376	**R**	RPOM	EN
80377	**R**	RPOM	EN
80378	**R**	RPOM	EN
80379	**R**	RPOW	OM
80380	**R**	RPOM	MA

80381–80395. Lot No. 30900 Wolverton 1977. Dia NS531. Converted from SK. Pressure ventilated. Fluorescent lighting. B5 bogies. xd. ETH 4X. 38t.

No.	Former No.		Type	Depot
80381	(25112)	**R**	RPOW	OM
80382	(25109)	**R**	RPOW	OM
80383	(25033)	**R**	RPOW	OM
80384	(25078)	**R**	RPOW	OM
80385	(25083)	**R**	RPOW	OM
80386	(25099)	**R**	RPOW	OM
80387	(25045)	**R**	RPOW	OM
80388	(25088)	**R**	RPOW	OM
80389	(25103)	**R**	RPOW	OM
80390	(25047)	**R**	RPOW	OM
80391	(25089)	**R**	RPOW	OM
80392	(25082)	**R**	RPOW	OM

80393 (25118) **R**	RPOW	OM		80395 (25056) **R**	RPOM	MA	
80394 (25156) **R**	RPOW	OM					

NT (POT) POST OFFICE STOWAGE VAN

Open vans used for stowage of mail bags in conjunction with POS. Various diagrams.

Lot No. 30488 Wolverton 1959. Dia. NT502. Originally built with nets for collecting mail bags in motion. Equipment now removed. B5 bogies. xd. ETH 3. 35t.

80400	RPOM	EN	80401**R**	RPOM	EN	80402**R** RPOM	EN

The following twelve vehicles were converted at York from BSK to lot 30143 (80403) and 30229 (80404–80414). No new lot number was issued. Dia. NT503 35 t. (*dia. NT501 BR2 bogies 38 t. ETH 3 (3X*).

80403 (34361)	RPOM	MA	80409 (35019) **R**	RPOW	OM
80404 (35014)	RPOM	MA	80410 (35001) **R**	RPOW	OM
80405 (35009)	RPOM	MA	80411 (35003) **R**	RPOW	BJ
80406 (35022) **R**	RPOM	MA	80412 (35002)	RPOM	MA
80407 (35015) **R**	RPOA	CA	80413 (35004) **R**	RPOW	BJ
80408 (35018) **R**	RPOA	CA	80414 (35005)	RPOM	MA

Lot No. 30781 York 1968. Dia. NT505. Pressure ventilated. B5 bogies. xd. ETH 4. 34t.

80415**R**	RPOM	EN	80419**R**	RPOM	EN	80422**R**	RPOE	HT
80416**R**	RPOM	EN	80420**R**	RPOE	HT	80423	RPOW	OM
80417**R**	RPOM	EN	80421**R**	RPOE	HT	80424**R**	RPOM	EN
80418**R**	RPOM	EN						

Lot No. 30840 York 1973. Dia. NT504. Pressure ventilated. fluorescent lighting. B5 bogies. xd. ETH 4X. 35t.

80425	RPOM	EN	80427**R**	RPOW	BJ	80429**R**	RPOM	EN
80426**R**	RPOW	BJ	80428**R**	RPOM	EN	80430**R**	RPOM	EN

Lot No. 30901 Wolverton 1977. converted from SK. Dia. NT521. Pressure ventilated. Fluorescent lighting. B5 bogies. xd. ETH 4X. 35t.

80431 (25104) **R**	RPOM	DY	80436 (25077)	RPOW	OM
80432 (25071) **R**	RPOE	HT	80437 (25068) **R**	RPOW	OM
80433 (25150) **R**	RPOE	HT	80438 (25139)	RPOM	DY
80434 (25119) **R**	RPOE	HT	80439 (25127) **R**	RPOM	DY
80435 (25117) **R**	RPOW	OM			

NU (BPOT) BRAKE POST OFFICE STOWAGE VAN

As NT but with brake. Mark 1.

80450–80455. Lot No. 30489 Wolverton 1959. Dia. NU501. BR2 bogies. vd. ETH 3X. 34t.
80456–80458. Lot No. 30782 York 1968. Dia. NU502. Pressure ventilated. B5 bogies. xd. ETH 4. 36t.

80450**R**	RPOW	OM	80451	RPOW	OM	80452**R**	RPOW	OM

80453 RPOW	OM	80455 RPOW	OM	80457**R** RPOM	EN
80454**R** RPOW	OM	80456**R** RPOM	EN	80458**R** RPOM	EN

NA/NC/ND (BG) GANGWAYED BRAKE VAN (90 mph)

These vans are built on short frames (57'). Load 10t. BR1 bogies. There are a number of variants:

NAV (Dia. NA501). BG steam heated.
NBV (Dia. NB501). High security letter mail van. Converted at WB from BG 1985. Gangways removed. vd. 84382/7/461/77 formerly renumbered 80460–3, but since given original number to prevent identification by potential train robbers!
NCV (Dia. NC501). Newspaper packing van. Fluorescent lighting. ETH 3X.
NDV or NDX (Dia ND501). BG dual heated. ETH 1 (1X*).
Braking and heating variations are shown in this section by the TOPS code.

Fitted with cage: 80735, 80965, 84152, 84330, 84437, 84508.

80501–80529. Lot No. 30009 Derby 1952–3. 31t.
80530–80535. Lot No. 30023 Eastleigh 1951. 31t.
80537–80565. Lot No. 30039 Derby 1954. 31t.
80570–80596. Lot No. 30040 Wolverton 1954–5. 32t.
80597–80671. Lot No. 30046 York 1954. 31.5t.
80672–80724. Lot No. 30136 Metro-Cammell 1955. 31.5t.
80726–80802. Lot No. 30140 BRCW 1955–6. 31.5t.
80803–80852. Lot No. 30144 Cravens 1955. 31.5t.
80855–80962. Lot No. 30162 Pressed Steel 1956–7. 32t.
80965–80999. Lot No. 30173 York 1956. 31.5t.

80501		RPXC	NDV*	CA
80512		RPXC	NDV*	CA
80513	**B**	RPPW	NCV*	BJ
80514		RPOW	NDV	BJ
80549		RPXC	NDV*	CA
80551		RPXC	NDV*	CA
80554		RPXC	NDV*	CA
80565		RPXC	NDV*	CA
80570		RPXC	NDV*	CA
80571		RPXC	NDV*	CA
80576		RPXC	NDV*	BJ
80580		RPXC	NDV*	CA
80582		RPXX	NDV*	BJ
80583		RPXC	NDV*	CA
80589		RPXC	NDV*	CA
80590		RPXC	NDV*	CA
80592		RPXX	NDV*	CA
80596		RPXC	NDV*	CA
80608		RPXC	NDV*	CA
80613	**B**	RPPA	NCV*	CA
80623		RPXC	NDV*	CA
80631	**B**	RPPA	NCV*	CA
80646		RPXC	NDV*	CA
80649	**B**	RPPA	NCV*	CA
80650		RPXC	NDV*	CA
80651		RPXX	NDV*	CA
80653		RPXC	NDV*	CA
80654		RPXC	NDV*	CA
80658		RPXC	NDV*	CA
80663		RPXX	NDV*	BJ
80664		RPXC	NDV*	CA
80668		RPXC	NDV*	BJ
80669		RPXX	NDV*	CA
80670		RPXC	NDV*	BJ
80671		RPXC	NDV*	CA
80674		RPXC	NDV*	CA
80680		RPXC	NDV*	CA
80682	**B**	RPPA	NCV	CA
80693		RPXC	NDV*	HT
80695		RPPA	NCV	CA
80699		RPXC	NDV	BJ
80700		RPXB	NDX*	CA
80701		RPXB	NDX	CA
80707		RPXC	NDV*	OM
80709		RPXC	NDV*	OM
80710		RPXC	NDV*	OM
80716		RPXC	NDV*	CA
80720	**B**	RPPW	NCV*	BJ

80728	RPXC	NDV*	CA
80730	RPXC	NDV	CA
80736	RPXC	NDV*	HT
80737	RPXB	NDX	CA
80742	RPXC	NDV*	HT
80748	RPXC	NDV*	BJ
80754	RPXC	NDV*	CA
80756	RPXC	NDV*	HT
80761	RPXX	NDV*	HT
80763	RPXC	NDV*	CA
80767 **B**	RPPW	NCV*	BJ
80768 **B**	RPPW	NCV*	BJ
80769	RPXC	NDV	BJ
80775 **B**	RPPW	NCV*	BJ
80780 **B**	RPPW	NCV*	BJ
80781	RPXB	NDX*	HT
80782 **B**	RPPW	NCV*	BJ
80789 **B**	RPPW	NCV	BJ
80790	RPXB	NDX	HT
80794	RPXC	NDV*	CA
80795	RPXC	NDV*	BJ
80804	RPXC	NDV*	CA
80805	RPXB	NDX*	CA
80806	RPXC	NDV*	CA
80810	RPXC	NDV*	CA
80820	RPXC	NDV*	CA
80824	RPXC	NDV*	CA
80831	RPXX	NDV*	CA
80834	RPXB	NDX*	CA
80835	RPXC	NDV	CA
80837	RPXC	NDV*	CA
80843	RPXB	NDX	CA
80850	RPXX	NDV*	BJ
80851	RPXC	NDV*	CA
80852	RPXC	NDV*	CA
80859	RPXB	NDX*	BJ
80861 **R**	RPXB	NDX*	HT
80868 **R**	RPXB	NDX*	EC
80885	RPXB	NDX*	CA
80886	RPXB	NDX*	BJ
80889	RPXB	NDX*	CA
80892	RPXB	NDX*	BJ
80893	RPXB	NDV	CA
80926	RPXB	NDX	HT
80945	RPXB	NDX	CA
80962	RPXB	NDX*	CA
80966 **B**	RPPA	NCV*	CA
80967	RPXC	NDV*	CA
80972	RPXX	NDV*	CA
80976	RPXC	NDV*	CA
80981	RPXX	NDV*	CA
80985	RPXC	NDV*	CA
80988	RPXB	NDV*	HT
80994 **B**	RPXC	NCV*	CA

NZ (DLV) DRIVING BRAKE VAN (110 mph)

Dia. NZ501. Mark 3B. Air conditioned. T4 bogies. ae. pg. ETH X.

Lot No. 31042 Derby 1988. t.

82101 **I**	IWCX	OY
82102 **I**	IWCX	OY
82103 **I**	IWCX	OY
82104 **I**	IWCX	OY
82105 **I**	IWCX	OY
82106 **I**	IWCX	OY
82107 **I**	IWCX	OY
82108 **I**	IWCX	OY
82109 **I**	IWCX	OY
82110 **I**	IWCX	OY
82111 **I**	IWCX	OY
82112 **I**	IWCX	OY
82113 **I**	IWCX	OY
82114 **I**	IWCX	OY
82115 **I**	IWCX	OY
82116 **I**	IWCX	OY
82117 **I**	IWCX	WB
82118 **I**	IWCX	WB
82119 **I**	IWCX	OY
82120 **I**	IWCX	OY
82121 **I**	IWCX	OY
82122 **I**	IWCX	OY
82123 **I**	IWCX	WB
82124 **I**	IWCX	OY
82125 **I**	IWCX	WB
82126 **I**	IWCX	WB
82127 **I**	IWCX	WB
82128 **I**	IWCX	WB
82129 **I**	IWCX	WB
82130 **I**	IWCX	WB
82131 **I**	IWCX	WB
82132		
82133		
82134		
82135		
82136		
82137		
82138		
82139		
82140		
82141		
82142		
82143		
82144		
82145		
82146		
82147		
82148		
82149		
82150		
82151		
82152		

NZ (DLV) DRIVING BRAKE VAN (140 mph)

Dia. NZ502. Mark 4. Air conditioned. Swiss-built (SIG) bogies. ae. pg. ETH 6X.

Lot No. 31043 Metro-Cammell 1988. 45.18 t.

82200	I	IECX	BN
82201	I	IECX	BN
82202	I	IECX	BN
82203	I	IECX	BN
82204	I	IECX	BN
82205			
82206			
82207			
82208			
82209			
82210			
82211			
82212			
82213			
82214			
82215			
82216			
82217			
82218			
82219			
82220			
82221			
82222			
82223			
82224			
82225			
82226			
82227			
82228			
82229			
82230			
82231			

NA/NC/ND (BG) GANGWAYED BRAKE VAN (90 mph)

Class continued from 80994. Note: All 84XXX vehicles were renumbered from the 81XXX series by adding 3000 to the original number.

84000–84014. Lot No. 30173 York 1956. 31.5t.
84015–84053. Lot No. 30224 Cravens 1956. 31.5t.
84055–84179. Lot No. 30228 Metro-Cammell 1957–8. 31.5t.
84180–84204. Lot No. 30234 Cravens 1956–7. 31.5t.
84205–84265. Lot No. 30163 Pressed Steel 1957. 31.5t.
84266–84312. Lot No. 30323 Pressed Steel 1957. 32t.
84313–84497. Lot No. 30400 Pressed Steel 1957–8. 32t.
84498–84572. Lot No. 30484 Pressed Steel 1958. 32t.
84573–84592. Lot No. 30715 Gloucester 1962. 31t.
84594–84612. Lot No. 30716 Gloucester 1962. 31t.
84613–84628. Lot No. 30725 Gloucester 1962–3. 31t.

84010		RPXB	NDX	CA
84012	**B**	RPPW	NCV*	BJ
84015		RPXC	NDV*	CA
84016		RPXC	NDV*	CA
84020		RPXC	NDV*	BJ
84024		RPXX	NDV*	CA
84026		RPXB	NDX	BJ
84029	**B**	RPPW	NCV	BJ
84036		RPXX	NDV*	CA
84037		RPXC	NDV*	CA
84041	**B**	RPPW	NCV	BJ
84044		RPXB	NDX	CA
84051		RPXB	NDX*	EN
84057		RPXC	NDV*	BJ
84065		RPXC	NDV*	CA
84082		RPXC	NDV*	CA
84086		RPXC	NDV	BJ
84098		RPXC	NDV	CA
84099		RPXB	NDX*	CA
84100	**B**	RPPA	NCV	CA
84104		RPXB	NDX	CA
84106		RPXC	NDV*	CA
84108		RPXC	NDV	BJ
84109		RPXC	NDV	CA
84112	**B**	RPPW	NCV	BJ
84114		RPXC	NDV	CA
84119	**B**	RPPA	NCV	CA
84121		RPXX	NDV	OM
84129	**B**	RPPA	NCV	CA
84133		RPXC	NDV	BJ
84134		RPXB	NDX	CA
84135	**B**	RPPW	NCV	BJ
84137		RPXC	NDV	CA
84143		RPXC	NDV	CA

84145		RPXC	NDV	CA
84146		RPXC	NDV	CA
84147		RPXC	NDV	CA
84151		RPXC	NDV	CA
84159		RPXC	NDV*	CA
84160		RPXC	NDV	BJ
84162		RPXC	NDV	CA
84181	**B**	RPPW	NCV*	BJ
84185		RPXC	NDV	BJ
84190		RPXC	NDV	OM
84191		RPXC	NDV*	CA
84193		RPXX	NDV*	CA
84194		RPXC	NDV	CA
84195		RPXC	NDV*	CA
84197		RPXC	NDV	BJ
84200		RPXB	NDX	BJ
84202		RPXC	NDV	BJ
84203		RPXC	NDV	CA
84204		RPXC	NDV	CA
84211		RPXB	NDX*	PC
84212		RPXB	NDX*	CA
84215		RPXB	NDX*	CA
84217		RPXB	NDX*	PC
84234		RPXB	NDX*	HT
84239		RPXB	NDX*	CA
84240		RPXB	NDX*	CA
84242		RPXB	NDX*	PC
84250		RPXB	NDX*	CA
84251		RPXB	NDX*	CA
84252		RPXB	NDX*	CA
84261		RPXB	NDX*	BJ
84262		RPXB	NDX*	CA
84273		RPXB	NDX	CA
84274		RPXB	NDX	CA
84279		RPXX	NDV	BJ
84280		RPXB	NDX	CA
84281		RPXC	NDV	BJ
84286		RPXC	NDV	CA
84287		RPXC	NDV	BJ
84289		RPXB	NDX	CA
84290		RPXC	NDV	BJ
84296		RPXC	NDV	BJ
84298		RPXC	NDV	BJ
84301		RPXC	NDV	BJ
84302		RPXC	NDV	BJ
84313		RPXB	NDX	CA
84318		RPXX	NDV	DY
84319		RPXC	NDV	BJ
84320		RPXB	NDX	PC
84322		RPXB	NDV	PC
84325		RPXB	NDX	CA
84331		RPXX	NDV	BJ
84332		RPXC	NDV	BJ
84339		RPXB	NDX	CA
84342		RPXC	NDV	BJ
84344		RPXC	NDV	BJ
84346		RPXB	NDV	CA
84347		RPXC	NDV	BJ
84350		RPXC	NDV	BJ
84351		RPXB	NDX	EC
84352		RPXB	NDX	CA
84354		RPXB	NDX	CA
84356		RPXC	NDV	BJ
84361		RPXC	NDV	BJ
84363		RPXB	NDX	CA
84364		RPXC	NDV	BJ
84366		RPXX	NDV	BJ
84370		RPXB	NDX*	CA
84371		RPXC	NDV	BJ
84373		RPXC	NDV	DY
84380		RPXB	NDX	CA
84382		RPOM	NBV	EN
84386		RPXC	NDV	BJ
84387		RPOM	NBV	EN
84388		RPXB	NDX	CA
84390		RPXC	NDV	DY
84395		RPXC	NDV	BJ
84399		RPXB	NDX	CA
84402		RPXX	NDV	BJ
84404		RPXB	NDX	BJ
84406		RPXC	NDV	BJ
84408		RPXC	NDV	BJ
84412		RPXX	NDV	BJ
84415		RPXC	NDV	BJ
84416		RPXC	NDV	BJ
84419		RPXB	NDX	BJ
84420		RPXB	NDX	CA
84421		RPXC	NDV	BJ
84422		RPXC	NDV	BJ
84427		RPXB	NDX	CA
84428		RPXC	NDV	BJ
84430		RPXB	NDX	CA
84434		RPXX	NDV	BJ
84435		RPXC	NDV	BJ
84436		RPXC	NDV	BJ
84439		RPXC	NDV	BJ
84445		RPXB	NDX	CA
84447		RPXC	NDV	BJ
84456		RPXB	NDX	CA
84458		RPXB	NDX	CA
84461		RPOM	NBV	EN
84464		RPXC	NDV	BJ
84467		RPOW	NDV	BJ
84469		RPXB	NDX	BJ

84472		RPXB	NDX	CA
84475		RPXB	NDX	BJ
84477		RPOM	NBV	EN
84499		RPXB	NDX	BJ
84501		RPXB	NDX	CA
84510	**R**	RPXB	NDX	PC
84512		RPXB	NDX	BJ
84515		RPXB	NDX	EN
84519		RPXC	NDV	BJ
84531		RPXB	NDX*	EN
84543		RPXX	NDV	BJ
84547		RPXC	NDV	BJ
84552		RPXC	NDV	BJ
84554		RPXX	NDV	BJ
84558		RPXC	NDV	OM
84577		RPXB	NDV	EC
84590		RPXB	NDX	CA
84619		RPXC	NDV	BJ
84626		RPXC	NDV	CA

NE/NH (BG) 100/110 mph GANGWAYED BRAKE VAN

As ND but rebogied with B4 bogies suitable for 100 mph–NE (110 mph with special maintenance–NH). d. ETH 1 (1X* and NHA). For lot numbers refer to original number series. Deduct 1.5t from weights. All NHA are a*pg.

Fitted with cages: 92094/6/7/102/8/15/25/48/51/65/88/90/5–7.

92000	(80723)	**I**	RPXM	a pg	OY
92001	(80855)	**I**	RPXM	NHA	WB
92002	(80858)	**I**	RPXM	x*to	WB
92003	(80864)	**I**	RPXM	x*to	WB
92004	(80867)	**I**	RPXM	x*to	WB
92005	(80876)	**I**	RPXM	x*to	WB
92006	(80878)	**I**	RPXM	NHA	WB
92007	(80880)	**I**	RPXZ	x*pg	EC
92008	(80895)	**I**	RPXM	NHA	WB
92009	(80897)	**I**	RPXZ	x*to	EC
92010	(80900)	**I**	RPXZ	x*to	EC
92011	(80903)	**I**	RPXM	a*pg	WB
92012	(80910)	**I**	RPXM	a*pg	WB
92013	(80916)	**I**	RPXZ	x*pg	EC
92014	(80923)	**I**	RPXZ	x*pg	PC
92015	(80927)	**I**	RPXZ	x*to	PC
92016	(80930)	**I**	RPXZ	x*pg	PC
92017	(80940)	**I**	RPXZ	x*to	EC
92018	(80941)		RPXZ	x*to	PC
92019	(80944)	**I**	RPXZ	x*pg	PC
92020	(80950)	**I**	RPXZ	x*pg	PC
92021	(80956)		RPXZ	x*to	EC
92022	(80958)		RPXM	x*pg	OY
92023	(80971)	**I**	RPXM	a*pg	WB
92025	(81023)	**I**	RPXM	NHA	WB
92026	(81060)	**I**	RPXM	NHA	WB
92027	(81061)	**I**	RPXM	NHA	WB
92028	(81064)	**I**	RPXM	NHA	WB
92029	(81077)	**I**	RPXM	NHA	WB
92031	(81102)	**I**	RPXM	NHA	WB
92032	(81117)	**I**	RPXM	NHA	WB
92033	(81123)	**I**	RPXM	NHA	WB
92034	(81142)	**I**	RPXM	NHA	WB
92035	(81150)	**I**	RPXM	a*pg	WB
92036	(81158)	**I**	RPXM	NHA	WB
92037	(81165)	**I**	RPXM	NHA	WB
92038	(81173)	**I**	RPXM	NHA	WB
92039	(81175)	**I**	RPXM	NHA	WB
92040	(81186)	**I**	RPXM	a pg	WB
92041	(81205)		RPXM	a to	WB
92042	(81207)		RPXZ	a to	PC
92043	(81208)	**R**	RPXM	a to	WB
92044	(81209)	**P**	RPXH	a to	HT
92045	(81210)		RPXZ	a to	PC
92046	(81214)	**I**	RPXM	NHA	WB
92047	(81216)		RPXZ	a to	PC
92048	(81218)	**I**	RPXM	NHA	WB
92049	(81219)		RPXZ	a to	PC
92050	(81220)	**I**	RPXM	a pg	OY
92051	(81221)		RPXZ	a to	EC
92052	(81222)		RPXM	a to	WB
92053	(81223)		RPXM	a to	WB
92054	(81224)		RPXZ	a to	PC
92055	(81225)	**I**	RPXM	a to	OY
92056	(81226)		RPXM	a to	WB
92057	(81227)	**I**	RPXM	a to	OY
92058	(81228)	**P**	RPXH	a to	HT
92059	(81229)		RPXZ	a to	PC
92060	(81230)		RPXM	a to	MA
92061	(81231)	**I**	RPXA	a	NC
92062	(81232)		RPXM	a to	WB
92063	(81233)	**I**	RPXM	a*pg	MA
92064	(81236)		RPXM	a to	MA
92065	(81237)		RPXZ	a	PC
92066	(81238)	**I**	RPXM	a to	WB
92067	(81243)	**I**	RPXZ	a to	PC
92068	(81244)		RPXZ	a to	PC
92069	(81245)		RPXM	a	WB

92070	(81246)	RPXM	a	WB
92071	(81249)	RPXM	a to	MA
92072	(81253)	RPXZ	a to	EC
92073	(81254)	RPXZ	a to	EC
92074	(81255)	RPXM	a to	MA
92075	(81256) **R**	RPXM	a to	MA
92076	(81257)	RPXM	a to	EN
92077	(81258)	RPXW	a to	LA
92078	(81259)	RPXM	a to	MA
92079	(81260)	RPXZ	a to	PC
92080	(81263) **I**	RPXZ	a to	PC
92081	(81264)	RPXM	a to	MA
92082	(81265)	RPXM	a to	EN
92083	(81266)	RPXM	a to	MA
92084	(81268)	RPXM	a to	EN
92086	(81282) **I**	RPXA	a to	NC
92087	(81283) **I**	RPXM	x*to	OY
92088	(81284) **I**	RPXA	a to	NC
92089	(81303) **P**	RPXH	a to	HT
92090	(81305)	RPXM	a to	EN
92091	(81308) **I**	RPXA	a to	NC
92092	(81309) **P**	RPXH	a to	HT
92093	(81312) **I**	RPXW	a*to	LA
92094	(81367) **I**	RPXZ	a to	PC
92095	(81375)	RPXM	a to	MA
92096	(81377)	RPXM	a to	WB
92097	(81378)	RPXZ	a to	PC
92098	(81381) **I**	RPXM	NHA	WB
92099	(81383) **I**	RPXM	x*pg	OY
92100	(81391)	RPXW	a to	LA
92101	(81392)	RPXM	a to	EN
92102	(81394)	RPXM	a to	EN
92103	(81400)	RPXW	x to	BJ
92104	(81401)	RPXM	a to	EN
92105	(81405)	RPXM	a to	EN
92106	(81409)	RPXA	x to	NC
92107	(81410)	RPXA	x to	CA
92108	(81411) **R**	RPXM	a to	EN
92109	(81413)	RPXM	a to	EN
92110	(81426)	RPXZ	x*to	EC
92111	(81432) **I**	RPXM	NHA	WB
92112	(81440)	RPXA	x to	CA
92113	(81442) **I**	RPXZ	a to	EC
92114	(81443) **I**	RPXM	NHA	WB
92115	(81444)	RPXM	a to	OY
92116	(81450) **I**	RPXM	a to	OY
92117	(81451)	RPXA	x	CA
92118	(81452) **I**	RPXM	a to	EN
92119	(81454)	RPXA	x to	CA
92120	(81455)	RPXA	x to	CA
92121	(81457)	RPXM	x*to	OY
92122	(81459)	RPXM	x*to	OY
92123	(81460)	RPXM	a to	EN
92124	(81465) **R**	RPXW	x	BJ
92125	(81470)	RPXM	a to	OY
92126	(81471) **I**	RPXM	a pg	OY
92127	(81473)	RPXM	x*pg	OY
92128	(81478) **I**	IECX	a to	NL
92129	(81479)	RPXM	a to	EN
92130	(81480)	RPXZ	a to	PC
92131	(81481)	RPXW	a to	LA
92132	(81482)	RPXM	a to	EN
92133	(81484)	RPXM	a to	EN
92134	(81485)	RPXM	a to	EN
92135	(81486)	RPXM	a to	EN
92136	(81487)	RPXW	a to	LA
92137	(81488) **P**	RPXH	a to	HT
92138	(81490)	RPXW	a to	LA
92139	(81491) **R**	RPXM	a to	MA
92140	(81492)	RPXM	a to	MA
92141	(81493)	RPXM	a to	EN
92142	(81494)	RPXM	a to	MA
92143	(81495)	RPXZ	a to	PC
92144	(81496)	RPXZ	a to	PC
92145	(81497)	RPXM	a to	EN
92146	(81498) **I**	RPXM	NHA	WB
92147	(81500)	RPXZ	a to	EC
92148	(81506)	RPXM	a to	EN
92149	(81509)	RPXA	x	CA
92150	(81514)	RPXW	x to	BJ
92151	(81516) **I**	RPXZ	a to	PC
92152	(81518)	RPXZ	a to	PC
92153	(81520) **I**	RPXM	x*to	OY
92154	(81522)	RPXA	x to	CA
92155	(81525) **I**	RPXM	a*pg	WB
92156	(81529)	RPXM	x*to	OY
92157	(81532) **I**	RPXM	x*pg	OY
92158	(81533)	RPXZ	a to	PC
92159	(81534) **I**	RPXM	NHA	WB
92160	(81537)	RPXA	x to	CA
92161	(81538) **I**	RPXM	a to	EN
92162	(81539)	RPXM	a to	EN
92163	(81540)	RPXM	a to	MA
92164	(81541)	RPXZ	a to	PC
92165	(81546)	RPXM	a to	EN
92166	(81550)	RPXM	a to	MA
92167	(81553) **I**	RPXM	a*pg	WB
92168	(81555)	RPXZ	a to	PC
92169	(81556)	RPXW	x	BJ
92170	(81559)	RPXW	x to	BJ
92171	(81560)	RPXA	x to	CA
92172	(81562) **I**	RPXM	a pg	OY
92173	(81565)	RPXA	x to	CA
92174	(81567) **I**	RPXM	NHA	WB

92175	(81568) I	RPXM	a pg	OY
92176	(81569)	RPXW	x to	BJ
92177	(81572)	RPXM	a to	MA
92178	(81574)	RPXW	x	BJ
92179	(81580) R	RPXA	x to	CA
92180	(81581)	RPXM	a to	EN
92181	(81582)	RPXW	x to	BJ
92182	(81585)	RPXW	x to	BJ
92183	(81588) I	RPXM	a*pg	WB
92184	(81589)	RPXM	a to	EN
92185	(81591) I	RPXZ	a to	PC
92186	(81595)	RPXM	a to	OY
92187	(81597) R	RPXM	a to	EN
92188	(81598)	RPXM	a to	EN
92189	(81599)	RPXW	x to	BJ
92190	(81600)	RPXM	a to	EN
92192	(81602)	RPXW	x to	BJ
92193	(81604) I	RPXM	a pg	OY
92194	(81606) I	RPXZ	a to	PC
92195	(81607)	RPXM	a to	EN
92196	(81609) I	RPXZ	a to	PC
92197	(81610) I	RPXZ	a to	PC
92198	(81613)	RPXM	a to	MA
92199	(81614)	RPXM	a to	EN
92200	(81615)	RPXA	x to	CA
92201	(81618)	RPXW	x to	BJ
92202	(81620) I	RPXZ	a to	PC
92203	(81621) I	RPXM	a to	OY
92204	(81622) P	RPXH	a to	HT
92205	(81623)	RPXM	a to	EN
92206	(81624)	RPXW	x	BJ
92207	(81627)	RPXW	x to	BJ
92208	(84403)	RPXW	x	BJ
92209	(80873)	RPXW	x	BJ
92210	(84466)	RPXW	x	BJ
92211	(81267)	RPXW	x	BJ
92212	(80937) R	RPXW	x	BJ
92213	(81321)	RPXW	x	BJ
92214	(84504)	RPXW	x	BJ
92215	(80911)	RPXW	x	BJ
92216	(81376)	RPXA	x	CA
92217	(80877) R	RPXA	x	CA
92218	(84314)	RPXW	x	BJ
92219	(84285)	RPXA	x	CA
92220	(80924)	RPXW	x	BJ
92221	(80888) R	RPXA	x	CA
92222	(80887)	RPXA	x	CA
92223	(80932)	RPXA	x	CA
92224	(84277) R	RPXW	x	BJ
92225	(80891)	RPXA	x	CA
92226	(80879)	RPXW	x	BJ
92227	(84337)	RPXW	x	BJ
92228	(80921) R	RPXW	x	BJ
92229	(80902)	RPXW	x	BJ
92230	(81423)	RPXW	x	BJ
92231	(80860) I	RPXW	x	BJ
92232	(80870)	RPXW	x	BJ
92233	(80890)	RPXW	x	BJ
92234	(84336)	RPXW	x	BJ
92235	(80908)	RPXW	x	BJ
92236	(80909) N	RPXW	x	OM
92237	(80738)	RPXA	x	NC
92238	(84563)	RPXW	x	BJ
92239	(81170) R	RPXW	x	BJ
92240	(80703)	RPXW	x	BJ
92241	(80943)	RPXW	x	BJ
92242	(80857)	RPXW	x	BJ
92243	(84489)	RPXW	x	BJ
92244	(84248)	RPXW	x	BJ
92245	(80939)	RPXW	x	BJ
92246	(80929)	RPXW	x	BJ
92247	(84536)	RPXH	x	HT
92248	(80935) P	RPXA	x	CA
92249	(84511)	RPXW	x	BJ
92250	(81235)	RPXH	x	HT
92251	(84425)	RPXA	x	NC
92252	(80959)	RPXW	x	BJ
92253	(80936)	RPXA	x	CA
92254	(80894)	RPXA	x	CA
92255	(80871)	RPXA	x	CA
92256	(84424)	RPXA	x	CA
92257	(80955)	RPXA	x	NC

NE (BG) 100 mph GANGWAYED BRAKE VAN

As ND but rebogied with Commonwealth bogies suitable for 100 mph. ETH 1 (1X*). For lot numbers refer to original number series. Add 1.5 t to weights to allow for the increased weight of the Commonwealth bogies.

92308	(80784)	RPXW	x*	BJ
92309	(84043)	RPXW	x*	BJ
92310	(84105)	RPXW	x	BJ
92311	(84453)	RPXW	x	BJ
92312	(84548)	RPXW	x	BJ
92313	(80992)	RPXH	x*	HT
92314	(80777)	RPXA	x*	CA
92315	(80848) N	RPXW	x*	BJ

92316	(80980)	RPXA	x*	CA
92317	(80836) **P**	RPXA	x*	CA
92318	(80847)	RPXA	x*	CA
92319	(84055) **N**	RPXW	x*	OM
92320	(84166)	RPXH	x	HT
92321	(84566)	RPXW	x	BJ
92322	(80771)	RPXW	x*	BJ
92323	(80832)	RPXH	x*	HT
92324	(84087) **N**	RPXW	x	OM
92325	(80791)	RPXA	x	CA
92326	(84270)	RPXA	x	CA
92327	(80842)	RPXA	x*	CA
92328	(80999) **R**	RPXH	x*	HT
92329	(84001)	RPXA	x*	CA
92330	(80995)	RPXA	x*	CA
92331	(84365)	RPXW	x	BJ
92332	(80845)	RPXA	x*	CA
92333	(80982)	RPXA	x*	CA
92334	(80983)	RPXA	x*	CA
92335	(80973)	RPXA	x*	CA
92336	(84045)	RPXW	x*	BJ
92337	(84140)	RPXW	x*	BJ
92338	(84524)	RPXW	x	BJ
92339	(84530) **R**	RPXW	x	BJ
92340	(84059)	RPXW	x*	OM
92341	(84316)	RPXW	x	BJ
92342	(84397)	RPXW	x	BJ
92343	(84505)	RPXW	x	BJ
92344	(84154)	RPXW	x*	BJ
92345	(84083)	RPXW	x*	BJ
92346	(84091)	RPXW	x	BJ
92347	(84326)	RPXW	x	BJ
92348	(84075)	RPXA	x*	CA
92349	(84178)	RPXW	x	BJ
92350	(84049)	RPXW	x*	BJ
92351	(84174)	RPXA	x	CA
92352	(84182)	RPXW	x	BJ
92353	(84323)	RPXH	x	HT
92354	(84353) **N**	RPXW	x	OM
92355	(84517)	RPXA	x	CA
92356	(84535)	RPXW	x	BJ
92357	(84136)	RPXW	x	BJ
92358	(84393)	RPXW	x	BJ
92359	(84275)	RPXW	x	BJ
92360	(84431)	RPXW	x	BJ
92361	(84463) **R**	RPXA	x	CA
92362	(84188)	RPXW	x	BJ
92363	(84294)	RPXW	x	BJ
92364	(84030) **R**	RPXA	x*	CA
92365	(84122)	RPXW	x	BJ
92366	(84551)	RPXH	x	HT
92367	(84293)	RPXA	x pt	CA
92368	(84334)	RPXA	x	CA
92369	(80960)	RPXA	x*	CA
92370	(84324)	RPXA	x	CA
92371	(80856)	RPXA	x*	CA
92372	(84362)	RPXA	x	CA
92373	(84528)	RPXA	x	CA
92374	(84317)	RPXA	x	CA
92375	(84335)	RPXA	x	CA
92376	(84374)	RPXA	x	CA
92377	(80928)	RPXH	x*	HT
92378	(84340)	RPXA	x	CA
92379	(80914)	RPXH	x*	HT
92380	(84247)	RPXW	x*	BJ
92381	(84476)	RPXA	x	CA
92382	(84561)	RPXA	x	CA
92383	(84429)	RPXA	x	CA

NJ/NK/NX (GUV) GENERAL UTILITY VAN

Mark 1. Short frames. Load 14t. All vehicles are through steam piped only (dia. NK501) unless otherwise stated. Electric wired vehicles or steam piped and electric wired vehicles are classified NJ501. Motorail vans are classified NX501 and are denoted by their TOPS code. All vehicles have BR mark 2 bogies except Motorail vans shown † which now have B5s. ETH 0 or 0X.

93078–93499. Lot No. 30417 Pressed Steel 1958–9. 30t.
93501–93519. Lot No. 30343 York 1957. 30t.
93521–93654. Lot No. 30403 York/Glasgow 1958–60. 30t.
93655–93834. Lot No. 30565 Pressed Steel 1959. 30t.
93835–93984. Lot No. 30616 Pressed Steel 1959–60. 30t.

93078	**B**	RPXD		BJ
93079	**B**	RPXD		CA
93081	**B**	RPXE	xy*	BJ
93084	**B**	RPXE	xy*	BJ
93086	**B**	RPXX		CA
93090	**B**	RPXD		CA
93091	**R**	RPXE	xy*	CA
93093	**B**	RPXD		CA

93100	**B**	RPXD		CA
93104	**B**	RPOC	xr*	PY
93105	**B**	RPXD		CA
93107	**B**	RPXD		BJ
93110	**B**	RPOC	xy*	PY
93112	**B**	RPXE	xy*	CA
93115	**B**	RPXD		CA
93117	**B**	RPXD		CA
93122	**R**	RPOC	xy*	PY
93126	**B**	RPOC	xy*	PY
93131	**B**	RPXD	vr*	CF
93134	**B**	RPOC	xy*	PY
93135	**B**	RPXD	vr*	BJ
93138	**B**	RPXD		CA
93140	**B**	RPXD		CA
93141	**B**	RPOC	xy*	PY
93142	**B**	RPXE	xy*	CA
93143	**B**	RPOC	xy*	PY
93160	**B**	RPOC	xy*	PY
93163	**B**	RPXD	vr*	BJ
93165	**B**	RPOC	xy*	PY
93166	**B**	RPXX	xy*	BJ
93167	**B**	RPOC	xy*	PY
93172	**B**	RPOC	xy*	PY
93174	**B**	RPOC	xy	PY
93176	**B**	RPXD	vr*	CA
93180	**B**	RPXD		CA
93185	**B**	RPXD	vr*	CA
93187	**B**	RPXD		CA
93191	**B**	RPXD		CA
93192	**B**	RPXE	xy*	CA
93193	**B**	RPXX		CA
93195	**B**	RPXD	vr*	CA
93205	**B**	RPOC	xy*	PY
93207	**B**	RPXD		CA
93210	**B**	RPOC	xy*	PY
93212	**B**	RPOC	xy*	PY
93214	**B**	RPXD		CA
93217	**B**	RPXD		CA
93234	**B**	RPXD		CA
93235	**B**	RPOC	xy*	PY
93236	**B**	RPXD	vr*	BJ
93239	**B**	RPXX		CA
93241	**B**	RPXD		CA
93249	**R**	RPXE	xy*	CA
93251	**B**	RPXD	vr*	BJ
93253	**B**	RPXX		BJ
93255	**B**	RPXE	xr*	BJ
93257	**B**	RPXD		BJ
93258	**B**	RPXD	vr*	CA
93262	**B**	RPOC	xy*	PY
93263	**B**	RPOC	xy*	PY
93265	**R**	RPOC	xy*	PY
93267	**R**	RPXD	vr*	CA
93269	**B**	RPXE	xy*	CA
93272	**B**	RPXD		BJ
93273	**B**	RPXD	xr*	CA
93277	**B**	RPOC	xy*	PY
93278	**B**	RPXD	vr*	CF
93284	**B**	RPXD	vr*	BJ
93287	**B**	RPXX		BJ
93291	**B**	RPXD		CA
93292	**B**	RPXD	vr*	BJ
93293	**B**	RPOC	xy*	PY
93295	**B**	RPXD		CA
93298	**B**	RPXX		CA
93300	**B**	RPXX		CA
93301	**B**	RPXX		CA
93304	**B**	RPXD		CA
93306	**B**	RPXD		CA
93307	**B**	RPXD		CA
93308	**B**	RPXX		CA
93319	**B**	RPXD		CA
93321	**B**	RPXD		CA
93323	**R**	RPXE	xy*	CA
93325	**B**	RPXD		CA
93329	**B**	RPXX		CA
93330	**B**	RPXD		CA
93336	**B**	RPXX		CA
93338	**B**	RPXD		CA
93339	**B**	RPXX		CA
93343	**B**	RPXX		CA
93345	**B**	RPXD	NXV	CA
93347	**B**	RPOC	xy*	PY
93348	**B**	RPXD		CA
93349		RPXD	vr	CA
93353	**B**	RPOC	xy*	PY
93356	**B**	RPXX		BJ
93359	**B**	RPXX		CA
93362	**B**	RPOC	xy*	PY
93367	**B**	RPXD	vr*	BJ
93370	**B**	RPXD	vr*	BJ
93374	**B**	RPXD		BJ
93375	**B**	RPXD	vr*	CF
93376	**B**	RPOC	xy*	PY
93379	**B**	RPXD	vr*	CF
93386	**B**	RPXD		BJ
93388	**B**	RPXD	vr*	CF
93390	**B**	RPXD		BJ
93391	**B**	RPXD		CA
93393	**B**	RPXE	xy*	BJ
93394	**B**	RPXD	vr*	BJ
93395	**R**	RPXE	xy*	CA
93396	**B**	RPOC	xy*	PY

93398	**B**	RPXD		CA
93401	**B**	RPXD	vr*	BJ
93406	**B**	RPXX		CA
93407	**B**	RPXE	xy*	CA
93411	**B**	RPXE	xy*	CA
93412	**B**	RPXX		CA
93413	**B**	RPXD		CA
93416	**B**	RPXE	xy*	CA
93417	**B**	RPXD		CA
93419	**B**	RPXD	vr*	BJ
93421	**B**	RPXD		BJ
93426	**B**	RPXE	xy	BJ
93429	**B**	RPXE	xy*	BJ
93431	**B**	RPXD	vr*	BJ
93446	**B**	RPXD	vr	BJ
93450	**B**	RPXD	vr*	BJ
93455	**B**	RPXD		CA
93458	**B**	RPXD		CA
93459	**B**	RPXD		CA
93462	**B**	RPOC	xy*	PY
93463	**B**	RPXD		CA
93464	**B**	RPXD		CA
93465	**B**	RPXX		CA
93468	**B**	RPXE	xy	CA
93474	**B**	RPXD	vr*	BJ
93475	**B**	RPXD		BJ
93476	**B**	RPXD		BJ
93478	**B**	RPXD	vr	BJ
93479	**R**	RPXD		CA
93482	**B**	RPXD		CA
93485	**B**	RPXE	xr*	CA
93490	**B**	RPXD	vr	CF
93495		RPXD	NXX†	BJ
93498	**B**	RPXD		CA
93499	**B**	RPOC	xy*	PY
93502	**B**	RPXX		CA
93504	**B**	RPXX		CA
93506	**B**	RPXX		BJ
93508	**B**	RPXD	vr*	BJ
93511	**B**	RPXD	vr*	CF
93512	**B**	RPXD		CF
93514	**B**	RPXD	vr*	CA
93515	**B**	RPXD	vr*	CF
93516	**B**	RPXD		CF
93518	**B**	RPXE	xy*	BJ
93519	**B**	RPXD		BJ
93521	**B**	RPPM	xy*	CA
93523	**B**	RPOC	xy	PY
93525	**B**	RPXD	vr*	BJ
93529	**B**	RPXD	vr*	CA
93533	**B**	RPXD		CA
93534	**B**	RPOC	xy*	PY
93536	**B**	RPXD	vr*	CF
93538	**B**	RPXD		CA
93539	**B**	RPOC	xy*	PY
93541	**B**	RPXD	vr*	CA
93542	**B**	RPXX		CA
93546	**B**	RPXD		CA
93549	**B**	RPXD		CA
93550	**B**	RPXD		CA
93554	**B**	RPXD		BJ
93556	**B**	RPXD	vr*	CF
93557	**B**	RPXD	vr*	BJ
93560	**B**	RPXE	xy*	BJ
93562	**B**	RPXD	vr*	BJ
93563	**B**	RPXD	vr*	BJ
93571	**B**	RPOC	xy*	PY
93573	**B**	RPXD		CA
93576	**B**	RPXE	xy*	CA
93578	**B**	RPXE	xy*	BJ
93579	**B**	RPXD		CA
93581	**B**	RPXE	xr*	CA
93585	**R**	RPXD	vr*	CA
93590	**B**	RPXE	xy*	CA
93595	**B**	RPXD		BJ
93596	**B**	RPXD	vr*	BJ
93600	**B**	RPXE	xy*	BJ
93601	**B**	RPXD		BJ
93604	**B**	RPXE	xy*	BJ
93606	**B**	RPOC	xy*	PY
93607	**B**	RPOC	xy*	PY
93610	**B**	RPOC	xy*	PY
93617	**B**	RPXD		CA
93618	**B**	RPXD		CA
93622	**B**	RPXD	vr*	CF
93623	**B**	RPXD		CA
93624	**B**	RPXD	vr*	CF
93630	**B**	RPXD	vr*	CF
93631	**B**	RPXD	vr*	CF
93633	**B**	RPXD	vr*	CF
93634	**B**	RPXD		CA
93635	**B**	RPXD		CA
93640	**B**	RPXD		CA
93643	**B**	RPXD	vr	BJ
93644	**B**	RPXD		CA
93646	**B**	IWRX		OM
93647	**B**	IWRX		OM
93648	**B**	RPXE	xy	BJ
93652	**B**	RPXX		CA
93656		RPXD		BJ
93660	**B**	RPXD	vr*	BJ
93663	**B**	RPXD		CA
93664	**B**	RPXD		CA
93667	**B**	RPXD		CA

93668	**B**	RPXE	xr*	BJ
93673	**B**	RPXE	xy*	BJ
93675	**B**	RPXE	xr	BJ
93677	**B**	RPXD	vr*	CF
93680	**B**	RPXD		CA
93681	**B**	RPXD		BJ
93683	**B**	RPXD	vr*	CA
93684	**B**	RPXX		CA
93687	**B**	RPXD		CA
93692	**B**	RPXE	xy*	BJ
93694	**B**	NSSX	xy*	EH
93696	**B**	RPXD		CA
93701	**B**	RPXD	vr*	BJ
93706	**B**	RPXD	vy	BJ
93710	**B**	RPXD	vy	CF
93711	**B**	RPXD	vy	CF
93713	**B**	RPXD	vy	CF
93714	**B**	RPXD	vy	CF
93715	**B**	RPXD	vy	CF
93717	**B**	RPXD	vy	CF
93720	**B**	RPXD	vy	CF
93722	**R**	RPXD	vy	CF
93723	**B**	RPXD	vy	CF
93725	**B**	RPXD	vy	BJ
93726	**B**	RPXD	vy	BJ
93727	**B**	RPXX	vy	BJ
93728	**B**	RPXD	vy	BJ
93756	**B**	ICCX		EC
93762	**B**	RPOC	xy*	PY
93764	**B**	RPXE	xy*	BJ
93765	**B**	RPXX		BJ
93767	**B**	RPXD		CA
93769	**B**	RPXX		CA
93775	**B**	RPOC	xy*	PY
93776	**B**	RPXD		CA
93779	**B**	RPXD		CA
93781	**B**	RPXX		CA
93782	**I**	IWRX		OM
93784	**B**	IWCX		EN
93788	**B**	RPXD		CA
93791	**B**	RPXD		CA
93792	**B**	RPXD		CA
93793	**B**	RPXX		CA
93795	**B**	RPXX		CA
93798	**B**	RPOC	xy*	PY
93800	**B**	RPXD		CA
93801	**B**	RPXD		CA
93807	**B**	RPXX		CA
93811	**B**	RPXD		CA
93815	**B**	RPXD		CA
93820	**B**	RPOC	xy*	PY
93822	**B**	RPXD		CA
93823	**B**	RPXD	vr	BJ
93824	**B**	RPXD		CA
93829	**B**	RPXD		CA
93830	**B**	RPXD	vr	BJ
93834	**B**	IWCX		EN
93836	**B**	RPOC	xr*	PY
93843	**B**	IWCX		EN
93844	**B**	RPXE	xy*	BJ
93846	**B**	RPXD		BJ
93847	**R**	RPXD	vr*	BJ
93849	**R**	RPXD	vr*	BJ
93851	**B**	RPXX		CA
93852	**B**	NSSX	xr*	EH
93853	**B**	RPXX		CA
93854	**B**	RPXD	vr*	BJ
93855	**B**	RPXD	vr*	BJ
93857	**B**	RPXD	vr*	BJ
93858	**B**	RPXD		CA
93859	**B**	RPXD	vr*	BJ
93860	**B**	RPXE	xr*	BJ
93863	**B**	IXXZ		BJ
93864	**B**	RPXE	xy*	BJ
93875	**B**	IWCX		EN
93877	**B**	RPXD		BJ
93880	**B**	ICCX		EC
93881	**B**	RPXD	vr*	BJ
93883	**B**	RPXD		BJ
93886	**B**	RPXD	vr*	CA
93887	**B**	RPXD	vr*	BJ
93888	**B**	RPXX		BJ
93889	**B**	RPXD	vr*	BJ
93893	**B**	RPXD	vr*	BJ
93896	**B**	RPXD		BJ
93897	**B**	RPXE	xr*	BJ
93901	**B**	RPXD		BJ
93904	**B**	RPXD		BJ
93905	**B**	RPXD	vr*	BJ
93907	**B**	RPXD		BJ
93910	**I**	IWCX		EN
93911	**B**	RPXD	vr*	BJ
93914	**B**	RPXE	xr*	BJ
93915	**B**	RPXD	vr*	CA
93919	**B**	RPXD		BJ
93922	**B**	RPXD		CA
93925	**B**	RPXX		CA
93927	**B**	RPXD		CA
93930	**B**	RPXX		BJ
93931	**B**	RPXD		BJ
93932	**B**	RPXD		BJ
93933	**B**	RPXX		BJ
93935	**B**	RPXD	vr*	BJ
93937	**I**	IWCX		EN

93942	**B**	RPXE	xr*	BJ
93943	**B**	RPXD		CA
93948	**B**	RPXD		BJ
93950	**B**	RPXD	vr*	BJ
93952	**B**	RPXD	vr*	BJ
93955	**B**	RPXD		BJ
93956	**B**	RPXE	xr*	CA
93957	**B**	RPXD		BJ
93958	**B**	IWCX	NXX	EN
93962	**B**	RPXD	vr*	CA
93969	**B**	RPXE	xy*	BJ
93973	**B**	RPXD	vr*	BJ
93976	**B**	RPXD		BJ
93978	**B**	IXXX		MA
93979	**B**	RPXD	vr*	BJ
93980	**B**	IXXX		MA
93982	**R**	RPXD	vr*	BJ

NLX NEWSPAPER VAN

Dia. NL501. Mark 1. Short frames (57′). Converted from GUV. Fluorescent lighting, toilets and gangways fitted. Load 14t. ETH 3X. Note: These vehicles were originally renumbered 85500–85534.

*–Fitted with roller bearing axleboxes.

Lot No. 30922 Wolverton or Doncaster 1977–8. xq. 33t (31.t B5).

94001	(86438)	**B**	RPPW	BR1*	BJ
94002	(86651)	**B**	RPPW	B5	BJ
94003	(86281)	**B**	RPPW	B5	BJ
94004	(86156)	**B**	RPPW	B5	BJ
94005	(86845)	**B**	RPPW	B5	BJ
94006	(86202)	**B**	RPPW	B5	BJ
94007	(86572)	**B**	RPPW	B5	OM
94008	(86838)	**B**	RPPW	B5	OM
94009	(86144)	**B**	RPPW	B5	OM
94010	(86151)	**B**	RPPW	BR1*	OM
94011	(86437)	**B**	RPPW	B5	OM
94012	(86082)	**B**	RPPW	B5	OM
94013	(86408)	**B**	RPPE	B5	HT
94015	(86484)	**B**	RPPM	B5	CA
94016	(86317)	**B**	RPPM	B5	CA
94017	(86467)	**B**	RPPM	B5	CA
94018	(86170)	**B**	RPPE	B5	HT
94019	(86098)	**R**	RPPM	B5	OM
94020	(86620)	**B**	RPPW	B5	BJ
94021	(86204)	**B**	RPPW	B5	OM
94022	(86718)	**B**	RPPE	B5	HT
94023	(86221)	**B**	RPPE	B5	CA
94024	(86106)	**B**	RPPE	B5	CA
94025	(86377)	**R**	RPPM	B5	OM
94026	(86703)	**B**	RPPE	B5	CA
94027	(86732)	**B**	RPPW	BR1*	BJ
94028	(86733)	**B**	RPPW	BR1*	BJ
94029	(86740)	**B**	RPPW	B5	OM
94030	(86746)	**B**	RPPW	B5	OM
94031	(86747)	**B**	RPPW	B5	OM
94032	(86730)	**B**	RPPW	B5	OM
94033	(86731)	**B**	RPPE	B5	CA
94034	(86200)	**B**	RPPW	B5	OM

NMV NEWSPAPER VAN

Dia. NM501/2. Mark 1. Standard GUVs modified as newspaper vans. vd or vd*. ETH 3 (3X*). For lot numbers see old number series. Formerly NLV.

94050	(93771)	**B**	RPPE	d*	CA
94051	(93708)	**B**	RPPE	e	CA
94052	(93709)	**B**	RPPE	e	CA
94053	(93716)	**B**	RPPE	e	CA
94056	(93804)	**B**	RPPE	e	CA
94058	(93530)	**B**	RPPE	d*	CA
94061	(93763)	**B**	RPPE	d*	CA
94062	(93803)	**B**	RPPE	d*	CA
94068	(93424)	**B**	RPPM	d*	CA
94071	(93544)	**B**	RPPM	d*	CA
94076	(93842)	**B**	RPPM	d*	CA
94077	(93862)	**B**	RPPM	d*	CA
94078	(93971)	**B**	RPPM	d*	CA

NCX NEWSPAPER VAN (100 mph)

Dia. NC501. BGs modified to carry newspapers. xe. ETH 3 (3X*). Commonwealth bogies. For lot Nos. see BG section. Add 2t to weight.

95200	(84019)	B	RPPE	*	CA
95201	(80875)		RPPW		BJ
95202	(80667)	B	RPPE	*	CA
95203	(80621)	B	RPPE	*	CA
95204	(80947)	B	RPPE	*	CA
95205	(80620)	B	RPPW	*	BJ
95206	(80561)		RPPW		BJ
95207	(80560)	R	RPPW	*	BJ
95208	(80660)	B	RPPW	*	BJ
95209	(84047)	B	RPPE		CA
95210	(80731)	B	RPPE		CA
95211	(80949)		RPPW		BJ
95212	(84179)	B	RPPE		CA
95213	(80826)	B	RPPE	*	CA
95214	(84360)	B	RPPM		EN
95215	(80643)	B	RPPW	*	BJ
95216	(84542)		RPPW		BJ
95217	(84385)	B	RPPM		EN
95218	(80675)	R	RPPW	*	BJ
95219	(80946)	B	RPPM		EN
95220	(80717)	B	RPPW	*	BJ
95221	(84153)		RPPW	*	BJ
95222	(80774)	B	RPPW		BJ
95223	(80933)		RPPW	*	BJ

NCV NEWSPAPER VAN (100 mph)

Dia. NC501. ve. Commonwealth bogies. For lot Nos. see BG section. Add 2t to weight.

95300	(80689)	B	RPPE	CA
95303	(80614)	B	RPPE	CA
95304	(84345)		RPPW	BJ
95305	(80863)	B	RPPE	HT
95307	(84124)	B	RPPE	HT
95308	(84161)	B	RPPE	HT
95310	(84292)	B	RPPW	BJ
95312	(80503)		RPPE	CA
95321	(80525)		RPPM	MA
95329	(84341)	B	RPPE	CA
95332	(84014)		RPPE	CA

NX (GUV) MOTORAIL VAN

Mark 1. Dia. NX501. Renumbered from 93XXX series. For details and lot numbers see 93XXX series. ETH 0(0X*).

96100	(93734)		IWCX	a*B5	EN
96101	(93741)		IWCX	a*B5	EN
96102	(93743)	I	IWCX	a*B5	EN
96103	(93744)		IWCX	a*B5	EN
96104	(93745)	B	IWCX	a*B5	EN
96110	(93738)	I	IWCX	a*C	EN
96111	(93742)	I	IWCX	a*C	EN
96112	(93750)	B	IWCX	a*C	EN
96130	(93736)		IWCX	a*C	EN
96131	(93737)		IWCX	a*C	EN
96132	(93754)	I	IWCX	a*C	EN
96133	(93685)		IWCX	a C	EN
96134	(93691)	B	IWCX	a C	EN
96135	(93755)	I	IWCX	a C	EN
96136	(93735)	I	IWCX	a C	EN
96137	(93748)		IWCX	a C	EN
96138	(93749)	I	ICCX	a C	EC
96139	(93751)	B	ICCX	a C	EC
96140	(93752)		ICCX	a C	EC
96141	(93753)	B	ICCX	a C	EC
96150	(93097)	B	IWCX	x*B5	EN
96151	(93286)		IWCX	x*B5	EN
96152	(93324)		IWCX	x*B5	EN
96153	(93327)	B	IWCX	x*B5	EN
96154	(93331)	I	IWCX	x*B5	EN
96155	(93334)		IWCX	x*B5	EN
96156	(93337)	B	IWCX	x*B5	EN
96157	(93344)		IWCX	x*B5	EN
96158	(93351)		IWCX	x*B5	EN
96159	(93355)		IWCX	x*B5	EN
96160	(93385)		IWCX	x*B5	EN
96161	(93443)	B	IWCX	x*B5	EN
96162	(93647)	I	IWCX	a*C	EN
96163	(93646)	I	IWCX	a*C	EN
96164	(93880)	I	IWCX	a*C	EC
96165	(93784)	I	IWCX	a*C	EC

96166	(93834)	**I**	IWCX	a*C	EN
96167	(93756)	**I**	IWCX	a*C	EC
96168	(93978)	**I**	IWCX	a*C	EN
96169	(93937)	**I**	IWCX	a*C	EN
96170	(93159)		IWCX	x*C	EM
96171	(93326)	**N**	IWCX	x*C	EN
96172	(93363)		IWCX	x*C	EN
96173	(93440)	**I**	IWCX	x*C	EN
96174	(93453)	**B**	IWRX	x*C	OM
96175	(93628)		IWCX	x*C	EN
96176	(93641)		IWCX	x*C	EN
96177	(93980)	**I**	IWCX	a*C	EN
96178	(93782)	**I**	IWCX	a*C	EN
96179	(93910)	**I**	IWCX	a*C	EN
96180	(93843)	**I**	IWCX	a*C	EN
96181	(93875)	**I**	IWCX	a*C	EN
96182	(93944)	**I**	IWCX	a*C	WN
96185	(93083)		IWRX	x*C	OM
96186	(93087)		IWRX	x*C	OM
96187	(93168)		IWRX	x*C	OM
96188	(93320)	**I**	IECX	x*C	BN
96189	(93447)	**I**	IECX	x*C	BN
96190	(93448)		IWRX	x*C	OM
96191	(93665)		IWCX	x*C	EN
96192	(93669)	**I**	IWCX	x*C	EN
96193	(93874)		IWCX	x*C	EN
96194	(93949)	**B**	IWCX	x*C	EN
96195	(93958)	**B**	IWCX	x*C	EN

NY EXHIBITION VAN

Various interiors. Converted from various vehicle types. Electric heating from shore supply. In some cases new lot numbers have been issued for conversions, but not always. All allocated to SA. Non-standard livery – varies according to job being undertaken.

New lot Nos.: 30841 (99602), 30842 (99620–5) and 30950/1/2 (99629/30/1).

99601	(34577)	**0**	ICHX	v	BR1	Wolverton 1955	NY503	Exhibition Coach.
99602	(21145)	**0**	ICHX	v	Mk4	Swindon 1972–3	NY502	Toilet Van. 8L.
99620	(34692)	**0**	ICHX	v	BR1	Swindon 1972–3	NY504	Exhibition Coach.
99621	(34697)	**0**	ICHX	x	BR1	Swindon 1972–3	NY503	Exhibition Coach.
99622	(34584)	**0**	ICHX	w	Mk4	Swindon 1972–3	NY503	Exhibition Coach.
99623	(34700)	**0**	ICHX	x	Mk4	Swindon 1972–3	NY503	Exhibition Coach.
99625	(34693)	**0**	ICHX	x	Mk4	Swindon 1972–3	NY503	Generator Van.
99629	(25972)	**0**	ICHX	x	BR1	Stewarts Lane 1979	NY515	Exhibition Coach.
99630	(34965)	**0**	ICHX	x	BR1	Stewarts Lane 1979	NY504	Exhibition Coach.
99631	(1018)	**0**	ICHX	x	C	Stewarts Lane 1979	NY517	Exhibition Coach.
99632	(4329)	**0**	ICHX	v	BR1	Doncaster 1981	NY522	Cinema/Lounge.
99633	(3827)	**0**	ICHX	v	BR1	Doncaster 1981	NY522	Cinema/Lounge.
99634	(25317)	**0**	ICHX	v	BR1	Doncaster 1981	NY520	Exhibition Coach.
99635	(25637)	**0**	ICHX	v	BR1	Doncaster 1981	NY520	Exhibition Coach.
99636	(25646)	**0**	ICHX	v	BR1	Doncaster 1981	NY526	Reception/Toilet.
99637	(15854)	**0**	ICHX	v	BR1	Doncaster 1981	NY521	Exhibition Coach.
99641	(1036)	**0**	ICHX	b	C	Stewarts Lane 1981	NY517	Exhibition Coach.
99642	(1040)	O	ICHX	b	C	Stewarts Lane 1981	NY517	Exhibition Coach.
99643	(4272)	O	ICHX	v	BR1	Doncaster 1981	NY522	Exhibition Coach.
99644	(3852)	O	ICHX	v	BR1	Doncaster 1981	NY522	Exhibition Coach.
99645	(1765)	O	ICHE	v	C	Salisbury 1981	NY523	Club Car.
99646	(1766)	O	ICHE	v	C	Salisbury 1981	NY524	Club Car.
99647	(34629)	**0**	ICHE	v	BR1	Salisbury 1981	NY504	Exhibition Coach.
99648	*	**0**	ICHX	x			NY525	Workshop/Store.
99649	(80502)	**0**	ICHE	v	BR1		NY524	Exhibition Coach.
99650	(80830)	**0**	ICHE	v	BR1		NY524	Exhibition Coach.
99657	(4569)	**0**	ICHE	v	BR1	Cathays 1985†	NY522	Exhibition Coach.
99658	(4592)	**0**	ICHE	v	BR1	Cathays 1985†	NY522	Exhibition Coach.
99659	(4518)	**0**	ICHE	v	BR1	Cathays 1985†	NY522	Exhibition Coach.
99661	(80510)	**0**	ICHE	v	BR1		NY524	Exhibition Coach.

 * Formerly BR ferry van 21 70 214 0144–3. Also carried GB 787017.

6. PRIVATELY OWNED COACHES

This list comprises privately owned coaches which are "plated" to run on BR. The BR PO number is carried on a yellow plate affixed to the solebar. Coaches have to be passed by BR each year and this is denoted by a white equilateral triangle painted on the solebar with the year painted in black. In the following list, the present number is shown if this is different from the original number. It should be noted that there are other vehicles with BR PO numbers which are not to be found in the following list. These are vehicles which are not at present registered for running.

Plated	*No.*		*Original No.*	*Present No.*	*Built*	*Type*
5407 Loco. Group, Steamtown, Carnforth, Lancs.						
PWDS	99030	vd	34666		1955	BR BSK
Midland Railway Centre, Swanwick Junction.						
MRC	99040	vs	E 21232		1961	BR BCK
Steamtown Railway Museum, Carnforth.						
SPG	99070	vd	E 35123		1957	BR BSK
SNG	99080	vs	E 21096		1956	BR BCK
Princess Elizabeth Loco. Society.						
PES	99090	vs	35131		1957	BR BSK
Birmingham Railway Museum, Tyseley (Standard Gauge Steam Trust).						
SGST	99103	vs	80972		1949	GWR saloon
SGST	99108	xdk	E 17018		1961	BR BFK
Pullman Rail. Based at Steamtown, Carnforth, but the trains are normally maintained at BN. FO set.						
PULR	99120	vd	M 21236	120	1961	BR BCK
PULR	99121	vd	M 3105	121	1962	BR FO
PULR	99122	vd	M 3106	122	1962	BR FO
PULR	99123	vd	M 3109	123	1962	BR FO
PULR	99124	vd	M 3110	124	1962	BR FO
PULR	99125	vd	M 3113	125	1962	BR FO
PULR	99126	vd	M 3116	126	1962	BR FO
PULR	99127	vd	M 3117	127	1962	BR FO
PULR	99128	vd	M 3130	128	1963	BR FO
PULR	99129	vd	SC 21272	129	1964	BR BCK
PULR	99130	vd	E 17013	130	1964	BR BFK
PULR	99131	vd	SC 1999	131	1945	LNER Saloon
PULR	99132	xd	1861	132	1962	BR RMB
The Travelling College, OM.						
TCL	99160	vd	18856		1961	BR SK
TCL	99161	vd	18871		1961	BR SK
TCL	99162	vd	18795		1961	BR SK

TCL	99163	vd	18994		1962	BR SK
TCL	99164	vd	18853		1961	BR SK
TCL	99165	vd	5034		1962	BR TSO
TCL	99166	vd	18776		1961	BR SK
TCL	99167	vd	21238		1961	BR BCK
TCL	99168	vd	18778		1961	BR SK
TCL	99169	vd	4941		1961	BR TSO
TCL	99170	vd	4957		1961	BR TSO
TCL	99171	vd	4921		1961	BR TSO

David Smith, Midland Railway Centre, Swanwick Junction.

WDS	99312	vd	35463		1963	BR BSK

Dinting Railway Centre.

DRC	99313	vd	35451		1963	BR BSK

Pullman Rail. Pullman set.

PULR	99335	xd	Car no. 335		1960	Pullman kitchen second
PULR	99347	xd	Car no. 347		1960	Pullman parlour second
PULR	99348	xd	Car no. 348		1960	Pullman parlour second
PULR	99349	xd	Car no. 349		1960	Pullman parlour second
PULR	99350	xd	Car no. 350		1960	Pullman parlour second
PULR	99351	xd	Car no. 351		1960	Pullman parlour second
PULR	99352	xd	Car no. 352		1960	Pullman parlour second
PULR	99353	xd	Car no. 353		1960	Pullman parlour second
PULR	99354	xd	HADRIAN BAR		1960	Pullman second bar
PULR	99355	xd	M 21249	355	1962	BR BCK
PULR	99356	vd	M 21245	356	1962	BR BCK
PULR	99357	xd	M 3112	357	1962	BR FO
PULR	99358	vd	M 3108	358	1962	BR FO

Humberside Loco. Preservation Group, Hull Dairycoates Depot.

HLPG	99420	vs	E 21214	21214	1958	BR BCK

Sea Containers Ltd., Stewarts Lane Depot, London. "Venice-Simplon-Orient Express".

VSOE	99530	ae	PERSEUS		1951	Pullman parlour first
VSOE	99531	ae	PHEONIX		1952	Pullman parlour first
VSOE	99532	ae	CYGNUS		1951	Pullman parlour first
VSOE	99533	ae	E 70741	7	1943	LNER BG
VSOE	99534	ae	IBIS		1925	Pullman kitchen first
VSOE	99535	ae	MINERVA		1928	Pullman kitchen first
VSOE	99536	ae	ZENA		1929	Pullman parlour first
VSOE	99537	ae	AUDREY		1932	Pullman kitchen first
VSOE	99538	ae	S 34991	9	1957	BR BSK
VSOE	99539	ae	IONE		1929	Pullman kitchen first
VSOE	99540	ae	S 3069		1955	BR FO
VSOE	99541	ae	LUCILLE		1929	Pullman
VSOE	99542	a	GB 889202		1959	Ferry Van

Train Tours Ltd. Stock kept at LL.

TRTS	99710	vd	M	18767		1961	BR SK
TRTS	99711	vd	M	18857		1961	BR SK
TRTS	99712	vd	M	18893		1962	BR SK
TRTS	99713	vd	M	19013		1962	BR SK
TRTS	99714	vd	M	7187		1961	BR CK
TRTS	99715	vd	M	35337		1962	BR BSK
TRTS	99716	vd	M	18808		1962	BR SK
TRTS	99717	vd	M	18837		1962	BR SK
TRTS	99718	vd	M	18862		1962	BR SK
TRTS	99719	vd	M	7191		1962	BR CK
TRTS	99720	vd	M	35461		1962	BR BSK

Keighley & Worth Valley Railway, Oxenhope.

COWS	99792	vd		17019		1961	BR BFK

Scottish Railway Preservation Society, BR Perth Depot. This stock will move to Bo'ness eventually.

SCR	99818	xd	SC	1730		1060	BR RBR
SCR	99819	vd	SC	4844		1959	BR TSO
SCR	99820	vd	SC	4871		1959	BR TSO
SCR	99821	xdk	SC	9227		1955	BR BSO

"The Royal Scotsman". Kept at EC.

THR	99052	vs		45018	41	1927	LMS saloon
THR	99880	vs		5159		1891	LNWR Dining Car
THR	99881	vs		807		1912	GNR Family Saloon
THR	99882	vs	E	19169		1963	BR SK
THR	99883	vs	E	2108		1960	BR SLF
THR	99884	vs	E	19208		1963	BR SK
THR	99885	vs		2110		1960	BR SLF
THR	99886	vs	E	35407		1963	BR BSK
THR	99887	vs	E	2127		1961	BR SLF

National Railway Museum, York.

NRMY	99950	vd	E	35362		1962	BR BSK

7. CODES

7.1. LIVERY CODES

All locomotives are blue and all coaching stock and multiple unit vehicles are blue and grey unless otherwise indicated.

B Plain Blue *(coaches and multiple units only)*
E LNER Tourist Green and Cream.
F New Railfreight (two-tone grey sides with sub-sector markings)
FO Old Railfreight (grey sides, yellow cabsides, red bufferbeams and full height BR logo)
FR As FO but with red solebar stripe
G Greater Manchester PTE (orange/brown)
I *Locos & DVTs.* New InterCity (white and dark grey with red stripe and Swallow motif)
IO *Locos.* Old InterCity (light grey and dark grey with red stripe and yellow lower cabsides.)
I *Coaches.* InterCity (white and dark grey with red stripe)
J Jaffa Cake (two-tone brown with orange stripe)
M Mainline. As 'IO' but without the yellow cabsides.
N Network South East (grey/white/red/white/blue/white).
O Other livery (non-Standard – refer to text)
P Provincial (grey/light blue/white/dark blue).
PE Provincial Express (buff/light grey/dark grey/light grey/buff, with dark blue, white and light blue stripes).
PM Provincial Midline. As 'P' but with grey and cream stripes.
PN Provincial Network NorthWest. As 'P' but grey and red stripe.
PO Provincial old livery (grey/dark blue/white/light blue).
PR Provincial railbus variant (dark blue/white/light blue).
PS Provincial Scotrail (light grey/light blue/white/dark grey)
R *Locos.* Revised blue (large numbers and full height BR logo)
R *Coaches.* Royal Mail Red
S Strathclyde PTE (orange and black)
T Tyne and Wear PTE (yellow and white)
Y West Yorkshire PTE (new) (red and cream)

7.2. SECTOR CODES

The following list sector codes for BR locos and rolling stock. The locomotive sector codes are also known as pool codes. The codes indicate the BR sub-sector which owns the vehicle.

Locomotives:

CYPD	Privately Owned Diesel Locos authorised for main line use.
DBMS	British Rail Maintenance Ltd. Shunting Locos.
DCAA	Departmental (Civil Engineer Anglia) Main Line Locos.
DCBS	Bletchley Depot Shunting Locos.
DCEA	Departmental (Civil Engineer Eastern) Main Line Locos.
DCEB	York Depot Shunting Locos.
DCHA	Departmental (Civil Engineer Scotrail) Main Line Locos.
DCHB	Perth Depot Shunting Locos.
DCLS	Longsight Depot Shunting Locos.
DCMA	Departmental (Civil Engineer London Midland) Main Line Locos
DCMB	Departmental (Area Civil Engineer Birmingham) Locos.
DCMC	Departmental (Area Civil Engineer Crewe) Locos.
DCML	Departmental (Area Civil Engineer Liverpool) Locos.
DCMM	Departmental (Area Civil Engineer Manchester) Locos.
DCMN	Departmental (Area Civil Engineer Nottingham) Locos.
DCMW	Departmental (Area Civil Engineer Watford) Locos.
DCQA	Departmental (Civil Engineer BRB) Main Line Locos.
DCSA	Departmental (Civil Engineer Southern) Main Line Locos.
DCSB	Selhurst & Stewarts Lane Depots Shunting Locos.
DCSC	Chart Leacon Depot Shunting Locos.
DCWA	Departmental (Civil Engineer Western) Main Line Locos.
DCWC	Departmental (Civil Engineer Western) Class 97/6 Locos.
DMEA	Departmental (M & E.E.) Main Line Locos.
DMEC	Departmental (M & E.E.) Class 97/7 Locos.
DOPA	Departmental (Operations) Class 97/8 Locos.
DRTC	Departmental (Research) Railway Technical Centre Locos.
DWCC	Departmental (Operating) Royal Train Locos.
FABT	Railfreight (Aggregates) Tinsley Locos.
FACM	Railfreight (Aggregates) Motherwell Locos.
FAGS	Railfreight (Aggregates) Stratford Locos.
FALY	Railfreight (Aggregates) Thornaby Locos.
FAME	Railfreight (Aggregates) Eastfield Locos.
FASB	Railfreight (Aggregates) Stewarts Lane Locos.
FAWK	Railfreight (Aggregates) Cardiff Canton Locos.
FAXN	Railfreight (Aggregates) Toton Locos.
FCCC	Railfreight (Trainload) Crewe Locos.
FCTY	Railfreight (Trainload) Thornaby Locos.
FEAN	Railfreight (PS Coal) Toton Class 58 (Nottinghamshire) Locos.
FEBN	Railfreight (PS Coal) Toton Class 58 (Yorkshire) Locos.
FECN	Railfreight (PS Coal) Toton Class 56 (Nottinghamshire) Locos.
FEDN	Railfreight (PS Coal) Toton Class 56 (Yorkshire) Locos.
FEEN	Railfreight (PS Coal) Toton Class 56 (North East) Locos.
FEFN	Railfreight (PS Coal) Toton Class 20 (Nottinghamshire) Locos.
FEGN	Railfreight (PS Coal) Toton Class 20 (North West) Locos.

FEKK Railfreight (PS Coal) Cardiff Canton Locos.
FEOE Railfreight (PS Coal) Eastfield Class 20 Locos.
FEPE Railfreight (PS Coal) Eastfield Class 26 & 37 Locos.
FGAT Railfreight (Distribution) Tinsley Class 47 Locos.
FGBC Railfreight (Distribution) Crewe Class 47 Locos.
FGCS Railfreight (Distribution) Stratford Class 47 Locos.
FGDS Railfreight (Distribution) Stratford Class 37 Locos.
FGET Railfreight (Distribution) Tinsley Class 37 Locos.
FGHC Railfreight (Distribution) Crewe Class 85/0 & 87/1 Locos.
FGII Railfreight (Distribution) Immingham Class 47/4 Locos.
FGJW Railfreight (Distribution) Class 86/5 Locos.
FGKG Railfreight (Distribution) Shields Class 81 Locos.
FGLC Railfreight (Distribution) Crewe Class 86/4 Locos.
FGOT Railfreight (Distribution) Tinsley Class 31 Locos.
FGPC Railfreight (Distribution) Crewe Class 86/6 Locos.
FGRC Railfreight (Distribution) Crewe Class 85/1 Locos.
FGQW Railfreight (Distribution) Willesden Class 90 Locos.
FGTE Railfreight (Distribution) Eastfield Class 20 & 37 Locos.
FGUV Railfreight (Distribution) Inverness Class 37 Locos.
FGWB Railfreight (Distribution) Stewarts Lane Class 33 Locos.
FGXC Railfreight (Distribution) Crewe Class 31 Locos.
FGYX Railfreight (Distribution) Stored Locos.
FHAC Railfreight (Coal) Crewe Class 31 Locos.
FHBK Railfreight (Coal) Cardiff Class 37 Locos.
FHTA Landore Depot Shunting Locos.
FHUB Blyth Cambois Depot Shunting Locos.
FHVW Springs Branch Depot Shunting Locos.
FHWN Toton Depot Shunting Locos.
FHZR Ayr Depot Shunting Locos.
FJLL Railfreight (Trainload) Laira Locos.
FMAK Railfreight (Metals) Cardiff Canton Class 37/0 & 47 Locos.
FMCK Railfreight (Metals) Cardiff Canton Class 37/9 Locos.
FMHK Railfreight (Metals) Cardiff Canton Class 37/7 Locos.
FMGM Railfreight (Metals) Motherwell Locos.
FMMT Railfreight (Metals) Tinsley Class 09 Locos.
FMSK Cardiff Canton Depot Shunting Locos.
FMTY Railfreight (Metals) Thornaby Locos.
FMYI Railfreight (Metals) Immingham Locos.
FMZA Thornaby Depot Shunting Locos.
FPAE Railfreight (Petroleum) Eastfield Locos.
FPBC Railfreight (Petroleum) Crewe Diesel Locos.
FPCI Railfreight (Petroleum) Immingham Locos.
FPEK Railfreight (Petroleum) Cardiff Canton Locos.
FPFS Railfreight (Petroleum) Stratford Locos.
FPZA Immingham Depot Shunting Locos.
FQCK Railfreight (Coal Distrib.) Cardiff Canton Class 37 Locos.
FSAE Eastleigh & Ryde Depots Shunting Locos.
FSBM Motherwell Depot Shunting Locos.
FSCG Grangemouth Depot Shunting Locos.
FSDH Haymarket Depot Shunting Locos.
FSEG Gateshead Depot Shunting Locos.
FSFN Neville Hill Depot Shunting Locos.

FSGB Bescot Depot Shunting Locos.
FSGH Holbeck Depot Shunting Locos.
FSHT Tinsley Depot Shunting Locos.
FSID Doncaster Depot Shunting Locos.
FSJS Stratford Depot Shunting Locos.
FSKC Cambridge Depot Shunting Locos.
FSLC Upperby Depot Shunting Locos.
FSMA Allerton Depot Shunting Locos.
FSNC Crewe Depot Shunting Locos.
FSPG Gloucester Depot Shunting Locos.
FSQA Bristol Bath Road Depot Shunting Locos.
FSSO Old Oak Common Shunting Locos.
IAAB Aberdeen Depot Shunting Locos.
IANA InterCity Anglia Services Locos.
IANB Norwich Crown Point Depot Shunting Locos.
IBRA InterCity Bristol Class 47/4 Locos.
ICCA InterCity Cross Country Services Locos.
ICCP InterCity Cross Country Services Class 43 Locos.
ICDA InterCity Crewe Class 47/4 Locos.
ICHA InterCity Charter Services Locos.
IEBA InterCity Euston–Birmingham Services Locos.
IECA InterCity East Coast Services Main Line Locos.
IECP InterCity East Coast Services Class 43 Locos.
IEDA InterCity Eastfield Class 47/4 Locos.
IISA Inverness Depot Shunting Locos.
ILRA InterCity Class 47/8 Locos.
IMLP InterCity Midland Line Services Class 43 Locos.
IPSA InterCity Push-Pull & Pullman Services Locos.
IVGA InterCity Gatwick Express Services Locos.
IWCA InterCity West Coast Services Locos.
IWNB Willesden Depot Shunting Locos.
IWRA InterCity Western Region Services Locos.
IWRB Laira Depot Shunting Locos.
IWRP InterCity Western Region Services Class 43 Locos.
NNEA Network South East Kings Lynn Services Locos.
NNWA Network South East London & N. Western Services ML Locos.
NSSA Network South East Solent & Sarum Class 47 & 50 Locos.
NSSB Network South East Solent & Sarum Class 33 & 73 Locos.
NSSC Network South East Solent & Sarum Services Shunting Locos.
NWRA Network South East Western Region Services Locos.
NWRB Old Oak Common & Reading Depots Shunting Locos.
NXXA Network South East Shunting Locos.
PCDA Provincial Crewe Class 47/4 Locos.
PCDB Provincial Crewe Class 31 Locos.
PCFA Provincial Cardiff Canton Locos.
PEDA Provincial Eastfield Locos.
PISA Provincial Inverness Locos.
RXLB Parcels Class 31 Locos.
RXLC Parcels Crewe Class 47 Locos.
RXLD Parcels Bristol & Old Oak Common Class 47 Locos.
RXLE Parcels AC Electric Locos.
RXLS Derby Depot Shunting Locos.

Rolling Stock:

DXXZ	Authorised for withdrawal for departmental use.
IANR	Inter-City – Anglia Inter-City. Refurbished.
IANX	Inter-City – Anglia Inter-City.
ICCX	Inter-City – Cross Country.
ICHA	Inter-City – Charter and Special Services (air refurbished).
ICHB	Inter-City – Charter and Special Services (vacuum refurbished).
ICHC	Inter-City – Charter and Special Services (vacuum not refurb).
ICHD	Inter-City – Charter and Special Services (air not refurb).
ICHH	Inter-City – Charter and Special Services (held for re-use).
ICHL	Inter-City – Charter and Special Services (for loan).
ICHM	Inter-City – Charter and Special Services (Manchester).
ICHS	Inter-City – Charter and Special Services (steam specials).
ICHV	Inter-City – Charter and Special Services (VIP).
ICHX	Inter-City – Charter and Special Services.
IECX	Inter-City – East Coast Main Line.
IMLX	Inter-City – Midland Main Line.
IVGX	Inter-City – Victoria–Gatwick.
IWCX	Inter-City – West Coast Main Line.
IWRX	Inter-City – Great Western Main Line.
IXXX	Inter-City – General.
IXXZ	Inter-City – Authorised for Withdrawal.
NGBX	Network South East – Gospel Oak–Barking.
NGEX	Network South East – Great Eastern.
NGNX	Network South East – Great Northern.
NKCX	Network South East – Kent Coast.
NKSX	Network South East – Kent Suburban.
NMLX	Network South East – Midland Line.
NMYX	Network South East – Marylebone.
NNEX	Network South East – North East London.
NNLX	Network South East – North London Line.
NNWX	Network South East – London and North Western.
NSBX	Network South East – Surrey and Berkshire Suburban.
NSLX	Network South East – South London.
NSSX	Network South East – Solent and Sarum.
NSXX	Network South East – Sussex.
NTSX	Network South East – London Tilbury & Southend Line.
NWRX	Network South East – Western Region.
NXXX	Network South East – General.
PXXX	Provincial – General.
PXXZ	Provincial – Authorised for Withdrawal.
REMD	Parcels – d.c. EMUs.
RNPA	Parcels – Newspapers (Anglia Region).
RNPE	Parcels – Newspapers (Eastern Region).
RNPM	Parcels – Newspapers (London Midland Region).
RNPW	Parcels – Newspapers (Western Region).
RPOA	Parcels – Post Office (Anglia Region).
RPOC	Parcels – Post Office central van fleet.
RPOE	Parcels – Post Office (Eastern Region).
RPOM	Parcels – Post Office (London Midland Region).
RPOW	Parcels – Post Office (Southern Region & Western Region).

RPOZ	Parcels – Post Office (Scottish Region).
RPPA	Parcels – Newspaper Vans (Anglia Region).
RPPE	Parcels – Newspaper Vans (Eastern Region).
RPPM	Parcels – Newspaper Vans (London Midland Region).
RPPW	Parcels – Newspaper Vans (Western Region).
RPXA	Parcels – General (100/110 mph BG. Anglia Region).
RPXB	Parcels – General (90 m.p.h. BG dual braked).
RPXC	Parcels – General (90 mph. BG vacuum braked).
RPXD	Parcels – General (GUV vacuum braked).
RPXE	Parcels – General (GUV air/dual braked).
RPXG	Parcels – General (Courier vehicles).
RPXH	Parcels – General (100/110 mph BG. Heaton).
RPXM	Parcels – General (100/110 mph. BG).
RPXN	Parcels – DPU cars (DPU).
RPXW	Parcels – General (100/110 mph BG. Western region).
RPXX	Parcels – General.
RPXZ	Parcels – General (100/110 mph. BG. Scottish Region).

Note: Provincial sector codes consist of the letter 'P' followed by the two letter depot code followed by a letter 'X'.

7.3. DEPOT CODES

AB Aberdeen Ferryhill TMD
AF Ashford Chart Leacon TMD
AN Allerton TMD (Liverpool)
AY Ayr TMD
BD Birkenhead North T&RSMD
BH Barrow Hill SP.
BI Brighton T&RSMD
BL Blyth Cambois SD
BM Bournemouth EMUD
BN Bounds Green T&RSMD (London)
BQ Bury EMUD
BR Bristol Bath Road TMD
BS Bescot TMD (Walsall)
BX Buxton TMD
BY Bletchley TMD
CA Cambridge T&RSMD
CC Clacton EMUD
CD Crewe TMD (D)
CE Crewe TMD (E)
CF Cardiff Canton T&RSMD
CH Chester DMUD
CJ Clapham Junction CSD
CK Corkerhill T&RSMD
CL Carlisle Upperby CWMD
DL Doncaster Major Depot
DR Doncaster TMD
DY Derby Etches Park T&RSMD
EC Craigentinny T&RSMD (Edbgh)
ED Eastfield TMD (Glasgow)
EH Eastleigh T&RSMD
EM East Ham EMUD (London)
EN Euston Downside CARMD
FR Fratton EMUD (Portsmouth)
FY Merehead (Foster Yeoman)
GD Gateshead TMD
GI Gillingham EMUD
GL Gloucester SD
GW Glasgow Shields TMD
HA Haymarket TMD (Edinburgh)
HE Hornsey TMD (London)
HO Holbeck TMD (Leeds)
HR Hall Road EMUD
HT Heaton T&RSMD (Newcastle)
IL Ilford T&RSMD (London)
IM Immingham TMD (S. Humberside)
IP Ipswich fuelling point
IS Inverness T&RSMD
KY Knottingley TMD
LA Laira T&RSMD (Plymouth)

LE Landore T&RSMD (Swansea)
LG Longsight TMD (E) (Manchester)
LL Liverpool Edge Hill CARMD
LO Longsight TMD (D) (Manchester)
LR Leicester SD
MA Manchester Longsight CARMD
ME Marylebone TMD (London)
ML Motherwell TMD
MO Muir of Ord CSD
NC Norwich Crown Point T&RSMD
NH Newton Heath DMUD (Manchester)
NL Neville Hill T&RSMD (Leeds)
OC Old Oak Common TMD (D)
OM Old Oak Common CARMD (London)
OO Old Oak Common (HST) TMD
OY Oxley CARMD (Wolverhampton)
PC Polmadie CARMD (Glasgow)
PE Peterborough Nene Sidings CSD
PH Perth SD
PM St. Phillips Marsh T&RSMD (Bristol)
PY Bristol Pylle Hill
RE Ramsgate T&RSMD
RG Reading TMD
RTC Railway Technical Centre (Derby)
RY Ryde (Isle of Wight) T&RSMD
SA Salisbury CARMD
[illegible] Stratford TMD (London)
[illegible] Slade Green T&RSMD
SH Strawberry Hill EMUD (London)
SL Stewarts Lane T&RSMD (London)
SR Stratford Major Depot
SU Selhurst TMD (London)
TE Thornaby TMD
TI Tinsley TMD (Sheffield)
TO Toton TMD (Notts)
TS Tyseley TMD (Birmingham)
WB Wembley Inter-City CARMD
WC Waterloo EMUD (Waterloo & City)
WD Wimbledon EMUD (London)
WN Willesden TMD (London)
YK York SD
ZC Crewe Works (BREL)
ZE Derby Locomotive Works (BREL)
ZG Eastleigh Major Depot
ZH Springburn Major Depot
ZN Wolverton Major Depot

OTHER CODES USED

CARMD	Carriage maintenance depot.
CSD	Carriage servicing depot.
SD	Servicing depot.
CWMD	Carriage and wagon mtce. depot
SRPS	Scottish Rly. P. S.
T&RSMD	Traction and rolling stock mtce depot.
TMD	Traction Maintenance Depot.
TMD (HST)	HST Maintenance Depot.
TMD (D)	Traction Maintenance Depot (Diesel).
TMD (E)	Traction Maintenance Depot (Electric).

8. OTHER RAILWAYS

8.1. BLACKPOOL & FLEETWOOD TRAMWAY

Depot: Rigby Road.

For the information in this section we are indebted to Mr. Ian McLoughlin of Blackpool. Note: Numbers in brackets are pre-1968 numbers.

ONE MAN CARS

Rebuilt 1972–76 from English Electric railcoaches built 1934–5. Radio fitted. 13 converted (1–13).
Seats: 48U.
Traction Motors: Two EE305 of 40 kW.

* Red & cream livery.

Note: First numbers in brackets are post 1968 numbers prior to conversion.

1 (616, 269)(U)
5 (609, 221)
8 (612, 265)
10 (614, 267)
11 (615, 268)
12* (611, 264) (U)

OPEN BOAT CARS

Built 1934–5 by English Electric. 12 built (225–236).
Seats: 56U.
Traction Motors: Two EE327 of 30 kW.

600* (225)
602† (227)
604§ (230)
605 (233)(U)
606b (235)
607 (236)

* On loan to Heaton Park Tramway, Manchester.
† Yellow and black livery.
§ Red and white livery.
b Blue & yellow livery.

REPLICA VANGUARD

Built 1987 on underframe of one man car No. 7.(619–282).
Seats: U.
Traction Motors: Two EE327 of 30 kW.

619

BRUSH RAILCOACHES

Built 1937 by Brush. 20 built (284–303).
Seats: 48U.
Traction Motors: Two EE305 of 40 kW. (EE327 of 30 kW*).

621 (284)
622* (285)
623 (286)
625 (288)
626 (289)
627 (290)
630 (293)
631 (294)
632 (295)

633 (296)	636 (299)	637 (300)
634 (297)		

CENTENARY CLASS

Built 1984–7. Body by East Lancs. Coachbuilders, Blackburn. One man operated. Radio fitted.
Seats: 52U.
Traction Motors: Two EE305 of 40 kW.

* Rebuilt from GEC car 651).

641	644	647
642	645	648*
643	646	

CORONATION CLASS

Built 1953 by Charles Roberts & Co. Resilient wheels. 25 built (304–328).
Seats: 56U.
Traction motors: Four Crompton-Parkinson 92 of 34 kW.

660 (324)

PROGRESS TWIN CARS

Motor cars (671–677) rebuilt 1958–60 from English Electric railcoaches.
Seats: 53U. **Traction Motors**: Two EE305 of 40 kW.
Driving trailers (681–687) built 1960 by Metro-Cammell.
Seats: 53U.

671+681 (281+T1)	674+684 (284+T4)	676+686 (286+T6)
672+682 (282+T2)	675+685 (285+T5)	677+687 (287+T7)
673+683 (283+T1)		

SINGLE CARS

Rebuilt 1958–60 from English Electric railcoaches. Originally ran with trailers.
Seats: 48U.
Traction Motors: Two EE305 of 40 kW.

678 (278)	679 (279)	680 (280)

"BALLOON" DOUBLE DECKERS

Built 1934–5 by English Electric. 700–712 were originally built with open tops, and 706 has now reverted to that condition.
Seats: 94U.
Traction Motors: Two EE305 of 40 kW.

700 (237)	703 (240)	707 (244)
701 (238)	704 (241)	708 (245)
702 (239)	706 (243)	709 (246)

710	(247)	716	(253)	721	(258)
711	(248)	717	(254)	722	(259)
712	(249)	718	(255)	723	(260)
713	(250)	719	(256)	724	(261)
715	(252)	720	(257)	726	(263)

ILLUMINATED CARS

732	(168)	Rocket	Seats: 47U
733	(209)	Santa Fe loco. & tender	Seats: 35U
734	(174)	Santa Fe coach	Seats: 60U
735	(222)	Hovertram	Seats: 99U
736	(170)	HMS Blackpool	Seats: 71U

WORKS CARS

259	(748, 624)	PW gang towing car.
260	(751, 628, 291)	Crane car and rail carrier.
749	(S)	Tower wagon trailer.
750		Cable drum trailer.
752	(2, 1)	Rail grinder and snowplough.
753	(143)	Overhead line maintenance car with generator.

JUBILEE CLASS DOUBLE DECKERS

Rebuilt 1979/82 from Balloon cars. Standard bus ends, thyristor control and stairs at each end. 761 has one door per side whereas 762 has two. Radio fitted.

Seats: 100U.

Traction Motors: Two EE305 of 40 kW.

761	(725, 262)	762	(714, 251)

PRESERVED CARS

Blackpool & Fleetwood 40	Box car. Bogie single decker built 1914
Blackpool 59	Dreadnought (Open top double decker) built 1902
Bolton 66	Bogie double-decker built 1901

No. 59 will be returned to the National Tramway Mmuseum, Crich at the end of the 1990 summer season.

In addition, certain other preserved cars may be loaned to Blackpool during 1989, e.g. Glasgow 1100, Sheffield 513.

DEMONSTRATION CAR

A six-axle articulated car built by Tatra of Czechoslovakia will be used in Blackpool during 1990.

8.2. DOCKLANDS LIGHT RAILWAY

This is a light rail line running in London's East End from Tower Gateway and Stratford to Island Gardens. It is being extended to Bank & Beckton, and eventually will run to Lewisham.

System: 750 V dc third rail (bottom contact).
Depot: Poplar.

SIX-AXLE ARTICULATED CARS — Bo–2–Bo

Built 1987 by Linke-Hofmann-Busch. 28.80 x 2.65 m. Inward folding plug doors. Chopper control. Scharfenberg Couplers.
Weight: 39 t. **Seats**: 84U.
Traction Motors: Two GEC of 185 kW.
Max. Speed: 80 km/h.
Electric Brake: Rheostatic.

01
02
03
05
05
06
07
08
09
10
11

SIX-AXLE ARTICULATED CARS — Bo–2–Bo

Built 1990 by BREL Ltd., York Works. 28.00 x 2.65 m. Inward Folding plug doors. Chopper control. Scharfenberg Couplers.
Weight: 39 t. **Seats**: 84U.
Traction Motors: Two GEC of 185 kW.
Max. Speed: 80 km/h.
Electric Brake: Rheostatic.

12
13
14
15
16
17
18
19
20
21

SIX-AXLE ARTICULATED CARS — Bo–2–Bo

On order from La Brugeoise et Nivelles, Belgium. 28.8 x 2.65 m. Sliding doors. End doors for staff use. Chopper control. Scharfenberg Couplers
Weight: t. **Seats**: 70U.
Traction Motors: Two Brush? .
Max. Speed: 80 km/h.
Electric Brake: Rheostatic.

22
23
24
25
26
27
28
29
30
31
32
33
34
35
36
37
38
39
40
41
42

8.3. STRATHCLYDE PTE UNDERGROUND

Depot: Broomloan.
4'0" gauge. 600 V dc third rail. Generally known as the "Subway".

SINGLE CARS

Built 1977–9 by Metro-Cammell. 12.58 x 2.34 m.
Seats: 36U.
Traction Motors: Four GEC G312AZ.

101	110	118	126
102	111	119	127
103	112	120	128
104	113	121	129
105	114	122	130
106	115	123	131
107	116	124	132
108	117	125	133
109			

8.4. TYNE & WEAR METRO

System: 1500 V d.c. overhead.
Depot: South Gosforth.

DIESEL WORKS LOCOMOTIVES 0–6–0

Built: 1977–8 by Brush Traction, Loughborough. BSI couplers.
Engine: Rolls-Royce DV8N of 427 hp.
Transmission: One Brush axle-hung traction motor fitted to the centre axle.
Max. Speed: 50 km/h.

WL 1
WL 2
WL 3
WL 4
WL 5

BATTERY/OVERHEAD ELECTRIC LOCOS

Built: 1989–90 by Hunslet, Leeds. BSI couplers.
Traction Motors: Hunslet-Greenbat T9-4P.
Weight: 26 t.

Three locos. Numbers not yet allocated.

One of the three new battery/overhead electric locomotives being delivered to the Tyne & Wear Metro from Hunslet. TWPTE

South Shields
M
4031
4031

SIX-AXLE ARTICULATED UNITS Bo–2–Bo

Built 1976, 1978–81 by Metro-Cammell. 27.80 x 2.65m. BSI couplers.
Weight: 39 t. **Seats**: 84U. **Traction Motors**: Two 187 kW.

4001
4002
4003
4004
4005
4006
4007
4008
4009
4010
4011
4012
4013
014
015
016
017
018
019
020
021
022
023
4024
4025
4026
4027
4028
4029
4030
4031
4032
4033
4034
4035
4036
4037
4038
4039
4040
4041
4042
4043
4044
4045
4046
4047
4048
4049
4050
4051
4052
4053
4054
4055
4056
4057
4058
4059
4060
4061
4062
4063
4064
4065
4066
4067
4068
4069
4070
4071
4072
4073
4074
4075
4076
4077
4078
4079
4080
4081
4082
4083
4084
4085
4086
4087
4088
4089
4090

ames:

)41	HARRY COWANS
)44	Director
)51	Times

ɔtes:

andard livery is yellow and white.
44 is painted yellow.
51 is painted claret.

ne & Wear Metro Car No. 4031. TWPTE

ont Cover: Class 90 No. 90020 in InterCity livery is seen hauling a rake of Mark 3 ɪches forming the down 'Lancashire Pullman', the 16.00 Euston–Blackpool on 06/09/
Hugh Ballantyne

ck Cover: Class 108 power twin 51561/51930 climbs towards Church Stretton on 08/89 with the 16.20 Central Wales Line service from Shrewsbury to Swansea.
Tom Clift

PLATFORM 5 PUBLISHING LTD–MAIL ORDER LIST

1. PLATFORM 5 AND METRO ENTERPRISES PUBLICATIONS

Today's Railways Review of the Year (Vol 2)	£11.95
Today's Railways Review of the Year (Vol 1)	£11.95
Light Rail Review 1	£5.95
Midland Railway Portrait	£12.95
Handbook of BR Steam Motive Power Depots Vol. 1 Southern England	£7.9
Handbook of BR Steam Motive Power Depots Vol. 2 Central England East Anglia & Wales	£8.95
Blackpool & Fleetwood by Tram	£7.50
The Railways of Winchester	£6.9
Motive Power Pocket Book Spring 1990	£1.5
Coaching Stock Pocket Book Spring 1990	£1.5
Diesel Unit Pocket Book Spring 1990	£1.5
Electric Unit Pocket Book Spring 1990	£1.5
Locomotives & Coaching Stock 1990	£5.9
Preserved Locomotives and Multiple Units 6th Ed.	£3.9
British Rail Depot Directory 2nd Edition	£4.2
Departmental Coaching Stock Fourth Edition	£4.9
Track Machines Third Edition	£3.7
DB/German Federal Railway 2nd Ed.	£6.9
ÖBB/Austrian Federal Railways 2nd Ed.	£6.9
Benelux Locomotives & Coaching Stock 2nd Ed.	£6.9
Irish Railways Traction & Travel 2nd Ed.	£6.9
SNCF/French National Railways **(including 1988 supplement)**	£4.7
Private Owner Wagons Vol. 1	£4.9
Locomotives & Coaching Stock 1985	£2.9
Locomotives & Coaching Stock 1986	£3.3
Locomotives & Coaching Stock 1987	£3.3
Locomotives & Coaching Stock 1988	£3.
Locomotives & Coaching Stock 1989	£4.

2. Foxline/Milepost etc.

Scenes from the Past 1 – Railways around Stockport	£4.
Scenes from the Past 2 – Railways around Buxton	£4.
Scenes from the Past 3 – Manchester Railway Termini	£4.
Scenes from the Past 4 – The Cambrian Coast Railway	£4.
Scenes from the Past 5 – Railways around Grimsby & Cleethorpes	£5.
Scenes from the Past 6 – Railways around Preston	£4.
Scenes from the Past 7 – The Railway from Buxton to Ambergate	£5.
Scenes from the Past 8 – Railways in and around Manchester	£5.
Scenes from the Past 8 – (Casebound version)	£12.
An illustrated History of the Cheshire Lines Committee	£10.
The Midland Railway Around Nottinghamshire Vol.1	£4.
Great Central east of Sheffield Vol. 1	£5.
A Nostalgic Look at Glasgow Trams since 1950	£3.
Edinburgh's Trams and Buses	£4.
Totley and the Tunnel	£3.
On the Trams	£4.
Canadian Trackside Guide	£7
BR Wagon Fleet Vols 3 & 4. Each	£3

Please send to Mail Order Department. Address as on title page. Please a
10% (UK) 20% (abroad) for postage & packing.